APOLLO IN THE
AGE OF AQUARIUS

Apollo in the Age of Aquarius

NEIL M. MAHER

HARVARD UNIVERSITY PRESS

Cambridge, Massachusetts, and London, England

2017

First Printing

Library of Congress Cataloging-in-Publication Data

Names: Maher, Neil M., 1964– author.
Title: Apollo in the Age of Aquarius / Neil M. Maher.
Description: Cambridge, Massachusetts : Harvard University Press, 2017. |
Includes bibliographical references and index.
Identifiers: LCCN 2016041938 | ISBN 9780674971998 (hardcover : alk. paper)
Subjects: LCSH: Project Apollo (U.S.) | United States. National Aeronautics and Space
Administration—History—20th century. | Astronautics—Social aspects—United States—
History—20th century. | Nineteen sixties. | Outer space—Exploration—United States—Public
opinion—History—20th century. | Science and state—United States—History—20th century.
Classification: LCC TL789.8.U6 A55355 2017 | DDC 629.45/4—dc23
LC record available at https://lccn.loc.gov/2016041938

For Stacy, Riggs, and Leif

Contents

Introduction: Launching the Sixties   *1*

1   Spaceship Earth: Civil Rights and NASA's War on Poverty   *11*

2   Shooting (from) the Moon: NASA, Nature, and the
New Left during the Vietnam War Era   *54*

3   Thinking Globally, Acting Locally: Cape Canaveral
and Whole Earth Environmentalism   *92*

4   Heavenly Bodies: "Manned Spaceflight" and
the Women's Movement   *137*

5   The New Right's Stuff: The Hippie Counterculture and
the Rise of the Conservative Crescent   *183*

Conclusion: Grounding the Space Race   *228*

Notes   *241*

Acknowledgments   *347*

Illustration Credits   *351*

Index   *353*

Introduction

Launching the Sixties

During an evening performance of *Hair* on June 4, 1970, just before the first-act curtain and soon after the entire cast paraded onto center stage buck naked, Apollo 13 astronauts James Lovell and John Swigert abruptly walked out of Broadway's Biltmore Theater. Lovell and Swigert, who just a few months earlier had named their lunar landing module *Aquarius* in honor of the musical, and who played the album from *Hair* during the troubled Apollo 13 mission, ducked out of their front-row mezzanine seats not because of the nudity or the play's depiction, during a drug-induced hallucination, of three astronauts shooting Catholic nuns with "ray guns." Rather, the astronauts left *Hair* to protest "Don't Put It Down," a scene that satirized American patriotism by using the Stars and Stripes as a blanket and then threatening to burn it at a be-in. "I don't like what you're doing to the flag," explained Swigert when he was stopped in the theater's lobby by a *Hair* press agent. "I don't like the way they wrapped the flag around that guy." Although the musical's spokesperson responded that the flag had been treated with respect and never touched the ground onstage, the astronauts remained unconvinced; they and their guests quickly exited the theater's lobby and headed uptown to Raffles, a private club in the chic Sherry-Netherland Hotel.[1]

The "*Hair* affair," reported the popular press the following day, symbolized a much deeper cultural divide emerging in 1960s America. On one side were crew-cut "squares," men and women residing across so-called Middle America, who, like the Apollo 13 astronauts, flew their flags more forcefully as the tumult of the decade wore on. President Richard Nixon called them "the forgotten Americans," a "silent majority" who paid their taxes, attended church on Sunday, and voiced alarm back in October of 1957 when the Soviet Union launched Sputnik, the world's first orbiting satellite. The success of the Apollo program, many in this generation believed, was a significant step toward catching up with the Russians in space and winning the Cold War. Not surprisingly, in mid-July of 1969 nearly one million of these Middle Americans made the pilgrimage to Florida's Cape Canaveral to witness the Apollo 11 moon shot. Many also would have most probably followed in Lovell's and Swigert's footsteps, slipping out of the Biltmore Theater after becoming uncomfortable during the first act of Broadway's self-proclaimed American Tribal Love-Rock Musical.[2]

On the other side of this cultural gap, according to the American press, were long-haired "hippies." They, like the characters in *Hair*, often refused to fly the flag at all and instead draped their half-naked bodies in red, white, and blue bell-bottoms and flowered shirts in open rebellion against their parents' patriotism. They were joined by what *Time* magazine in 1969 called "liberals," "radicals," "blacks," and "many of the young," who similarly questioned the direction of mainstream political culture and fought instead for civil rights, an end to the Vietnam War, a cleaner environment, and women's liberation.[3] For these groups on the left, the race to the moon had not only dangerously heated up the Cold War between the United States and the Soviet Union, but also distracted the nation from a host of problems closer to home. As a result, rather than heading south to the National Aeronautics and Space Administration's (NASA) Kennedy Space Center for Apollo 11's launch, just three weeks later many from this camp joined approximately 400,000 youths who drove, hitchhiked, and finally walked through standstill traffic to a farm near Woodstock in upstate New York for a three-day outdoor music festival. Many, no doubt, would also have felt quite comfortable returning to their seats for the second act of *Hair*.

Taking its cue from the summer of 1969, when astronauts first set foot on the moon while America's youth danced at Woodstock, *Apollo in the Age of Aquarius* tells the story of the shared history of the space race and the social and political movements of the 1960s era. It argues not only that these two historical phenomena were mutually dependent on one another for popular and political support, but also that they represented two factions in a national debate regarding the present and future course of the United States. Was America the sort of nation that would spend $24 billion to land men on the lunar surface, or a country more interested in using such resources to solve political problems such as racism, war, pollution, and sexism right in its own backyard? This controversy over national purpose, moreover, ultimately transformed, albeit unevenly, both the space race and these grassroots movements. While shared opposition to space exploration nurtured these movements during the 1960s, making NASA a somewhat reluctant driver of broader social change, in the 1970s civil rights leaders, peace activists, environmentalists, feminists, and those in the counterculture successfully pressured the space agency to reorient some of its technology away from outer space toward our bright blue planet. As a result, Apollo, in partnership with what the cast of *Hair* was calling the "Age of Aquarius," brought both the space race and American politics back down to Earth.

The summer of 1969 was also a pivotal moment in this shared history. Prior to the moon landing, activists from all of these movements censured the space race for diverting attention from more pressing issues affecting African Americans, draft-age youths, citizens living near pollution, women, and countercultural dissidents. The Apollo program, with its end goal of reaching the moon and its novel perspective of the entire planet seen from outer space, aggravated this concern that an emerging global gaze was obscuring the needs of particular groups of people on Earth.[4] Administrators at NASA ignored such grassroots opposition during the 1960s because the majority of Americans supported Apollo as a means of counteracting the Soviets. Yet as it became clear that astronauts Neil Armstrong and Buzz Aldrin would beat the Russians to the lunar surface, a palpable "NASA fatigue" set in nationwide, resulting in plummeting popular support and dwindling congressional dollars for space exploration.[5] Space agency officials spent much of the 1970s trying to recover both support and funding by illustrating to these grassroots activists, as well as to politicians in Washington, DC, the benefits of space exploration for those back on Earth. By reacting to such special interest groups, Apollo not only encouraged the identity politics of the 1960s but also the perception, widely accepted during the following decade, that American society was becoming disaggregated, fragmented, and fractured.[6] *Apollo in the Age of Aquarius* traces the rise and fall of this global perspective within an increasingly divided America.

I had not originally set out to write such a book. The seeds of a very different approach took root one day while I was reading *Of a Fire on the Moon*, Norman Mailer's rambling firsthand account of the Apollo 11 launch during that summer of 1969. After describing in mind-numbing detail

NASA's thirty-six-story Saturn rocket, the Apollo spacecraft perched on top, and the pad where both would launch toward the moon, the Pulitzer Prize–winning author, who throughout the book goes by the name "Aquarius" to distinguish himself from the space agency's straitlaced astronauts, turned his attention to the surrounding Cape Canaveral environment. "Through all that water and flat swampy waste of wilderness, through all the evening din of crickets, cicadas, beetles, bees, mosquitoes, grasshoppers and wasps," he wrote, "a million people began to foregather on all the beaches and available islands and causeways and bridges and promontories which would give clear view of the flight."[7] As an environmental historian, I was hooked. Sending men from Earth to the moon, I was surprised to learn, involved not only technology but wild nature as well.

As a historian of twentieth-century America, I also knew that Mailer's literary linking of high technology with lowly nature at the Apollo 11 launch was not an isolated event. The post–World War II era in particular was a period of deep cultural ambivalence regarding the impact of modern technology and science on the natural environment. Beginning with the dropping of two atomic bombs on Japan and continuing with the nuclear arms race into the early 1960s, Americans increasingly questioned the belief, widespread during the first half of the twentieth century, that technology signified progress. Books such as *Hiroshima*, John Hersey's 1946 best seller that described the effects of the atomic blast on six residents of that city, and Rachel Carson's *Silent Spring*, which in 1962 exposed the dangers of chemical pesticides, were only the most popular expressions of this widespread belief that technology run amok could harm both humans and their environment. This anxiety reached its peak in 1969 when a pair of environmental disasters—the spoiling of Santa Barbara's pristine beaches by an oil spill and the incomprehensible burning of Cleveland's Cuyahoga River due to ignited pollution—each made national front-page news.

Yet as California's white sands turned sticky black and Ohio's waterways blazed toxic red, an alternative view arose that embraced, rather than rejected, certain types of technologies as solutions to these same problems. E. F. Schumacher's "appropriate technology" movement and Murray Bookchin's idea of "ecotechnologies" were just two examples of the belief that alternative technologies, if used correctly, could cure environmental ills.[8] Buckminster Fuller, an eco-designer who popularized the term "Spaceship Earth" after viewing one of NASA's photographs of our home planet, promoted similar ideas. Such thinkers countered the understandable technological pessimism of Hersey and Carson by promoting the notion that "green" technologies such as windmills, thermal windows, and solar panels could help create a more socially egalitarian and eco-friendly world. Nuclear energy, synthetic fertilizers, and even the birth control pill sparked similar cultural debate over the possibilities and pitfalls of modern technology.

The space race was yet another example, and I initially intended to use it as a road map for exploring this ambivalence regarding nature and technology during the 1960s era. As a result, my early research focused on five features of the Apollo program where nature and technology unexpectedly converged. The book would open with a discussion of John F. Kennedy's announcement, at the 1960 Democratic presidential convention in Los Angeles, of a New Frontier in science and technology that portrayed outer space as the ultimate wilderness. It would then examine the construction of Apollo space capsules that were designed, by both engineers and ecologists, to mimic enclosed ecosystems that recycled air, water, and waste. I would next turn to simulators, which NASA developed to acclimate astronaut bodies to foreign space environments, as well as to spacesuits that protected those bodies by transporting a livable Earth environment into space. The book would end by exploring two of the most obvious examples of this convergence: the capturing, by Apollo spacecraft, of images of Earth from space that became icons for the environmental movement, and the collection, by orbiting satellites, of global data that allowed environmental scientists to track ecological changes back on Earth. In its early incarnation, in other words, this book was an environmental and technological history of the space race.

Yet as I made my own pilgrimage to archives in Cape Canaveral and Houston as well as at the National Air and Space Museum and NASA Headquarters in Washington, DC, I began to lose faith in my original approach. First of all, by focusing on links to nature both in space and back on Earth, my environmental history of the space race would exacerbate a serious problem already plaguing this field. To put it bluntly, too many environmental historians continue to write about narrow topics such as parks, pollution, and environmental policy that appeal mostly to fellow environmental historians. While such scholarship built an intellectual foundation during the field's early years back in the 1970s and 1980s, its continuance today tends to separate environmental history from bigger, broader, more mainstream historical events, movements, and ideas. Women's history once suffered a similar fate.[9] As a result, instead of conceptualizing nature as a central category of historical analysis, historians from other subdisciplines tend to "add nature and stir" when describing a moment in time with obvious links to the natural world. Almost every U.S. history textbook, for example, makes room for nature during the colonial period, when everyday life seemed closer to the ground, so to speak, but then shunts it aside thereafter. For environmental history to be taken more seriously by all historians, practitioners need to write fewer histories on, for instance, Yosemite National Park, smog, and the Clean Water Act and more environmental histories of slavery, the Great Depression, and the women's movement.[10] My original idea for this book, I soon realized, fell into this trap.

It also refrained from addressing a second ailment afflicting the history of technology. For decades scholars in this field have debated whether or not technologies drove history: whether, for instance, the cotton gin made inevitable the expansion of slavery across the South and into the American West. Since the 1980s the field as a whole has thankfully shifted from emphasizing technology's role in determining history to an approach that makes room for other factors, from cultural beliefs to political practices to economic relationships, in guiding how technologies shape the past.[11] Most space historians, unfortunately, have not followed suit. This subfield has "largely been insulated from the paradigmatic revolution in the history of technology," argues space historian Asif Siddiqi of this cultural turn from determinism. Space historians, he warns, must "engage much more actively with the mainstream American history community" in order to move the field "from a fledgling subdiscipline to one that is vibrant, mature, and complex."[12] *Apollo in the Age of Aquarius* as originally conceived did not connect space technology to this mainstream American history.

The third problem with my original approach was that it failed to incorporate the politics of what is often called "the 1960s era." Seeping beyond the neat decadal boundaries of the 1960s proper, this long 1960s, which many scholars agree ran from the late 1950s through the 1970s, represented one of the most politically tumultuous periods of the post–Civil War era.[13] Civil rights marches, anti-war sit-ins, demonstrations by environmentalists and feminists, and even the protest songs of the counterculture became common everyday experiences. They also kept surfacing, intermittently, during my early research for the book. The "green" politics of the 1970s are only the most obvious example. While the burgeoning environmental movement during the late 1960s openly criticized Apollo for causing pollution and distracting politicians from cleaning it up, in the 1970s NASA responded to such political pressure, and the decrease in congressional funding that accompanied it, by "spinning off" some of its space technology for environmental purposes on Earth. A similar political give-and-take occurred between the space agency and the other grassroots movements of the period.

Addressing this concern became even more difficult for me because the politics of the 1960s era were so fluid, fuzzy, and difficult to pin down. At the highest level, the country seemed politically schizophrenic, swinging back and forth from the presidencies of Kennedy and Johnson on the left during the 1960s to those of Nixon and later Ronald Reagan, separated by the brief four-year term of Democrat Jimmy Carter, on the right during the 1970s and early 1980s. The political landscape from the bottom up was equally confusing, crowded with movements that were both distinct yet overlapping. Martin Luther King Jr. and Ralph Abernathy were synonymous with civil rights but also spoke out publicly against the Vietnam War. The environmentalist Stewart Brand promoted ecological stewardship, yet participated in the counterculture aboard the Merry Pranksters' psychedelic school bus, Further, which toured the country administering acid tests. Even feminists, including Gloria Steinem and Betty Friedan, took a stand on issues beyond equality for women, such as civil rights. And, of course, grassroots activists from all of these movements headed to upstate New York during the summer of 1969 to listen to Joan Baez, Janis Joplin, Jefferson Airplane, and dozens of other counterculture musicians. The political terrain of the 1960s era, in other words, much like the rain-soaked fields of Woodstock, was wonderfully muddied.

So, too, were the politics of space exploration. While the competition between the United States and the USSR to land humans on the moon was straightforward enough, the means of doing so in each country was less so. In the Soviet Union the space program remained a classified military effort; failures were rarely acknowledged publicly and accomplishments were announced only after assured success. In response to such secrecy, in July of 1958 President Eisenhower created NASA as a distinctly civilian agency, albeit with clandestine ties to the U.S. military. Yet while more transparent, NASA was far from monolithic. During the so-called Apollo era, which, like its long 1960s counterpart, encompassed most of the 1960s through the mid-1970s, numerous factions vied with one another to publicly define both space exploration and NASA's place within it.[14] These included a handful of administrators, from James Webb and Thomas Paine to George

Low and James Fletcher; three dozen or so astronauts, including not only Apollo's Lovell, Swigert, Armstrong, and Aldrin but also Project Mercury's John Glenn and Scott Carpenter as well as Gemini's Frank Borman and Pete Conrad; thousands of engineers and scientists, such as the famed German rocketeer Wernher von Braun and the climate-change expert Jim Hansen; and tens of thousands of private contractors scattered across the nation. Such competing interests help to explain the public apology by NASA administrators to the cast of *Hair* the day after Apollo 13's astronauts walked out in mid-performance.[15] Much like the numerous movements of the 1960s era, the space agency itself was a complex organization with discrete yet coinciding actors.

Such problems with my original approach raised a simple, yet important question: Could I push my narrow history of the space race, with its attention to space technology and earthbound nature, toward more mainstream American history by actually incorporating the turbulent politics of the 1960s era? Scholars have asked similar questions for quite some time. For answers, environmental historians have turned to the concept of "enviro-technologies" to investigate the intersections between nature and technology, while historians of technology have embraced the idea of "technopolitics" to expose the often hidden, and always unequal, power relationships embedded within technological artifacts and technological systems.[16] Political ecologists have instead focused on the interrelationship between political power and environmental change.[17] Yet few writers, and even fewer historians, have moved beyond these hybrid approaches to explore how the trio of nature, technology, and politics together shape the past.[18] *Apollo in the Age of Aquarius*, I decided, must do just this.

So I went back on the road to a very different set of archives. I looked south to Atlanta's Martin Luther King Jr. Center for Nonviolent Social Change in search of material connecting Ralph Abernathy and the Southern Christian Leadership Conference to the space race. I visited New York University's Tamiment Library to seek links between peace activists from the New Left, such as those involved in Students for a Democratic Society, and space technology being used by the military in Vietnam. I scoured speeches by environmentalists including the Sierra Club's David Brower, pamphlets and posters from the first Earth Day in 1970, and the pages of Stewart Brand's *Whole Earth Catalog*, which in 1968 began promoting eco-friendly technologies to "hippie environmental spacemen," for any and all references to Apollo.[19] I headed north, as well, to the Radcliffe Institute's

Schlesinger Library at Harvard for evidence of the troubled relationship between the National Organization for Women and high-level administrators at NASA. I even paid a visit to the Bethel Woods Collection, better known as the Woodstock Museum archive, in upstate New York for countercultural references to the moon landing during the summer of 1969.

The result is a very different book. Rather than merely explaining the unexpected ecology of NASA's space capsules, *Apollo in the Age of Aquarius* now begins with civil rights leaders in the 1960s protesting against such technology and with space agency engineers' subsequent attempts to retool it in order to help African Americans in inner cities. The book then turns to the Vietnam War, when NASA's covert development of aerospace hardware to help the U.S. military wage "ecocide" across Southeast Asia sparked opposition by peace activists and college students from the New Left. *Apollo in the Age of Aquarius* next demonstrates how early opposition to space exploration by both environmental activists and environmental scientists compelled NASA in the 1970s both to preserve land locally at Cape Canaveral and to develop satellites and computer models that made possible scientific research on the global environment. Additionally, instead of simply examining environmental simulators on Earth and astronaut suits in space, the book now illustrates how NASA administrators used these same technologies to deny female bodies a place in space and how second-wave feminists in turn organized successfully against such body politics. Finally, *Apollo in the Age of Aquarius* ends with *Hair*, or more precisely with the cultural battle between hippies, who opposed the space race through art, music, and lifestyle choices, and Nixon's silent majority, who supported Apollo as a counter to the counterculture. The book, reincarnated, is thus less an environmental and technological history of the space race and more a political history of the 1960s era, albeit one that focuses on the unforeseen relationship between space technology and the natural environment both in space and back on Earth.

Which brings us back to the "*Hair* affair." It, too, brought together the space race and the grassroots movements of the 1960s era. Not only did crew-cut astronauts attend the "hippie" musical in June of 1970, but the play itself frequently commingled NASA's mission to the moon and each of the social and political struggles taking place back on Earth. In "Walking in Space," for instance, one of the most important scenes in the second act, the play's "tribal members" smoke marijuana laced with LSD before taking off "on a rocket to the fourth dimension." During their trip "from mainline to moonville," an African American character hallucinates the murder of whites under a "Black Power" banner. A draft dodger imagines skydiving into the jungles of Vietnam, while a feminist proclaims to the audience, "I certainly don't want to be a housewife with kids."[20] Later the play's main protagonist, claiming to be an extraterrestrial, sings longingly of his pollution-free home planet "where the air is pure, the river waters crystal bright." While Swigert and Lovell missed all of these Act II references linking the space race to the era's grassroots movements, President Lyndon Johnson did not. "Space was the platform from which the social revolution of the 1960s was launched," he argued less than a year after the "Hair affair." "If we could send a man to the Moon," he added, we should be able to solve other problems such as racism and poverty in inner cities, war in Southeast Asia, pollution, gender discrimination, and an increasingly disillusioned youth culture.[21] This book follows Johnson's lead by illustrating how the Apollo program launched not only rockets but also the political and social struggles of the Age of Aquarius.

1

Spaceship Earth

Civil Rights and NASA's War on Poverty

In the early afternoon of July 15, 1969, the day before the Apollo 11 launch, NASA's top administrator, Thomas O. Paine, and the Southern Christian Leadership Conference's (SCLC) president, the Reverend Ralph Abernathy, faced one another across a wide field just outside the western entrance to the Kennedy Space Center in Cape Canaveral, Florida. It was hot, humid, and misting. Paine had arrived with several fellow space agency administrators, all dressed in gray suits, white button-down shirts, and thin black ties only to find Abernathy, who had organized the event, already waiting with twenty-five poor African American families from the Deep South; four scruffy mules pulling two rickety wagons; and, much to Paine's dismay, dozens of newspaper reporters and several television news crews. When Abernathy's group, all of whom were members of the SCLC's Poor People's Campaign, began slowly marching hand-in-hand singing "We Shall Overcome," Paine and his entourage walked forward to meet them in the middle of the open field. As one reporter from the *Washington Post* explained of the gathering, "in the distance thunder rolled across the sky and, had we looked over our shoulders, Apollo 11 could have been seen on its launch pad."[1]

With four mules as a backdrop and the Apollo 11 spaceship off in the distance, Abernathy raised a microphone and explained to the crowd of more than one hundred that the Poor People's Campaign did not travel from across the South all the way to this field at the Cape to protest the space program. "On the eve of man's noblest venture, I am profoundly moved by the nation's achievements in space and the heroism of the three men embarking for the moon," he explained. Rather, they were demonstrating against the country's "distorted sense of national priorities."[2] One-fifth of the nation was without adequate food, clothing, medical care, and shelter, Abernathy argued, yet little had been done to solve these social problems. "My people," he continued with a nod toward the African American men, women, and children gathered around him, "are Americans too, with no homeland but America and we must improve their lot." The reverend then turned directly to face Paine. "I want NASA scientists and engineers and technicians," he stated emphatically, "to find ways to use their skills to tackle the problems we face in society."[3]

As Paine approached the microphone, reporters later noted that the NASA administrator seemed sincerely sympathetic. "If we could solve the problems of poverty in the United States by *not* pushing the button to launch men to the moon tomorrow," he explained, referencing Apollo 11 over his shoulder in the distance, "then we would *not* push that button." The sophisticated technological advances resulting in the Apollo spacecraft, he added, are "child's play compared to the tremendously difficult human problems with which you and your people are concerned."[4] Paine also suggested to the Poor People's Campaign that a successful launch the following day, and a triumphant moon landing over the weekend, might encourage Americans to help fight poverty back on Earth.[5] He then promised them VIP passes to the launch, which in the past few days had drawn one million tourists to Cape Canaveral. "I hope you will hitch your mule wagons to our rockets," Paine concluded, and use "the space program as a spur to the nation to tackle problems boldly in other areas."[6]

Paine's and Abernathy's conflicting views of the Apollo 11 launch were quite understandable, considering the divergent life paths that led them to that damp field during the summer of 1969. Paine had grown up in Berkeley, California, the son of a Navy commodore, and he studied engineering at the Ivy League's Brown University before receiving a Ph.D. in physical metallurgy in 1949 from Stanford. During the following decade he served as a research associate and manager at some of the most powerful technology firms in America, including General Electric, before being tapped in 1968 by President Lyndon Johnson to head NASA.[7] Like many NASA administrators, Paine hailed from the world of high tech. Abernathy, on the other hand, felt more at home on the ground; he was born on his family's farm in Linden, Alabama; wrote his master's thesis in sociology about a local grassroots social movement; and in 1948 decided to become a Baptist minister.

It was while serving as the senior pastor of the First Baptist Church in Montgomery that Abernathy met Martin Luther King Jr., with whom he worked on a host of civil rights efforts, including the Montgomery bus boycott of 1955, the creation of the SCLC in 1957, and the Poor People's Campaign in May 1968, which was an attempt to promote economic justice through a march on, and month-long occupation of, Washington, DC.[8] While Abernathy had run the campaign, which ultimately took place one month after King's assassination, out of a hastily built shantytown housing thousands of demonstrators on the capital's National Mall, Paine was working tirelessly on Apollo, most probably in air-conditioned comfort just a few blocks away in NASA's new gleaming headquarters.[9]

The meeting of the space agency's chief administrator and the successor to Martin Luther King during the Apollo 11 launch thus illustrates the shared history of the space race and race relations during the late 1960s. Scholars since that time have analyzed numerous aspects of America's civil rights struggles, focusing not only on the efforts of leaders and grassroots participants but also on divisions within the movement as well as its international reach during the height of the Cold War.[10] An equally rich literature exists detailing the often-successful borrowing of civil rights strategies honed during the 1950s and 1960s—from marches to boycotts to sit-ins— by subsequent social change activists, including those advocating for peace in Vietnam, a clean environment, women's equality, and countercultural lifestyles.[11] Less understood, however, is the cross-fertilization between the black freedom struggles of the 1960s era and what was arguably the most popular national concern of the period—the race against the Soviet Union to land a man on the moon.[12]

Abernathy's Poor People's protest at the Kennedy Space Center also highlights the key roles played by technology and nature in this somewhat surprising relationship between racial politics and the space race.[13] On that humid summer day in July of 1969 both Paine and Abernathy consciously and repeatedly referenced in their public remarks the thirty-six-story Apollo 11 rocket ship standing visibly on the horizon. While Paine and other NASA officials present that day, along with the majority of citizens across the United States watching the Apollo countdown that week, viewed the spaceship as a technological wonder, Abernathy and the poor families accompanying him to the Cape saw it instead as a political blunder that left unresolved more pressing problems back on Earth.[14] In an effort to publicize such criticisms, SCLC organizers brought to the Apollo 11 protest the

same four mules the movement had used the year before during the campaign's occupation of Washington, DC.[15] Abernathy's four mules, immediately recognizable as icons of rural, agricultural backwardness, were the
perfect foil to Paine's bright, white rocket.[16]

What follows traces three shifts in this unexpected link between rockets
and mules, NASA administrators and civil rights leaders, racial inequality
and the space race. During the early 1960s, space agency officials paid little
attention to criticism of Apollo levied by the civil rights movement and focused instead on developing technology that could safely transport men to
the moon. By the middle of the decade, however, such disregard became
impossible as African Americans held sit-ins, marches, boycotts, and demonstrations against such space technology, along with its high price tag, for
distracting the nation from poverty and a variety of environmental problems affecting inner cities. When such opposition weakened overall popular
support for space exploration in the late 1960s, NASA responded by "spinning off" several of its technologies for applications in programs related to
President Lyndon Johnson's Great Society and its War on Poverty. While
space agency officials proved enormously successful at promoting such efforts to the public, this retooled hardware ultimately failed to significantly
address the urban problems plaguing African American neighborhoods.
Civil rights activists were successful, though, in using their protests against
Apollo and the space race to broadcast the movement's concerns regarding
racial inequality to a more global audience.

Housing NASA's Astronauts

With NASA's 360-foot-tall rocket ship looming over his shoulder, Abernathy had good reason to complain about America's "distorted sense of national priorities." He understood that the space agency had dedicated
nearly the entire 1960s to building the vehicle, which was actually composed of two quite different technologies. The three-stage Saturn V booster
rocket, which with its eleven engines weighed in at nearly six million
pounds, transported astronauts through space at a maximum speed of
17,400 miles per hour. The Apollo spacecraft, which the *New York Times*
called "the home-away-from-home of the three American moon men," sat
atop the Saturn V and itself comprised three sections: a command module
that served as the control center during the mission, a lunar module that
transported two of the three Apollo astronauts to and from lunar orbit to

the moon's surface, and finally a service module that housed the space-craft's technical system.[17] All told, the space vehicle was taller than the Statue of Liberty, with which it was often compared.[18] As Abernathy and his Poor People's Campaign also knew, the space agency had spent more than $25 billion of taxpayers' money designing, testing, and building this space vehicle, making it the single most expensive piece of technology constructed up to that time.

Much of this federal money went toward the research and development of technologies that protected astronauts from the life-threatening hazards of the space environment. Such practice was not new. Back in 1953, in an article titled "Man's Survival in Space," *Colliers* magazine explained that "in a region so unlike the environment we've always known, there is only one way to protect life: bring our environment with us."[19] Technicians at NASA agreed, and by the late 1960s they had developed a slew of technologies that maintained fresh air and water in space and that protected astronauts from the harmful effects of radiation, meteorites, and burning and freezing temperatures. The Apollo spaceship "is a major engineering feat," reported the *Huntsville Times* in December of 1968. "Besides the intricate rocketry, there is the capsule which provides a miniature world for the astronauts to live in."[20] *The Miami News* concurred, reporting soon after the successful Apollo 11 moon landing that "man has walked on another celestial body than his own for the first time, carrying his earth environment with him."[21]

To transport a small portion of the Earth's environment into outer space, the space agency relied not only on tens of thousands of its own engineers and those of its private contractors throughout the country, but, perhaps more surprisingly, also on a small army of ecologists. While scientists had been involved in constructing artificial living environments for submarines and underground atomic shelters immediately after World War II, ecologists' interest in studying space environments began in earnest in late August of 1962 at the annual meeting of the American Institute of Biological Sciences. The meeting, held that year at Oregon State University, included a symposium titled "Space Biology: Ecological Aspects" that was sponsored by, among other organizations, the Ecological Society of America, the Botanical Society of America, and the American Society of Zoologists. After introductory remarks by University of Texas biologist Jack Myers, who drew stark contrasts between life support systems designed by ecologists versus engineers, nearly a dozen biological scientists presented their research on the use of closed ecological systems to process astronaut metabolisms involving oxygen exchange, water regeneration, food production, and waste disposal.[22] Myers's well-known colleague, the ecologist Howard Odum, presented his theory that conventional engineering approaches that relied solely on "pipes, electric circuits, fuel, and maintenance parts sent from Earth" would be less energy efficient, and ultimately more unstable, than a life support system based on a "complex multispecies climax ecosystem" that naturally recycled gases, water, and waste while simultaneously providing food for astronauts.[23] Symposium participants made similar arguments, supporting them with elaborate flow diagrams depicting algae, bacteria, plankton, fish, and mammals as possible ecological components of what they called "bioregenerative" life support systems for space capsules.[24]

In 1963 NASA officially began supporting this ecological research by cosponsoring, along with the Office of Naval Research, three successive annual conferences at Princeton University on human ecology in spaceflight. It was at the first Princeton meeting that Howard Odum's equally well-known brother, the University of Georgia ecologist Eugene Odum, organized a panel discussion regarding "regenerative systems" for space capsules, during which he explained the theories presented by his brother the previous year.[25] "Why not take advantage of the fact that nature has worked on systems for years?" asked Eugene Odum of his fellow conference participants, who included an impressive mix of nearly two dozen biologists and physiologists from NASA, the aerospace industry, and academia. "I think it is very obvious," Odum continued, "that a space system will have to combine natural components."[26] Along with sponsoring similar conferences, which were intended to share knowledge about life support systems for outer space, NASA also began directly underwriting such research. By the end of 1966, for instance, the space agency had spent $1 million to fund approximately one hundred scientists and engineers investigating closed ecological systems for space capsules.[27]

During the summer of 1964 Eugene Odum became one of the earliest recipients of this type of funding from NASA. In order to test his brother's hypotheses, he and the ecologist Robert Beyer, a former student of Howard Odum who had recently taken a faculty position in Eugene's ecology department at the University of Georgia, applied for and received a multiyear grant from NASA to run laboratory experiments that compared two different ecological life support systems for possible use in space capsules. To do this, Odum and Beyer first built their own "space capsules" at the University of Georgia; they designed and constructed two walk-in "growth chambers" that automatically controlled both temperature and humidity. Inside one the ecologists tracked the growth of a two-species ecosystem based on algae and an imagined astronaut, while in the other they monitored an ecosystem comprising multiple species, including protozoa, fungi, crustaceans, several types of algae, and ten different kinds of bacteria. Rows of flasks in each chamber housed samples of these different microcosms, and fluorescent lights provided energy. Over the next several months Odum and Beyer measured the growth rate of each microcosm's biomass, the exchange rates of each regarding the conversion of carbon dioxide to oxygen, and both system's chlorophyll levels. They even dried the organic matter from the two-species and multispecies ecosystems and fed it to laboratory mice in an effort to determine the potential nutritional values for future astronauts. Finally, in order to assess each system's stability within a simulated space environment, Odum and Beyers irradiated samples of biomass from each chamber and measured disruptions in growth. "One of the primary objectives of the research under contract," explained the University of Georgia scientists of their NASA work, "is to test the hypothesis that true stability in a life support system is obtainable only after the system becomes adjusted to boundary conditions (i.e., outside environment) by the process of ecological succession."[28]

In 1966, midway through conducting their NASA experiments, Odum, Beyers, and Dennis Cooke, one of their graduate students, presented their preliminary findings at the "Bioregenerative Systems Conference" in Washington, DC, which was cosponsored by NASA and the American Institute of Biological Sciences. The three scientists first dismissed a completely engineered life support system for long-duration space exploration on the grounds that it would necessitate the storage of enormous quantities of fuel, food, oxygen, and water. Odum and his colleagues next shared their laboratory data, which indicated that a multispecies closed ecosystem, once it reached a climax state, was both more energy efficient and more stable as a space capsule life support system than its two-species counterpart. As important, this mature ecosystem would provide a more diverse, more easily digestible, and more nutritious food source for astronauts than any algae harvested from the two-species system.[29] "It is the mature ecosystem which we propose as the theoretical basis for the development of life-support systems" in space, the ecologists argued in their talk.

While promoting climax ecosystems for space travel, Odum and his colleagues nevertheless understood that nature alone would not get astronauts to the moon and beyond. Because their ecological life support system would theoretically necessitate approximately two acres of growing room per astronaut, the ecologists welcomed technology into their designs. "Practicality dictates some compromise," they admitted. "Since such a system will necessarily be larger," they added, "we need to determine which processes can be satisfactorily supplemented or replaced by mechanical or chemical devices."[30] According to Odum's NASA research, then, a new form of what he called "ecological engineering," involving both nature and technology, was necessary for a stable and efficient space capsule life support system.[31]

After listening intently to the presentation, conference attendees from academia, NASA, and the aerospace industry praised Odum and Beyer's theoretical approach before attacking it as highly impractical. Two acres per astronaut was far too large an area to transport into space, even in the distant future, they argued, while the cost of testing multiple species for efficiency and stability in orbit would be prohibitive. Most of these criticisms, however, cohered around a single argument: technology alone could solve all of NASA's life support problems. "We are dealing with a rather special situation here of a man in a space capsule," explained a plant physiologist from the University of Maryland during the panel's question-and-answer discussion. By "trading off stability of an ecosystem to engineering, we are able to maintain just one species as a matter of fact, for a reasonably long time." Others agreed. "If the spacecraft were large enough, he could take corned beef sandwiches and oxygen," joked a biochemist from the National Institutes of Health. Because it is not, NASA should simply supplement that which astronauts cannot carry. "I wonder if this necessarily involves us in an ecosystem or just a continuous supply of food and some exchange of atmosphere to maintain him," the biochemist asked rhetorically. "Spaceflight," stated one Martin Marietta Corporation scientist in an attempt to end the discussion once and for all, "is not interested in the ecosystem."[32]

Partly due to similar professional conversations among scientists, engineers, and NASA administrators during the mid-1960s, the space agency not only halted funding for Odum and Beyers's research before its third and final year, but engineers working on Apollo also ignored the potential role of natural ecosystems when designing astronaut life support systems and instead embraced a purely technological approach.[33] The result was NASA's Environmental Control Subsystem, built by the Grumman Aircraft Engineering Corporation, which mechanically regulated natural resources, including air and water, while reprocessing waste and monitoring climate for the Apollo space capsule.[34] For instance, rather than rely on natural ecosystems to create breathable air, Grumman stored pure oxygen in tanks during the mission.[35] For potable water, engineers used Apollo fuel-cell power generators that, when producing electricity from a chemical reaction involving the combination of hydrogen and oxygen, made water as a by-product.[36] "It's like carrying your own well with you on a desert journey," proclaimed one space scientist of the fuel cells in 1965.[37] Technology likewise managed onboard waste: a bed of lithium dioxide transformed carbon dioxide into breathable oxygen, charcoal filters absorbed odors, "debris traps" captured particulate matter such as undigested food and dust floating inside the cabin, and the astronauts' bodily waste was either jettisoned overboard, disinfected and collected in specially designed bags, or, in later missions such as Skylab, disposed of through pneumatic toilets that transported excretions to an onboard sealed containment system.[38]

The final major component of NASA's Environmental Control Subsystem, the Heat Transport Section, managed the climate inside the Apollo space capsule. The heating and cooling technology was composed of two closed loops filled with an antifreeze-like solution of 65 percent water and 35 percent ethylene glycol that circulated through two heat exchangers to regulate onboard temperatures. When cooled, the water-glycol solution circulated inside the spacecraft to lower the temperatures of the cabin's atmosphere and the ship's electronic equipment and batteries. After warming during this process, the water-glycol mixture then flowed through tubes running along the exterior of the spacecraft, where it radiated its heat into space and became cool once again. The mixture was then piped back inside the spacecraft, where it once more reduced the temperature of the ship's air and electronics equipment.[39] Such confidence in technology's ability to regulate air, water, waste, and climate in space was not new; as far back as 1959 the *New York Herald Tribune* argued that "man can survive and function efficiently in outer space if he is provided with a hermetically sealed self-contained compartment or space cabin, i.e. a miniature terrestrial ecology."[40] Yet it wasn't until the mid-1960s that NASA's Environmental Control Subsystem became the technological version of this tiny land-based ecosystem.

Although the Odums and their fellow ecologists failed to convince NASA of the practicality of bioregenerative life support systems, they nevertheless successfully promoted the idea that spaceships and ecosystems had much in common by publishing their research in popular journals, including *American Biology Teacher*; in numerous reader-friendly monographs written for nonscientific audiences; and in best-selling ecology textbooks.[41] "The problems of man's survival in an artificial spacecraft," explained Eugene Odum's *Fundamentals of Ecology*, "are the same problems involved in his continued survival on his earth spaceship."[42] James Lovelock, who worked as a scientific consultant for NASA throughout the 1960s, argued similarly through his Gaia hypothesis, which posited that planet Earth was a synergistic, self-regulating, holistic system.[43] Such thinking regarding "Spaceship Earth" quickly seeped from scientific circles to more mainstream media before inundating the general public. "NASA has spent billions to provide the moonbound astronauts with the ecology of the 'good earth'—pure air, pure water, and careful disposal of waste," explained *Newsweek* magazine in July of 1969. "Each spaceship, in effect, is a model and a reminder of what earth should be like."[44] Readers encountering similar statements in *Time*, the *New York Times*, and other mainstream publications embraced the insight.[45] "Ecologists dote on the concept that we are all passengers aboard space ship earth," explained one *Chicago Tribune* letter to the editor, which went on to argue that Apollo space cabins could serve as laboratories for the stewardship of planet Earth. Because of this "no true conflict exists between space and welfare," the *Tribune* reader concluded. "In the larger sense the development of space is our best welfare program."[46]

Space Capsules and the Urban Environment

Back at Cape Canaveral, in the shadow of the astronauts' $24 billion "home away from home," Ralph Abernathy could not have disagreed more.[47] Such sentiment was prominently displayed early in the morning of July 15, 1969, several hours before his Poor People's Campaign protested the Apollo 11 launch, when Reverend Abernathy held a prayer service at St. Paul's in the heavily African American section of West Cocoa, Florida. Crowded into the narrow, red-brick church were hundreds of African American parishioners as well as a throng of newspaper reporters and television crews. After Abernathy led the congregation in prayer, Hosea Williams, another SCLC leader and a co-organizer of the Poor People's Campaign, laid out the goals of the Cape Canaveral demonstration. "We don't want to be obstructive here, we don't want to irritate," he explained to both the congregants and the news media. It was simply a matter of priorities, he added. "We are spending billions of dollars to explore outer space but if Americans spent that same amount of money feeding the poor and hungry, then poverty and hunger would be gone from the face of America today."[48] The space race, Williams stated emphatically to dozens of reporters, was exacerbating the problems of racial discrimination and poverty in the United States.

Since the dawn of the space age, civil rights leaders had been making similar arguments, initially focusing their criticism on discriminatory hiring practices within NASA and among its tens of thousands of private contractors nationwide. Employment data supported such claims. Throughout the 1960s, when the space agency's workforce fluctuated between more than 18,000 in 1961 and nearly 34,000 at its peak in 1965, the proportion of African Americans employed at NASA hovered consistently around a paltry 3 percent, with most minorities relegated to low-wage positions as janitors and secretaries.[49] Rates of employment for minorities were equally grim, or worse, for NASA's private contractors.[50] While space agency administrators continually blamed such low numbers on the difficulty of finding technically qualified African American scientists, engineers, and managers who were willing to move to the Deep South, where many of NASA's centers and contractors were located, civil rights leaders rejected such claims and censured NASA even more for failing to train African American astronauts.[51] Such criticism reached its apex in 1965 when the space agency dropped its first black astronaut candidate, Air Force Captain Edward J. Dwight, from its training program. *Ebony, Jet,* and the *Chicago Defender* broke stories strongly suggesting that racism forced Dwight from the program, while the National Association for the Advancement of Colored People (NAACP) demanded that the Department of Defense investigate the cause of Dwight's dismissal.[52] "A sincere and dedicated young American who was willing to join with others in risking his life for advancing our space program," argued the NAACP in a public letter to the Department of Defense in 1965, "has been the victim of a shameful combination of prejudice and unfair tactics."[53]

Civil rights leaders also worried that the space race was distracting the nation from efforts for racial equality outside the space agency. Such concerns first surfaced in October 1957 when the launch of Sputnik nudged from national headlines attempts by civil rights activists to integrate Central High School in Little Rock, Arkansas; in a successful effort to embarrass the United States by bringing Central High back into the news cycle, Soviet press agencies publicized worldwide the exact date and time that the famous satellite would fly directly over the racially divided town.[54] John Lewis, chairman of the Student Nonviolent Coordinating Committee (SNCC), became similarly concerned when news of the first U.S. manned spaceflight in May 1961 quickly upstaged press coverage of violence against Freedom Riders occurring at a segregated Greyhound bus terminal in Rock Hill, South Carolina. Although brief reports of the racially motivated beatings appeared in morning newspapers the following day, Lewis doubted that "many Americans paid attention to that little story" because most headlines were still celebrating the successful May 5 flight of Alan Shepard aboard Project Mercury's Freedom 7. "I had no idea that history was being made up in space while I was being beaten in Rock Hill," concluded Lewis.[55] Even the third civil rights march from Selma to Montgomery, often considered the high-water mark of the early civil rights movement, was partially eclipsed on national television by the space race. "Immediately following the events in Alabama the network switched to Cape Kennedy, Florida," explained the New York Times on March 26, 1965, for live coverage of a Gemini 3 press conference with Virgil Grissom and John Young, "the two astronauts," the Times concluded somewhat sarcastically, "who orbited the earth three times while the marchers covered a few miles."[56]

Civil rights leaders were particularly concerned that NASA and the space race were diverting public attention as well as federal funding from programs aimed at improving the urban environment.[57] Dr. Martin Luther King Jr., for instance, accompanied by fellow SCLC leaders Andrew Young and Walter Fauntroy, expressed such anxieties on December 15, 1966, in testimony before a Senate subcommittee investigating the plight of inner cities. "There is a striking absurdity in committing billions to reach the moon where no people live, and from which none presently can benefit, while the densely populated slums are allocated minuscule appropriations," King explained to the assembled members of Congress. "With the continuation of these strange values in a few years we can be assured that we will set a man on the moon and with an adequate telescope he will be able to see the slums on earth with their intensified congestion, decay, and turbulence."[58] The more militant wing of the civil rights movement was even more scathing. "The white man," argued Malcolm X in his 1965 auto-biography, "is now solving the problems of sending men exploring into outer space." Yet when focusing on social problems, he continued, espe-cially in urban areas, "the white man's working intelligence is hobbled."[59] As the *Christian Science Monitor* concluded later that summer, "spokesmen for varying shades of opinion among America's black citizenry, stretching all the way from the mild moderation of a Roy Wilkins or a Whitney Young to the bitter and revolutionary outlook of an Eldridge Cleaver and a Stokely Carmichael," are drawing "unflattering contrast between what has been done on the moon and what has not been done at home."[60]

Although a wide variety of civil rights activists criticized NASA for redirecting financial resources from the fight to improve the urban envi-ronment, they were primarily concerned with one problem in particular: unhealthy and unaffordable living conditions. "Some of those people watched men walk on the crater-pocked moonscape on television sets in rundown tenements in the ghettos of America," explained the National Urban League's executive director, Whitney Young, in the *Washington Daily News* one week after the Apollo 11 launch. "They watched, while plaster peeled from the ceilings and rats scratched in the walls."[61] Congres-sional budget allocations during the late 1960s only fueled such critiques. In 1966, for instance, Congress made cuts to the federal budget for housing and urban development that were seventeen times larger than cuts for NASA.[62] More militant civil rights activists were outraged. "America spends over 300 million dollars a year in a race for outer space, but cannot afford to spend 300 dollars a month to insure a decent living space for her Black communi-ties," wrote the activist Bernice Jones in *Black Panther* magazine just two weeks after the Apollo 11 moon landing. Because of such misplaced alloca-tions, Jones argued, "the *environments* in which Black people have been forced to live have cultivated a new nature, that of an angry people." "Hell will break loose," she predicted, if Washington kept funding space explora-tion at the expense of cleaning up run-down, polluted, and unhealthy urban living conditions. "WE WANT DECENT HOUSING," she warned at the end of her editorial, "OR ELSE!"[63]

Hell had already broken loose across urban America during the mid- to late 1960s, at the exact moment when NASA was intensifying its efforts to beat the Russians to the moon. During this period, as the civil rights movement shifted course geographically from the rural South to the urban

North, and in the process refocused its priorities from fighting racial discrimination in public spaces to highlighting impoverished and unhealthy environments in the nation's ghettos, activists' tactics also veered from the nonviolence espoused by King toward the more militant response promoted by black power advocates such as Malcolm X and Bernice Jones.[64] Partly as a result, between 1964 and 1968 more than 300 riots convulsed African American neighborhoods in more than 250 American cities.[65] Poor living conditions were one of the most important causes of this unrest.[66] "Inadequate housing" and other forms of economic deprivation, argued Martin Luther King Jr., soon after the August 1965 riots in Los Angeles's Watts neighborhood, "give birth to tragic expressions of violence." As a result, King concluded, the Watts riots "are *environmental* and not racial."[67] African American residents of Detroit and Newark agreed. When the *Detroit Free Press* polled black residents after that city's 1967 riot they listed "poor housing," second only to police brutality, as the major cause for the upheaval, while a New Jersey governor's survey found that African American residents in Newark cited "bad housing conditions" as the top cause of the riots in their neighborhoods that same summer.[68]

While residents of Los Angeles, Detroit, Newark, and hundreds of other cities across America took to the streets to protest polluted and unhealthy living conditions in America's cities, and perhaps NASA's role in exacerbating them, less militant civil rights leaders such as Ralph Abernathy began consciously employing tried-and-true civil rights strategies used during the 1950s and early 1960s to publicly critique the space race in the late 1960s. For instance, even before Abernathy's Poor People's Campaign had packed up their mules and left Cape Canaveral, other civil rights leaders had organized a sit-in not at a Woolworth's counter in Greensboro, North Carolina, but rather underneath a full-size mock-up of the Apollo lunar landing module in Houston, Texas. Orchestrated by the National Welfare Rights Organization, a civil rights group created in 1966 to advocate on behalf of welfare recipients, the sit-in took place just outside NASA's Mission Control, at what was then known as the Manned Spacecraft Center. The protest involved seventy-five African American protestors waving signs demanding a halt to future moon missions until the federal government began cleaning up cities and providing for affordable and healthy urban housing.[69] The following year more than a dozen civil rights protesters returned to Houston to demonstrate once again under the same model of the lunar lander, where President Richard Nixon was presenting the Medal of Freedom to NASA's flight control team for rescuing the Apollo 13 astronauts. As sheriff's officers escorted the protesters away from the ceremony site, newspaper reporters covering the event noted that the demonstrators raised their fists in a black power salute.[70]

Along with sitting in, civil rights activists also took to their feet and organized marches to highlight the discrepancy between federal funds spent on America's space effort versus those used to better the urban environment. In a protest intended to mirror the Selma-to-Montgomery demonstrations of 1965, civil rights activists from the SCLC and the Daytona Beach Citizens Coordinating Committee organized a three-day, seventy-five-mile "March against Moon Rocks" from Daytona Beach, Florida, to Cape Canaveral to coincide with the January 31, 1971, Apollo 14 launch. Among the 200 protesters were thirty black maids from Daytona who earned $35 a week cleaning $50-a-day hotel rooms. Marchers held rallies along the route, stayed overnight in the homes of local black citizens, and ended the march with a prayer service in the town of Titusville near the launch site. "Our country is spending $30 billion," explained one march organizer, "to get some moon rocks for Vice President Spiro Agnew to hand out to heads of state." Another participant couched this practice of gifting moon rocks to foreign countries in order to curry diplomatic favor, which President Nixon initiated, in less diplomatic terms. "America is sending lazy white boys to the moon because all they're doing is looking for moon rocks," he complained. "If there were work to be done, they'd send a nigger." At the end of their march, the protestors demanded that federal dollars instead be redirected from exploring outer space to helping "the plight of the working poor" in cities back on Earth.[71]

Following a long tradition, civil rights leaders also organized boycotts to protest against the space race and to raise awareness of urban poverty and environmental degradation in minority neighborhoods. Instead of boycotting buses in Montgomery, however, former SNCC chairman and future Washington, DC, mayor Marion Barry used his position as a co-founder of Pride, Inc., a program that retrained unemployed black men in the nation's capital, to encourage the country's black community to ignore President Nixon's National Day of Participation. The proclaimed federal holiday on July 21, 1969, gave federal employees the day off, and it encouraged the private sector and the nation's schools to do the same, so that as many Americans as possible could celebrate Neil Armstrong's and Buzz

Aldrin's moonwalk. "Why should blacks rejoice when two white Americans land on the moon, when white America's money and technology have not yet even reached the inner city?" asked Barry at a press conference in the heart of Washington's ghetto. "Why should blacks celebrate Monday as a national holiday when President Nixon didn't feel that Dr. Martin Luther King's assassination deserved to be observed?" While boycotting the civic holiday, Barry argued, African Americans should volunteer their time with community groups aiding in the environmental cleanup and financial rejuvenation of their own urban neighborhoods.[72]

More informal boycotts took place as well. On the day of the Apollo 11 moon landing, for example, black patrons at the Metropole Bar on the corner of Twenty-Third Street and Michigan Avenue in Chicago tuned the television instead to a local baseball game. In New York, 50,000 mostly African American music lovers flocked to the Harlem Cultural Festival, where the announcement of the lunar touchdown brought forth a chorus of boos from the audience.[73] The following day, when the *New Yorker* dispersed reporters throughout the city to record people's reactions to the Apollo 11 moonwalk, they also found residents in Harlem disinterested if not openly hostile. Inside the Lincoln Bar, at the corner of Lenox Avenue and 135th Street, explained the magazine's "Talk of the Town" feature, "scarcely anyone was looking or listening to the television coverage when Armstrong placed his foot on the moon."[74] "There ain't no brothers in the program," explained an African American bar patron, who added that he was rooting for the "Mets over the Moon." "The whole thing uses money that should be spent right here" in the ghetto, he added. Three days after splashdown the *New York Times* reported similar findings, concluding that "many black Americans found ways in recent days to ignore the Apollo 11 moon shot, an effort, they say, that ignored them."[75]

Finally, in yet another attempt to draw national attention to deplorable living conditions in the nation's cities, civil rights leaders also organized numerous acts of nonviolent civil disobedience during the public celebrations for returning Apollo astronauts. On the West Coast such demonstrations began soon after Armstrong, Aldrin, and Collins had returned safely to Earth. On August 13, 1969, several hundred activists participated in a boisterous rally organized by the SCLC and other civil rights groups at Los Angeles's Century Plaza Hotel, where President Nixon was bestowing the Medal of Freedom on the three Apollo astronauts at a lavish state dinner. The guest list of more than 1,500 included forty-four of the fifty state governors, diplomats from eighty-five nations, several Supreme Court Justices, Hollywood actors and comedians, and Charles Lindbergh. While diners sipped champagne inside the hotel's grand ballroom, the demonstrators outside in the parking lot dipped defiantly in the Century Plaza's fountain and distributed leaflets declaring that although Americans "have moved forward in space we have repeatedly lagged behind in areas of civil rights" such as poverty, hunger, and "intolerable ghetto conditions."[76] When the dinner ended, reported the *New York Times*, and "the tuxedos and sequins rolled past in their long cars," civil rights protesters again "silently raised their fists."[77]

The most elaborate of such nonviolent demonstrations occurred on March 8, 1971, during New York City's ticker-tape parade for Apollo 14. After leaving their hotel, the St. Regis, at noon in a sixteen-car caravan flanked by twenty-four police motorcycles blaring their sirens, astronauts Alan Shepard, Edgar Mitchell, and Stuart Roosa waved to millions gathered along Fifth Avenue from their convertible under a shower of confetti. When the parade reached Broadway and West Third Street, however, the party was abruptly interrupted. There, outside the Broadway Central Hotel, which housed 300 welfare families, twenty African American protestors blocked traffic by pushing baby strollers into the street and waving signs reading "White Astronauts Fly to the Moon While Black Children Die in Welfare Hotels." Although police officers soon cleared the motorcade route, an hour later at City Hall when Mayor John Lindsay presented the city's Gold Medal to the Apollo 14 astronauts, the hotel's residents resumed the demonstration by waving additional placards reading "Welfare Hotels—The New-Style Concentration Camps." Although Lindsay broke from his prepared remarks to suggest that the technology used to transport Apollo 14 to the moon "can also help to bring our cities back," the protestors remained unconvinced. The activists' continual chant of "crumbs for the children and millions for the moon," reported the *New York Times*, "provided a steady obbligato to the prayers and speeches" in honor of the astronauts.[78]

NASA's Ghetto Problem

Whether they involved marches with mules at Cape Canaveral, sit-ins under replicas of lunar landing modules in Houston, boycotts of moonwalk holidays in the nation's capital, or nonviolent demonstrations during lavish state dinners and ticker-tape parades for returning astronauts in Los Angeles and Manhattan, these civil rights protests against NASA began appearing in the mainstream media. This coverage, in turn, popularized the notion that the space race was diverting federal funds and national attention from a decaying city environment and unhealthy urban housing. Mainstream African American magazines and newspapers were among the first to promote this idea through articles with titles such as "Moon Probe Laudable—But Blacks Need Help," "Moonshot Missed Most Blacks," and "'Only Way I Could Get on the Moon Is to Be the First Janitor There.'"[79] "Especially to the nation's black poor, watching on unpaid-for television sets in shacks and slums," explained *Ebony* in September of 1970, "the first moon landing in July of last year was viewed cynically as one small step for 'The Man,' and probably a giant step in the wrong direction for mankind."[80] Four years later the magazine was even more forthright, arguing that the federal government should spend millions of man-hours and billions of dollars on "an earth probe rather than a space probe." Such a mission, *Ebony* concluded, could beam back photographs of minority children who are "poorly clothed and live in homes that are a disgrace to the 20th Century."[81]

The black press also depicted African American opposition to the space race visually, through a wide variety of political cartoons. The week before the Apollo 11 launch, for instance, the *Chicago Defender* published "What about the Space between Races of Man," which portrayed a black hand labeled "Humanity on Earth" reaching out, longingly, toward a stream of exhaust, which took the form of billowy dollar signs, trailing a small Apollo spacecraft orbiting the moon in the distance. African American irritation regarding Apollo is also suggested by the body language of the cartoon's Uncle Sam, who protectively straddles the rising dollar signs while turning his face, in apparent annoyance, to the outstretched hand (see Figure 1.1).[82] Three months later, during the Apollo 12 mission, the same newspaper depicted African American indifference toward the space race by running "An Expensive Rerun," another editorial cartoon that included a black man napping, head tipped back with mouth wide open as if in mid-snore, through television coverage of the second lunar landing.[83] During the late 1960s and early 1970s the *Chicago Defender* alone ran nearly a dozen similar cartoons illustrating both the anger and apathy of African Americans toward the space race.[84]

1.1   Chester Commodore, "What about the Space between Races of Man," *Chicago Defender*, July 12, 1969.

While such media coverage was widespread, its resonance within the African American community paled in comparison to that of the poet Gil Scott-Heron's "Whitey on the Moon." Recorded to music in a Harlem nightclub in 1970 and later that year included on Scott-Heron's debut album, *Small Talk at 125th and Lenox*, the spoken-word performance piece was a scathing indictment of NASA's role in exacerbating environmental injustice in urban America. The poem begins with African drums and Scott-Heron intoning, in a proto-rap cadence, "A rat done bit my sister Nell / With Whitey on the moon." He adds, "Her face and arms begin to swell / And Whitey's on the moon." Scott-Heron then runs through a litany of urban ailments—from a lack of hot water, electric lights, and working toilets in inner-city apartments to drug addiction and high medical bills associated with life in the unhealthy American ghetto—and tethers them to the high cost of placing white men on the lunar surface.[85] "I think I'll send these doctor bills," he concludes of the financial burden associated with living in the inner city, "Airmail special (to Whitey on the moon)."[86] Scott-Heron's poem, which proved immensely popular in the inner-city neighborhoods of New York, Los Angeles, and Detroit, set to music the same criticism leveled against the space race by the black press. It also reverberated in "Transported," an illustration by the Pulitzer Prize–winning cartoonist Herblock, in which "Whitey" sat quite literally on the moon attentively watching coverage of Apollo 11 while the dark clouds of "war," "poverty," and "prejudice" roiled the Earth below (see Figure 1.2).[87]

As Herblock's editorial cartoon indicates, African American critiques of Apollo were soon covered by the mainstream media. Whereas during the early 1960s the *New York Times* had bragged that the high-tech Apollo space capsule was the perfect "home-away-from-home of the three American moon men," by the late 1960s, in editorial after editorial, it openly contrasted Apollo's cozy astronaut accommodations with the dilapidated housing in inner cities. Four days after Armstrong, Aldrin, and Collins returned to Earth, for instance, the *Times*'s John Hamilton wrote an editorial titled "Meanwhile, Back on Earth . . ." to highlight the paltry annual budget of the Department of Housing and Urban Development (HUD).[88] That same month, the *Minneapolis Tribune*, the *Washington Post*, and *Newsweek* also used editorials about the moon landing to call for the replacement of Project Apollo with what one editor called "project low-cost housing."[89] The *Minneapolis Tribune*'s Richard Wilson put it most succinctly in his editorial "Will It Be Mars—or Housing?" After noting that HUD Secretary George Romney responded to Apollo 11 by advocating publicly for the reallocation of NASA funds for urban housing programs, Wilson argued "the issue is coming down toward the simple question of placing men on Mars in the century or building the houses the nation needs and solving the problems of poverty."[90]

1.2   Herblock, "Transported," *Washington Post*, July 18, 1969.

The mainstream American press, like its African American counterpart, supplemented these written critiques with political cartoons that also visually contrasted the Apollo 11 moon landing with poor urban housing conditions. In the *Milwaukee Journal*'s "On to Mars!" an African American family sits on their stoop, surrounded by overflowing garbage cans, a leaking fire hydrant, dangling electric lines, and a torn overhead awning as Vice President Spiro Agnew parades by holding a placard topped by dollar signs that proclaims his support, which he had announced publicly

'There's a cat who's really out of this world!'

1.3   Bill Sanders, "On to Mars!," *Milwaukee Journal*, July 20, 1969.

at the Apollo 11 launch, for additional funding for exploration of the red planet.[91] "There's a cat who's really out of this world!" complains the cartoon's father, who holds a newspaper with the headline "Urban Crisis" splashed across its front page (see Figure 1.3).[92] The Louisville, Kentucky, *Courier-Journal*'s "American Know-How" was no less subtle, replacing Scott-Heron's sister Nell with an obviously malnourished African American boy covered in rags in a run-down inner-city tenement; plaster peels from the walls, a drooping rod hangs torn curtains, and a rat scampers at the boy's feet while he stares, with toy rocket ship in hand, through a cracked window at a bright full moon (see Figure 1.4).[93]

1.4   Hugh Haynie, "American Know-How," *Courier-Journal* (Louisville, KY), July 17, 1969.

As civil rights activists protested against the space race for diverting attention from the degraded inner-city environment, and the national media in turn publicized such opposition through coverage, editorials, and cartoons, many Americans began to question their support for NASA. This was most obvious within the African American community; a Gallup survey conducted in late July 1969, just days after Apollo 11 returned to Earth, found that blacks opposed Agnew's call for a mission to Mars by a three-to-one margin.[94] Yet similar doubts also started seeping throughout the broader American public. Again according to Gallup, which conducted more than thirty national polls between 1965 and 1975 on the American space program, the high-water mark for public support for Apollo occurred in October 1968, when the ten-day journey of Apollo 7 convinced Americans that the United States would beat the Soviets to the moon.[95] From then throughout the 1970s, public concern with domestic issues worked to decrease popular support for NASA. Worry about polluted, poverty-stricken, unhealthy cities was one such issue. "With all the poverty, crime, and urban decay that we have on this planet," argued one respondent from Lansing, Michigan, in the same July 1969 Gallup survey, "I see no reason why we should use all of our resources to get to a planet where life probably does not exist."[96] Paul Bunge, a senior at Aurora Central High School, agreed, noting in a February 1969 *Denver Post* article that "slum dwellers, as they watch Apollo 8 circle the earth, may justifiably ask why more money cannot be spent to provide decent housing."[97] The space agency, both Bunge and Gallup seemed to suggest, had a ghetto problem.

This dip in public support for NASA, due in part to opposition from civil rights activists, forced politicians at both the local and national levels to rethink their own positions on the space race. Charles Grigsby, a newly elected member of Boston's South End Urban Renewal Committee, criticized the space program during the summer of 1969 for drawing funds away from the urban housing crisis, as did New York Congressman Ed Koch the following spring. "I cannot justify approving moneys to find out whether or not there is some microbe on Mars, when in fact I know there are rats in Harlem apartments," argued Koch, who at the time served as a member of the House Committee on Science and Astronautics.[98] Yet the most abrupt political about-face regarding NASA involved Massachusetts Senator Ted Kennedy, whose own brother had launched the space race in 1961 by challenging the nation to land a man on the moon by decade's end. Although a staunch supporter of Apollo during the early to mid-1960s, the younger Kennedy reversed his thinking as the decade wore on and civil rights activists took to the streets, as well as to Cape Canaveral and Houston, to demand a reallocation of space funding to improve the urban environment.

Senator Kennedy went public with this political realignment in a speech on May 19, 1969, at the dedication ceremony for Clark University's new

Robert H. Goddard Library. During the celebration, which named the library in honor of the NASA physicist who launched the nation's first liquid-fueled rocket, Kennedy addressed a crowd of more than 3,000 that included the astronaut Buzz Aldrin and the NASA rocketeer Wernher von Braun. "I believe in the space program," he admitted at the outset of his speech, "but we also have other national goals." These included, according to the senator, "the abolition of poverty" and "a decent house for every American." When Kennedy then called for the diversion of a substantial portion of the space budget to solve more pressing problems on Earth, the crowd erupted in applause. "Our challenge today," he concluded, "is to use the same techniques and the same discipline" of the Apollo mission "to lower the cost of production for home building."[99] According to those present, Aldrin and von Braun, much like the Uncle Sam in the *Chicago Defender*'s cartoon, were visibly annoyed.

Ted Kennedy, Ed Koch, and their colleagues in Washington, DC, not only spoke publicly about their own growing unease regarding NASA funding, they also began acting on it. In early July 1970, less than one year after the success of Apollo 11, liberal senators fought simultaneously to slash $710 million from NASA's $3.2 billion proposed budget for 1971 while increasing by $700 million funding for urban renewal and other city programs.[100] "Why do we have to keep going to the moon," asked Wisconsin Democrat William Proxmire of his colleagues on the Senate floor, when "Man has immediate needs on earth that need taking care of?" Although failing in a close vote to reduce NASA's proposed budget, the seven-hour floor debate over what the *Washington Post* called "national priorities" marked the beginning of a steep reduction in federal funding for the American space program.[101] Between 1970 when this congressional debate took place and 1972, NASA's federal budget decreased from $3.7 billion to $3.3 billion annually, a reduction of more than 20 percent after accounting for inflation.[102] "The cutbacks already made have effected a tremendous reduction in our national space capability," complained NASA Administrator Thomas Paine during the Senate deliberations. "Any cuts below that would have a devastating effect on this program."[103]

NASA's War on Poverty

Back at the Apollo 11 launch, in another VIP viewing section not far from the one occupied by Abernathy and the members of his Poor People's

Campaign, sat former president Lyndon Johnson and his wife, Lady Bird. The couple had arrived on an Air Force jet the day before to attend a luncheon in honor of NASA's second chief administrator, James Webb, who, after leading the space agency through most of the 1960s, had stepped down several months before the Apollo 11 liftoff. On his way into that luncheon at the Cape Kennedy Hilton, while just up the road Abernathy was driving his mules towards his Kennedy Space Center protest, Johnson stopped to sign autographs for cheering children in bathing suits. He then headed inside the hotel, chuckled at a gag newspaper reporting that he and NASA's ex-chief were flying together to the moon, and strode up to a podium to address his thirty-seven luncheon guests. "If we can lead the world to the moon," he began as the audience dined on crab salad and sherbet, "we can lead them to peace and bountiful prosperity at home." By the end of his seven-minute speech, however, Johnson's confidence in such proclamations wavered. Sounding more like Reverend Abernathy, who at that moment stood in a damp field imploring NASA's new administrator to tackle poverty and discrimination back on Earth, Johnson warned his audience that while the space program had made "a valiant and very proud record for our country," the nation nevertheless has "much yet to do" in education, poverty, and other areas affecting America's cities.[104]

Johnson's lunchtime praise for NASA was understandable because he, like Webb, had spent the better part of the last decade building up America's space program. As the Democratic majority leader in the Senate, he not only convened public hearings of the Preparedness Subcommittee immediately after the October 1957 launch of Sputnik to emphasize the dangers of falling further behind the Russians in space, but the following year, as chairman of the Special Committee on Space and Aeronautics, he also oversaw the creation of NASA. Such commitment deepened in 1961, when John F. Kennedy made his new vice president head of the country's space council, and it endured long after Kennedy's assassination as Johnson worked tirelessly to achieve his predecessor's goal of landing a man on the moon by the end of the decade. "There were those men in our government who ten years ago fought to guarantee America's role in space," Johnson had stated proudly a few months before taking the podium at the Cape Kennedy Hilton. "I am proud that I have stood with the space effort from its first days—and I am so glad to see it now flower in this most marvelous achievement."[105]

Yet Johnson's concern regarding the nation's poor, also expressed while his guests dined and sipped wine, was equally comprehensible because he had worked similarly hard to fight poverty and racial discrimination, especially that experienced by African Americans. Although active on such issues throughout his political career, it was on January 8, 1964, in his first State of the Union address, that President Johnson declared an "unconditional war on poverty in America."[106] A flurry of federal legislation followed, resulting in passage of the Economic Opportunity and Civil Rights acts in 1964 as well as the creation of dozens of antipoverty programs, including Medicare and Medicaid, Head Start, Upward Bound, and a domestic version of the Peace Corps. Johnson fought especially hard to build Great Society welfare programs to help urban minorities. The Economic Opportunity Act, for instance, established the Community Action Program to promote economic self-sufficiency in poor neighborhoods, especially those in cities, while HUD, which Johnson created the following year, initiated the Model Cities Program to fund urban renewal and redevelopment. Johnson capped such efforts in 1968 by successfully pushing for passage of the Federal Fair Housing Act, which outlawed discrimination in the housing industry.[107]

Johnson's luncheon speech at the Cape Kennedy Hilton thus expressed many of the same concerns being voiced by Abernathy at the nearby Kennedy Space Center. Johnson, too, worried that the space effort was diverting national resources away from the battle against racial discrimination, urban poverty, and a degraded city environment. Such anxiety had first surfaced for Johnson during the early 1960s, when civil rights leaders began criticizing the space program for its discriminatory hiring practices. Johnson responded in 1962 by arguing publicly that the space age necessitated "recruiting the best talent regardless of race."[108] When such statements failed to alter significantly NASA's hiring practices or fully appease civil rights activists, Johnson turned to federal legislation. Although his Civil Rights Act of 1964, which made it illegal to discriminate on the basis of race, exempted federal agencies, Johnson's executive order the following year prohibited racial discrimination in the workplace by the federal government. Due in part to similar prodding by the president, in 1966 Webb established at NASA both an equal employment policy, which exceeded the federal law's requirements, as well as an Equal Opportunity Officer to implement it.[109]

When such efforts to integrate the space agency's workforce continued to stall during the late 1960s, Johnson turned next to the Congressional Budget Office.[110] In 1967, for instance, during deliberations over a proposed tax bill, the president passively accepted congressional cuts to NASA's budget in a conscious effort to stave off reductions to his War on Poverty initiatives.[111] Between then and the summer of 1969, however, when Johnson toasted Webb and then went off with Lady Bird to watch the Apollo 11 launch, he came to realize that simply moving tax dollars from the space agency back to programs aimed at helping the American city, or, for that matter, promoting through public statements and national policy nondiscriminatory hiring practices at NASA, would not by themselves reconcile the goals of the space race with those of the civil rights movement.

But aerospace technology might. "The space program," explained Webb to President Johnson back in November of 1964, because of its technological innovations, "lies in your first area of building the great society."[112] Johnson agreed, and the following year his administration began encouraging NASA and its contractors to retool their aerospace technologies in order to fight poverty.[113] The cornerstone of such efforts was NASA's Technology Utilization Program, which collected information regarding such innovations and disseminated it to industries, academic institutions, and the American public. While the program reached engineers, scientists, and managers within industry through a flood of specialized publications, including contractors' reports, invention disclosures, and what it called Tech Briefs, in 1970 it began reaching out as well to the general public through a glossy magazine called *Spinoff*, which it distributed to politicians, the news media, public libraries, and individual citizens interested in new developments in aerospace technology. By 1971 NASA's Technology Utilization Program had disseminated 700,000 documents to the public at large, and just three years later it was providing services to more than 4,300 companies.[114]

The Johnson administration was particularly interested in having NASA's spinoff program explore how aerospace technology might combat problems affecting the urban environment. "The techniques that are going to put a man on the Moon," explained Johnson's vice president, Hubert Humphrey, in a 1967 *Aerospace Technology* editorial, "are going to be exactly the techniques that we are going to need to clean up our cities."[115] To foster such techniques, beginning in the mid-1960s NASA sponsored research, including a book-length study by General Electric titled *Application of Aerospace Technologies to Urban Community Problems*; forged partnerships, such as those with the International City Management Association; and hosted a series of conferences that explored the possibilities of using space technology within cities.[116] At NASA's 1963 "Space, Science, and Urban Life" conference, for instance, which was also supported by the Ford Foundation and organized in cooperation with the University of California and the City of Oakland, more than 150 city managers, aerospace executives, and NASA administrators, engineers, and scientists convened to discuss how systems management techniques involving space technology could be applied to urban settings. "It is imperative that we depart from time-worn traditions and concepts," explained an Oakland city administrator, Wayne Thompson, in his opening remarks for the conference, "and adopt space-age techniques to cope with the problems of our space-age cities." Webb, who also spoke at the conference, agreed, noting that "if we can overcome the problems of water and air pollution and sewage disposal for a trio of men in space, we may learn much that is valuable about the same problems for our cities."[117]

During the early 1970s the space agency tried to put Webb's proclamation into practice. "Pollution is the greatest threat to our water supplies," explained a NASA study on the application of space technologies to urban problems, "due both to industrialization and to population growth and the constant dumping of industrial wastes and raw sewage into streams and rivers."[118] In response, administrators at NASA's Langley Research Center in Virginia, which had developed technology to recycle wastewater onboard the Apollo spacecraft, awarded a contract in 1972 to General Electric to design, fabricate, and test a similar water-treatment system for domestic use.[119] Engineers at NASA likewise considered retooling Apollo fuel cells to increase supplies of clean water for urban areas.[120] Along similar lines, in the mid-1970s the space agency directed two Landsat satellites to collect data above the western United States in order to monitor snowpack in mountainous areas to help, as NASA engineers explained, in "estimating future water supplies for major western cities."[121]

The space agency made similar efforts to address urban air pollution. After noting that approximately 90 percent of the country's urban population lived in neighborhoods with poor air quality, and arguing that such an environment dramatically increased city dwellers' incidences of respiratory ailments, such as emphysema, bronchitis, asthma, and lung cancer, NASA administrators once again promoted the conversion of space capsule technology as a possible solution. "NASA improvements and developments in the field of artificial atmospheres can be effectively utilized for creating clean atmospheric environments within buildings," argued one of the space agency's early reports. "Extension of this concept from space ships and space living quarters indicates that it should be possible to provide atmospheres for entire cities."[122] By 1971 engineers at NASA's Lewis Research Center in Cleveland had begun investigating such possibilities by establishing an Environmental Research Office that soon developed techniques for sensing trace elements and compounds for that city's Division of Air Pollution Control.[123] These efforts went national three years later when NASA teamed up with the Environmental Protection Agency (EPA), established in 1970, to create an "air pollution detection program" that reconfigured instruments originally developed to identify contaminants in space capsule interiors to measure instead airborne formaldehyde, a primary indicator of car exhaust and a major component of urban air pollution. Engineers and scientists at the Houston space center complemented such research by converting devices that had been used to monitor spacecraft air turbulence into "laser Doppler technology" that helped the EPA measure the flow rate and particulate size of urban smokestack emissions. By the end of 1974 the environmental protection agencies of Philadelphia and New York City were assessing NASA's Doppler technology for possible use in tracking the air quality in and around their cities.[124]

Finally, NASA administrators initiated several programs aimed at tackling urban waste. "In many U.S. cities, sewage treatment plants are dumping raw sewage into lakes, rivers and streams," explained *Spinoff* magazine, "because the facilities were not designed to handle as much sewer flow."[125] To help correct this problem, in the mid-1970s NASA engineers and scientists began retooling technology that had been developed at the Jet Propulsion Laboratory, primarily to insulate rocket engines, for use instead in its Activated Carbon Wastewater Treatment System, which burned solid sewage to make activated carbon that in turn filtered more sewage. In 1977 the space agency evaluated the carbon wastewater system in a one-million-gallon-per-day sewage-treatment plant near Los Angeles, and later that year it installed another waste-treatment system, also derived from spacecraft waste technology, in a 600-unit apartment complex in Jacksonville, Florida.[126] Two years later, technicians from the Marshall Space Flight

Center in Huntsville, Alabama, applied electronic circuitry and data-collection principles, acquired while developing the Apollo spacecraft, when designing a high-tech sewer monitor that could, in theory, more efficiently track solid waste removal in cities.[127]

Space Age Urban Renewal

Such attempts by NASA to improve the urban environment were not restricted to improving water quality, alleviating air pollution, and enhancing sewage disposal. In part because civil rights leaders had focused much of their condemnation on NASA's role in diverting federal funds and national attention from the urban housing crisis, the space agency's Technology Utilization Program also sought to improve urban living conditions. Thinking about such problems began at NASA as early as 1966, when the space agency teamed up with HUD and invited fifty scientists, urban specialists, and government officials to Woods Hole, Massachusetts, for a three-week summer workshop titled "Science and Urban Development." In a report published the following year, workshop participants argued that although "urban housing is more complex than a rocket and the city is subject to more perturbations than the moon," if NASA and HUD worked together the nation's housing crisis "can be attacked in the same logical way we have gone about exploring the universe."[128]

Beginning in 1968 HUD took several steps to "work together" with the space agency. It established its own technology transfer department, called the Office of Research Technology, to identify useful innovations for the private housing industry, and the following year it tapped Harold Finger, who had worked as a NASA engineer for more than a decade, to manage it. "In his new role at HUD," explained HUD Secretary George Romney of Finger, "he will have the challenging assignment of adapting space age techniques to the production of more efficient housing."[129] This synergy between the two agencies continued two months later when Thomas Paine, who during the early 1960s had himself consulted for HUD on what he called "slum housing" problems, became head of NASA and immediately held meetings with Romney and Finger regarding how the space agency and the housing department could collaborate.[130]

One result of those meetings was the creation during the spring of 1972 of NASA's Urban Systems Project Office at the Manned Spaceflight Center in Houston. The new office, which was a cooperative effort among the space

agency, HUD, and the Atomic Energy Commission, was charged with researching, developing, and testing aerospace technologies that could more
efficiently provide utilities for housing complexes. According to Ted Hays,
the manager of the new urban office, for nearly a decade NASA engineers
had been providing astronauts with a comfortable living environment in
space through the development of independent, integrated life support
utilities, the end result of which was the Apollo space capsule's Environmental Control Subsystem. This space technology, Hays argued, now
had practical applications on Earth. "We put it all together for 33 hours in
the Mercury Project; we put it all together for 14 days in the Gemini and
Apollo Programs," he explained, concluding that the space agency would
now, in the early 1970s, "put it all together" not only for poor residents
in low-income apartment buildings but also for entire urban renewal
neighborhoods.[131]

The space agency's Urban Systems Project Office was just one component
of Operation Breakthrough, a much broader housing initiative begun by
HUD's Office of Research and Technology during the Apollo 11 summer
of 1969. Created in response to the housing crisis and the associated urban
riots of the mid- to late 1960s, the program promoted the research and
development of mass-produced, prefabricated components, including pre-
assembled windows and doors, precast concrete-wall sections, and even
standardized kitchen cabinets and appliances, in an overall effort to lower
housing costs specifically for low-income groups.[132] "Now that we've landed
on the moon," Secretary Romney explained while introducing Operation
Breakthrough to a room full of reporters in Washington, DC, "we should
revise and reverse our priorities. Housing is a key means of dealing with
domestic problems and should have top priority."[133] To accomplish this,
HUD awarded contracts to twenty-two private manufacturers, including
six aerospace firms such as TRW, General Electric, and Grumman, the last
of which built Apollo's Environmental Control Subsystem during the early
1960s, to construct prototype housing models at ten demonstration sites
across the country.[134] While these prototypes were expected to encourage
additional factory-built construction by the private housing industry, Operation Breakthrough itself hoped to build 5,000 low-cost prefabricated
housing units between 1969 and 1974.[135]

Within Operation Breakthrough NASA was responsible for determining
how aerospace technology might be "spun off" to lower the costs of utili-

ties such as electric power, clean water, healthy air, and waste disposal for HUD's housing prototypes. To undertake such research NASA transferred approximately thirty-five of its engineers and technicians with "life support" expertise, gained while developing Apollo's Environmental Control Subsystem, to work on Operation Breakthrough. "A space suit engineer became a water and liquid specialist; an expendables analysis engineer became a solid waste management engineer; a structural engineer became the subsystems office manager," explained one reporter covering the new space agency project.[136] The result was NASA's Modular Integrated Utility System (MIUS), a proposed on-site physical plant that could reduce energy consumption, and thus utility costs for consumers, by more efficiently using natural resources in the provisioning of electric power, residential water purification, air conditioning and space heating, and solid and liquid waste processing.[137] "The NASA aerospace team," proclaimed the space agency's deputy administrator, George Low, in 1972, "has extensive capabilities in systems engineering, environmental control, water supply, waste management, materials structures, habitability, and electrical power generation and distribution which are directly applicable to this group effort."[138] According to Low, who managed the MIUS program, NASA planned to use these capabilities to help solve the urban housing crisis.

The space agency's Urban Systems Project Office began its MIUS research the following year when it awarded the United Aircraft Corporation, which back in the 1960s had developed the Apollo spacesuit and fuel cell, a contract to build a test laboratory at the Johnson Space Center in Houston. Similar to NASA's Environmental Control Subsystem, which employed heat exchangers to regulate temperature within the Apollo space capsule, this new MIUS Integration and Subsystem Test laboratory, or MIST lab for short, relied on an on-site "total energy" power plant that captured thermal energy generated by electricity production and used it for space heating, to generate hot water, and to power absorption chillers for air conditioning.[139] Solid waste at the lab was either burned in an incinerator, which met all EPA standards, or treated by filtration, chlorination, and reverse osmosis, resulting in purified water that, much like the coolant used in Apollo capsules, was then pumped back through the system to cool the lab's electrical circuits.[140] Tests conducted at the MIST lab in 1974 concluded that while MIUS technology applied to a twelve-story apartment complex could reduce energy use by 38 percent, for an urban development

1.5   Aerial view of Operation Breakthrough site in Jersey City, New Jersey, circa mid-1970s.

area of 110,000 people it could not only reduce energy use by a similar amount but might also lower water consumption by 17 percent, wastewater effluent by 27 percent, and the volume of solid waste by an astounding 80 percent.[141] As important, it could alleviate local air pollution by requiring less burning of fossil fuels. "MIUS is a spinoff from the environmental system used in manned spacecraft," explained one reporter of the new NASA technology, "in which over half the waste is recovered and used for heating air and water, air conditioning, and liquid waste treatment."[142]

After a year of testing at the MIST lab, in 1974 NASA's Urban Systems Project Office began overseeing the installation of its MIUS technology in the field at a HUD Operation Breakthrough demonstration site in Jersey City, New Jersey. The six-acre complex, called Summit Plaza, consisted of one commercial building, a local public school with a heated pool, a central power plant, and four residential apartment buildings comprising 485 units rented out to low-income residents at federally subsidized rates (see Figure 1.5).[143] As with most of the other nine Breakthrough test sites located across the country, HUD officials consciously situated Summit Plaza within a minority urban renewal neighborhood, in this case the St. John's Urban Renewal Area in downtown Jersey City, in an effort to aid economic revitalization. "Our objective," explained a HUD spokesperson, Alfred

Perry, of the Jersey City project, is not only a technological breakthrough but also "a new life style" for the surrounding community.[144] To support such efforts, federal officials at HUD worked closely with the local Jersey City Redevelopment Agency in designing the Summit Plaza complex, and they awarded contracts to local private developers experienced in mass-produced and prefabricated housing techniques.[145]

At Summit Plaza the space agency's MIUS technology, much like the Environmental Control Subsystem from which it was derived, regulated the "climate," air, and waste not of Apollo astronauts in space capsules but of low-income residents in their apartments. To regulate temperatures throughout the complex, HUD replicated the MIST lab's total energy system, which in Jersey City comprised five 600-kilowatt diesel engine generators. Thermal heat released during the generation of electricity was recovered by specially designed water jackets surrounding the engines and then pumped through heat exchangers, much like those used in Apollo's Environmental Control Subsystem, for space heating and hot water in Summit Plaza's seven buildings and heated swimming pool. During warmer seasons the captured thermal energy was piped through a primary hot-water loop, much like the coolant loops developed for the Apollo space capsule, and used to drive absorption chillers that produced cold water for the complex's air conditioning system.[146] "We don't have an energy crisis here," exclaimed one Summit Plaza tenant in the late 1970s. "Our system's so effective," added the complex's property manager, "that most of us keep the heat turned off for much of the day."[147] While they were quite conscious of these financial savings, most Summit Plaza residents remained ignorant of the technological transfer that had made its way from Apollo's Environmental Control Subsystem to NASA's MIUS program to their low-income housing project in Jersey City, New Jersey.

Engineers also relied on NASA's MIUS technology in their efforts to reduce urban air pollution in and around Summit Plaza. While HUD was unable to install air filters such as those used on Apollo spaceships, which removed odors and particulate matter from the astronauts' atmosphere, the total energy system collected exhaust from the five diesel generators, redirected it into a common duct, diluted the exhaust with air, and then propelled it higher above Jersey City with a stack dilution fan. In 1977 HUD administrators evaluated such efforts by funding an air-quality assessment, undertaken by the Oak Ridge National Laboratory, of Summit Plaza's total energy plant. According to the laboratory's findings, the housing complex produced concentrations of carbon monoxide, methane, and sulfur dioxide that were indistinguishable from ambient concentrations of these same pollutants from other sources in Jersey City.[148] In other words, while NASA's MIUS technology did not result in atmospheric conditions comparable to the inside of an Apollo space capsule, the highly efficient total energy system caused significantly less air pollution than would have been generated by a traditional power plant.

Finally, HUD engineers turned to MIUS to more efficiently dispose of Summit Plaza's solid waste. Based on waste-management technology developed at NASA's MIST lab, the pneumatic trash-collection system installed at Summit Plaza was the first of its kind in the nation.[149] After residents of the complex deposited garbage into chutes in each apartment building, the pneumatic underground network used air to transport the refuse to a centrally located containment area, much as vacuums removed astronaut waste from space capsules, where it was compacted and transported from the site. "These systems," explained one scientific assessment of the MIST lab's pneumatic waste-management technology, "permit automatic, rapid transport of refuse from several locations near points of generation to a central storage location." Although the NASA study recommended that HUD incorporate incinerators into its Operation Breakthrough waste-management system so that collected garbage could be burned to produce additional electricity, such a facility was never built in Jersey City due to high costs and environmental fears regarding increased air pollution.[150]

Similar to the MIUS experiments performed at NASA's MIST lab, which engineers continually assessed through an elaborate array of electronic monitors, the total energy plant at the Jersey City Breakthrough site also included a data-acquisition system. Developed by the National Bureau of Standards and installed in Summit Plaza in April of 1975, the data-gathering technology included 200 sensors for monitoring approximately 225 power-plant variables at five-minute intervals for a period of three years.[151] According to a 1977 HUD report based on these data, the Jersey City total energy system used 17 percent less fuel annually for a total savings of 160,000 gallons of oil. The report also indicated that if technicians properly adjusted the plant's absorption chillers, and made several minor modifications to the overall system, the Jersey City total energy plant could consume 24 percent less fuel oil annually.[152]

Administrators at NASA undertook a nationwide public relations campaign to promote the space agency's efforts to spin off aerospace technologies that would improve the urban environment. They published the proceedings from their conferences on space technology and urban life, circulated hundreds of news releases describing how Apollo spacecraft technologies had been redesigned to treat urban wastewater, and encouraged coverage, in mainstream publications from the *New York Times* to *Popular Science*, of NASA's attempt to both monitor urban air pollution and reduce it across America's cities.[153] Space agency administrators also participated in a nationwide tour, kicked off by HUD during the summer of Apollo 11, to promote NASA's role in trying to alleviate the urban housing crisis. "Crisscrossing the country with tub-thumping evangelism," explained *Newsweek*, administrators at the housing department "sought to enlist the support of governors, mayors, local builders, labor leaders, Rotary Clubs and major corporations."[154] They also sought to convince African Americans. For instance, HUD Secretary George Romney, with former NASA administrator Harold Finger at his side, not only pitched Operation Breakthrough to the National Housing Producers Association, which was a coalition of black lawyers, architects, builders, and mortgage brokers formed by the NAACP, but he also announced at the NAACP's annual meeting that year that HUD had awarded the civil rights organization nearly $200,000 to encourage minority contractors to participate in Operation Breakthrough construction projects.[155] At many of these promotional stops, which continued into the early 1970s, HUD officials referenced both concurrent Apollo missions as well as NASA's role in converting aerospace technology for use in the Operation Breakthrough program in an effort to gain wider publicity for these housing efforts.[156]

The African American media understandably took the bait, reporting favorably on NASA's space age approach to urban problems. New communities in urban neighborhoods "may be designed in the next few years much like space stations," argued the *Atlanta Daily World* in 1973, "with environmental systems that waste precious little energy and use 38 percent less fuel to operate." The African American newspaper went on to describe in detail NASA's Operation Breakthrough partnership with HUD, the creation of the space agency's Urban Systems Project Office, the construction of the MIST lab at the Johnson Space Center in Houston, and the development, by NASA engineers who had worked on the Apollo spacecraft's

Environmental Control Subsystem, of the MIUS "total energy" system used in the Summit Plaza low-income housing project. "Unlike conventional methods which waste most of the energy," the newspaper explained to its readers, "all of the excess heat would be funneled off for other uses."[157] Similar coverage appeared in the *Chicago Defender* and *Jet* magazine, the latter of which argued that this pioneering program by HUD and NASA "could reduce current costs and allow more black families to own their own new homes."[158]

Such coverage understandably convinced some civil rights leaders to soften their stance on NASA and instead praise the space agency for improving the urban environment. During the summer of 1969, for instance, at the fifty-ninth annual conference of the National Urban League, participants publicly commended Operation Breakthrough's new method of low-cost housing. "The plan of George Romney," explained one Urban League leader during the conference's luncheon, might very well result in "improved environmental quality at a cost that will bring the homes within reach of all families."[159] The space agency, many in the audience understood, had developed technologies during the Apollo program that were now being used in Operation Breakthrough. Thus while civil rights leaders during the mid- to late 1960s had consciously used the space race to emphasize the degraded state of the urban environment, especially poor housing conditions, by the mid-1970s a few, at least, had begun looking to NASA for possible solutions.

Space Age Civil Rights

Back at the Kennedy Space Center in the early afternoon of July 15, 1969, Abernathy and the Poor People's Campaign participants began wrapping up their protest of the Apollo 11 launch. It was hotter and more humid than it had been an hour earlier when they began the demonstration. Before concluding the rally, Abernathy again faced Paine and reiterated his demand that the space agency use its technology to tackle poverty back on Earth. "I am here to demonstrate with poor people in a symbolic way," stated Abernathy while referencing the rocket over his shoulder, "against the tragic and inexcusable gulf that exists between America's technological abilities and our social injustices." He then held up a noose and gestured toward Paine's neck. When the director of America's space program

1.6 SCLC president Ralph Abernathy "hanging" NASA administrator Thomas Paine, *Jet*, July 31, 1969.

smiled uncomfortably and then slowly bent forward, Abernathy carefully slipped the rope, which carried a small tag reading "I Helped Hang Poverty," over Paine's head in an effort to encourage NASA's leader to support the campaign. Photographers covering the protest, such as the pair seen in the background of a *Jet* magazine photo, swarmed to capture the moment (see Figure 1.6).[160]

This dramatic conclusion to the Poor People's Campaign protest at the Kennedy Space Center symbolizes the dysfunctional relationship between the civil rights movement and the space race during the 1960s era. Abernathy's noose, much like the marches, sit-ins, boycotts, and demonstrations by civil rights activists against Apollo during the late 1960s, forced NASA administrators to bend to the movement's demands. In an attempt to appease such criticism, and to resurrect public support as well as congressional funding for space exploration, during the early 1970s NASA engineers and scientists reoriented some of their space technology toward more pressing urban problems, especially those affecting African American neighborhoods, including air and water pollution, poor sewage and garbage disposal, and unhealthy living conditions. The civil rights movement, in other words, had begun bringing the space race back down to Earth.

Yet as both civil rights activists and black urban residents quickly understood, NASA's space technology ultimately failed to improve significantly the inner-city environment. The space agency, for example, funded several companies that *studied,* and convened multiple conferences that *discussed,* the possibility of retooling aerospace technologies to solve a variety of city problems. Administrators at NASA *considered* redesigning Apollo fuel cells as a source of urban drinking water but were never able to develop a cost-effective prototype, while the cities of Philadelphia and New York merely *assessed,* rather than implemented, the air-pollution monitor that engineers had reappropriated from Apollo space capsules. Scientists *researched* sewage-treatment systems derived from NASA's Environmental Control Subsystem, but then tested them in only two sewage-treatment plants, one for an apartment building with only 600 units. Perhaps most telling, although in the mid-1970s NASA had announced, with much fanfare, its desire to apply MIUS technology to poor inner-city communities that each had a population of more than 100,000, sparking the *New York Times* to praise it and Operation Breakthrough as a "symbol of opportunity and hope for the renewal program," the space agency installed its "total energy" plant in only a single low-income housing project in Jersey City, New Jersey.[161] NASA's war on poverty, in other words, although well intentioned, was more performative than permanent, more public relations wishful thinking than a technological dream come true.[162]

This perhaps explains why Paine, soon after he was "hung" by Abernathy's noose, offered several members of the Poor People's Campaign VIP passes to the Apollo 11 liftoff the following day. As the head of NASA undoubtedly hoped, the SCLC activists were transformed after watching the enormous rocket blaze into the sky, at least temporarily. "There's a great deal of joy and pride" in that spectacle, Abernathy told reporters immediately after the launch. "For that particular moment and second," he admitted, "I really forgot that we have so many hungry people in the United States of America."[163] The space agency's efforts to improve the urban environment relied on a similar technological sublime, one that offered up not rocket launches but instead the promise of space age air-pollution monitors, water-treatment plants, and energy-efficient heating and cooling systems for low-income housing projects. The Nixon administration, which occupied the Oval Office during Apollo 11, similarly understood that space technology could mask, whitewash, one could even use the term "scrub," racial problems closer to home.[164] "Warm congratulations on handling a very delicate situation with consummate skill," wrote Nixon's most trusted advisor on domestic issues, Peter Flanigan, to Paine after learning of Abernathy's protest. "The fact that I, as one of your guests, didn't even realize this problem existed," he contended, "indicates how successful you were." After complimenting NASA's chief on his ability to erase the "problem" of the Poor People's Campaign from the spectacle of the Apollo 11 launch, Flanigan concluded his memo with a hint of irony. "Now that you've handled the problem of getting to the moon," he joked with Paine, "maybe you'd like to tackle those of urban affairs."[165]

Yet while NASA failed to alter significantly the everyday lives of urban blacks, the space race nevertheless transformed civil rights by shining an extremely bright public relations spotlight on the movement. In the Kennedy Space Center VIP viewing area, for instance, as they waited for hours for the Apollo 11 countdown, Abernathy and the families joining him consciously worked the captive audience. "This is really holy ground," he stated during an impromptu address to the politicians, foreign dignitaries, movie stars, and, most important, the dozens of reporters who were all waiting for the rocket to roar into the air. "But it will be even more holy once we feed the hungry and care for the sick and provide for those who do not have homes."[166] Such media coverage continued later that day, when many of the same photographers who had captured Abernathy hanging a noose around Paine's neck also framed images of Poor People's Campaign protestors touring the Kennedy Space Center with signs reading "Billion$ for Space Pennies for the Hungry" while NASA's rockets loomed in the background (see Figure 1.7).[167] As *Jet* magazine argued soon after the protest, no other civil rights demonstration organized by Abernathy "has reached more people and served a role as his attendance at the launching of Apollo 11."[168]

Such media exposure of civil rights activism, much like the coverage of Apollo itself, also went global. News stories reporting on Abernathy's Cape Canaveral protest, for example, appeared in newspapers and magazines from Kingston, Jamaica, to Tokyo, Japan, from far up in Canada to down under in Australia, and all across Europe.[169] United Press International even captured a Poor People's Campaign organizer, Hosea Williams, during the demonstration using a black power salute directly underneath two NASA rockets, and it circulated the image worldwide to the newswire's

Elaine Tomlin
With sign "billions for space, pennies for the hungry," Mrs.
Mattie Gray, lame daughter Jackie, 8, demonstrate at NASA site.

7

1.7   Mrs. Mattie Gray and her daughter, Jackie, demonstrating with
the Poor People's Campaign at the Kennedy Space Center, *Jet*, July 31,
1969.

more than 6,000 media subscribers (see Figure 1.8).[170] The same could be
said of civil rights sit-ins, marches, boycotts, and nonviolent demonstrations
against the space agency. Such coverage was especially intense in the Soviet Union, which used Abernathy's demonstration, and others like it
during the early 1970s, to dampen global enthusiasm for America's space
achievements. "What can the spaceship, this miracle of ultramodern technology, and a cart hitched to a pair of mules have in common?" asked one

1.8 SCLC leader Hosea Williams in the Kennedy Space Center Rocket Garden during the Poor People's Campaign demonstration at the Kennedy Space Center, *Boston Globe*, July 16, 1969.

reporter in *Izvestiya*, Moscow's daily newspaper, while Apollo 11 was still orbiting the moon. The reporter then quoted Paine's suggestion that Abernathy hitch his mules to NASA's rocket and concluded by noting that the phrase "only brought a sarcastic smile to the lips of the poor people's leader."[171]

2

Shooting (from) the Moon

NASA, Nature, and the New Left during the Vietnam War Era

Just before midnight on July 23, 1980, in the high desert of Kazakhstan, the Soviet Union launched Soyuz 37 from the Baikonur Cosmodrome. Lifting off from "Russia's Cape Canaveral" were Colonel Victor Gorbatko, a seasoned Soviet cosmonaut, and Colonel Pham Tuan, who, on this mission, became the first Asian to travel into outer space. Six days later, after docking in Earth's orbit with the Soviets' Salyut 6 space station, cosmonaut Tuan proudly addressed his fellow countrymen and -women on Vietnamese national television, and reached much of the rest of Southeast Asia through a live, translated satellite feed. "The Vietnamese revolution has always enjoyed the great assistance of the Soviet Union," he explained while floating 200 miles straight up. Tuan then became quite specific regarding the type of aid bestowed on Vietnam by the Russians. "I would like to convey my heartfelt thanks," he stated, to the Soviet "scientists [and] engineers . . . who have taken part in preparation for our flight." Asia's first cosmonaut then ended his televised broadcast from space by holding up a scale model of the Salyut 6–Soyuz 37 orbital space complex for all of Vietnam to see.[1]

The deployment of space science and technology as Cold War propaganda did not take off with the Soyuz 37 mission during the summer of 1980. Beginning with the launch of Sputnik in 1957 up through the Apollo 11 landing more than a decade later, both the United States and the Soviet Union continually publicized space feats as a means of promoting their competing political systems.[2] This practice of linking a country's scientific and technological prowess to its national identity and international prestige was also far from new; it harked back at least to the seventeenth century, was reinforced in the European context by the Enlightenment, and became even more widespread during the Industrial Revolution. Yet it was during the Cold War that this relationship between a nation's scientific and technological capabilities on the one hand, and its political ideology on the other, became most interconnected.[3] By holding up a scale model of the Salyut–Soyuz space complex on Vietnamese national television, Pham Tuan was in effect broadcasting the benefits of Soviet-style communism to the people of Southeast Asia.

In his televised broadcast from space Colonel Tuan also praised his own country's natural environment, which he passed over more than 142 times during his nearly eight days in orbit. "I was deeply moved each time I flew over Vietnam and saw our beautiful homeland," he beamed from space after watching the lush green landscape race by below.[4] Vietnamese on the ground also experienced this historic spaceflight through the nature all around them. "Everyone in our country—on the rivers, in the mountains, on the seas, in the forests, in the rice fields," explained Radio Hanoi in a national broadcast on the day of the launch, "is happy."[5] This use of nature to promote national purpose also had deeper historical roots.[6] Yet while such joy was obviously patriotic, Tuan and his fellow Vietnamese were also excited for more pragmatic reasons. Experiments conducted during the mission, announced the Vietnamese government to the public just days before liftoff, "include some very important and interesting ones to study natural resources of the country from space."[7] The Vietnamese people thus understood, perhaps more than their Russian counterparts, that although superpower technologies such as spaceships were difficult if not impossible to develop from the ground up *within* the developing world, they might nevertheless be put to good use on the ground *across* it.[8]

Perhaps most telling, however, was Tuan's use during his broadcast of both space technology and earthbound nature to criticize American aggression during the Vietnam War. Not only did he carry with him into orbit a copy of his country's Declaration of Independence as well as the will of Ho Chi Minh, who led Vietnam's independence movement, but Tuan also directly referenced the war, along with the Sino-Vietnamese border skirmish of 1979, in his televised comments. "The Vietnamese people have not been able to enjoy genuine peace due to the war threats made by U.S. imperialists and the Chinese expansionists against the peoples of Vietnam, Laos, and Kampuchea," he explained. "I wish," he continued, "that our children may live under a clear sky never again obscured by the clouds of war."[9] While such sentiments also had much deeper roots, Tuan's criticism of American and Chinese intervention in Southeast Asia placed such linkages involving technology, nature, and imperialism squarely, and quite publicly, into its Cold War context.[10]

Administrators at NASA, who no doubt scoured transcripts of Tuan's broadcast, must have cringed since they had been trying for the past two decades to distance their agency, along with its technology, from the war in Southeast Asia. Perhaps because of such efforts, writers rarely mention NASA or the space race when chronicling American involvement in the Vietnam War.[11] Nor do scholars of the New Left. They distinguish this grassroots movement from the Marxist-oriented Old Left of the 1930s, define it loosely as an association of mostly white college student groups, and trace its activism from an immediate fight for free speech on university campuses in the early 1960s to protests and demonstrations in the 1970s aimed at a broader set of concerns, the most important of which was opposition to the Vietnam War.[12] Nowhere in this history does NASA or the space race appear, much to the relief of space agency officials. Thus, while the relationship between the space race and the fight for racial equality during the 1960s era might have been unexpected, the connections discussed from orbit by Pham Tuan regarding space exploration and the military conflict that ravaged his homeland remain, for the most part, unknown.

Uncovering this history must begin in the mid-1960s, when NASA officials ignored demonstrations initiated by college students from the New Left who had become increasingly alarmed by the space agency's research on technologies for the Vietnam War. During this period, landing men on the moon was just too popular for administrators to take seriously a few thousand coeds protesting against Apollo. Yet as such anti-war activism migrated beyond university campuses during the early 1970s, just as civil rights concerns regarding the space race spread outside the African American community, NASA officials were again forced to take notice. At first they tried to conceal from the public the particulars of the agency's involvement in the war effort, and when that failed they halted research on several military technologies. Then NASA began funding technological alternatives that could help the U.S. government wage a more subtle type of imperial warfare, on both the ideological and environmental fronts, both in Vietnam and across the rest of the developing world. While this about-face by NASA was never broadcast in Pham Tuan's orbiting news conference, it was nevertheless evident in the preparations for his groundbreaking spaceflight, which began not in the dry desert of Kazakhstan in 1980, but rather in the tropical jungles of Southeast Asia during the Vietnam War.

Vietnam's Natural Ally

Both the Soviet Union and the Communist Party of Vietnam chose Colonel Pham Tuan for the Soyuz 37 spaceflight because on the evening of December 27, 1972, at the height of the Vietnam War, he became the first Vietnamese fighter pilot to shoot down an American B-52 bomber. As publicized worldwide during the Soyuz 37 mission by both TASS, the official Soviet press agency, and *Quon Doi*, the Vietnamese military newspaper, on that night back in 1972 the deputy wing commander flew his Soviet-made MiG-21 behind a squadron of U.S. Air Force B-52s, maneuvered one of the high-flying bombers into his sights, and fired. The American jet burst into flames and crashed. This bravery, reported Soviet and Vietnamese sources, had been stoked days earlier by a pep talk given in person by North Vietnamese Prime Minister Pham Van Dong, who encouraged Tuan's unit to "completely smash Nixon's strategic blitz."[13] Although the U.S. military at the time denied the incident, for his aerial feat Tuan became a national hero, received Vietnam's highest military and civilian decoration, the Gold Star medal, and leap-frogged, eight years later, to the front of the line of Vietnamese cosmonauts.[14]

Tuan had honed his skills as a fighter pilot during years of special training for flying airplanes at night, a talent he used successfully in December 1972. This education began seven years earlier when he enrolled in the Vietnamese Air Force and accepted an assignment in the Soviet Union for three years of flight school, during which he reportedly received the highest grades in his class and did extremely well in the practical training of day and night flying. After graduating in May 1968 with a commendation proclaiming him a "strong-willed and very promising pilot," Tuan returned to Vietnam, where he joined the Air Force's "Red Star" regiment, which specialized in night combat missions. By 1972 he had become deputy commander of the regiment, and soon thereafter commander.[15] Yet while Colonel Tuan was forced to travel all the way to the Soviet Union to learn how to use the dark of night to his advantage in flight, his experiences nevertheless reflected the more local Vietnamese practice of using the natural environment to one's own advantage during military conflict.

During the First Indochina War, which took place from 1946 through the summer of 1954, Vietnamese communists had turned to the Mekong Delta's water environment for military assistance by navigating along unmapped backcountry creeks, by scheduling the transportation of soldiers and materials in coordination with beneficial tides, and by setting ambushes along strategic bends in channels and canals where French vessels would be most vulnerable.[16] In 1965 when French gunboats in Vietnam gave way to U.S. airplanes and helicopters, Vietnamese communists, now under the banner of the National Liberation Front (NLF), began relying to an even greater extent on local nature for military advantage. Now it was the dense Vietnamese jungle, rather than the country's winding waterways, that were the NLF's primary weapon against the U.S. military's increasingly sophisticated war technology. "The woods were being used to conceal armed bands of Viet Cong," lamented an *Air Force Times* journalist in May of 1966. Trees, grasses, and especially the thick jungle canopy were all employed to camouflage their positions.[17]

In many instances Vietnamese communists hid in plain sight, directly underneath dense jungle foliage. The extremely thick canopy of the Boi Loi forest, for instance, located approximately twenty-six miles northwest of Saigon and extending to within a few miles of the Cambodian border, hid communist guerrillas from aerial view so well that these fighters were able, over the course of a decade, to develop the forty-eight-square-mile area into one of the NLF's most important military operations centers in South Vietnam.[18] When the jungle itself proved inadequate the NLF's military, the People's Liberation Armed Forces (PLAF) burrowed beneath it by constructing thousands of miles of caves, tunnels, trenches, and other underground passageways for the surreptitious transportation of troops and munitions. The most elaborate of these underground networks were the Cu Chi tunnels, which ran from the outskirts of Saigon for approximately seventy-five miles to the Cambodian border.[19] Finally, in an effort to augment the jungle's cloaking capabilities, many guerrillas moving through the Vietnamese forest did so after dusk, using the night much as Pham Tuan did during his flight in December 1972. "Enemy battalions that break into small units to escape discovery in the daylight," wrote the Pulitzer Prize–winning photojournalist Horst Faas from Saigon in 1966, "come together under the concealment of the night and strike in force." The Ho Chi Minh Trail, a nearly 10,000-mile-long network of truck roads and paths for both bicycles and foot traffic, was a case in point; since daylight afforded U.S. bomber pilots better visibility deep within the jungle, nearly all PLAF movement along the Trail connecting North and South Vietnam occurred after sundown.[20] "The night," Faas concluded, "is an ally" of the Vietnamese communists.[21]

America's Visibility Problem

The Vietnamese communists' practice of using their country's jungle as a wartime defense made it increasingly difficult for the U.S. military to "see" clearly in Southeast Asia.[22] Magazine and newspaper articles from the early to mid-1960s, with titles such as "Uncovering Charlie," "The Invisible Foe," and "U.S. Airborne Device Sniffs for Foe under Jungle Canopy," all speak to this growing visibility problem as well as to an increasingly desperate reliance by the American military on other senses to pinpoint PLAF soldiers.[23] "The key to defeating the guerrillas is finding them," explained Lieutenant Colonel Stanley D. Fair in a September 1963 issue of *Army* magazine. Fair went on to lament the U.S. military's inability to do just this—see and locate the enemy. "They move by stealth," he wrote of Vietnamese guerrillas. "If only there were some magic way to clear away the trees and brush," he concluded, "there would be less difficulty in finding the VC and defeating them."[24]

Long before Fair arrived in Southeast Asia, the French had brought their own "magic" to "clear away the trees and brush" in Vietnam. These efforts involved technologies such as land surveys and dredges, the latter of which cut deeper and wider canals throughout the Mekong Delta in order to enhance the movement and thus the visibility of French war vessels patrolling the region.[25] When the U.S. military replaced French forces in the mid-1960s it doubled down on this approach. For example, to prevent the PLAF from using dense foliage as camouflage, the U.S. Army in 1967 unleashed more than 150 giant "jungle-eating" bulldozers across South Vietnam that had, by 1971, scraped bare 750,000 acres of jungle, 9,000 of which lay in the Boi Loi forest.[26] Communists hiding beneath the jungle in the NLF's widespread tunnel network were also targeted by U.S. military technology; in January 1966 the U.S. Air Force initiated Operation Crimp, a carpet-bombing campaign aimed at destroying the Cu Chi tunnel system.[27] Finally, to highlight guerrillas moving through Vietnam's jungle in the dark, the U.S.

military brightened the forest by dropping flares from planes at night and also by spraying herbicides, such as Agent Orange, during the day to allow more natural light to penetrate the foliage. This effort, which began in 1961 with Operation Ranch Hand and ultimately defoliated an area the size of Massachusetts, focused much of its spraying on the PLAF's main transportation route between North and South Vietnam.[28] "We're going to turn the Ho Chi Minh Trail brown," bragged one U.S. military advisor in Saigon.[29] "Jungle trails are being denuded," reiterated an American reporter, "to expose convoys of men and materials" to aerial reconnaissance.[30]

Although the American government dedicated hundreds of millions of dollars' worth of technology to make communists in Vietnam's jungles more visible, the U.S. military admitted publicly and frequently that such efforts proved ineffective. "In spite of the advantages of mobility, manpower, equipment, and firepower," warned one U.S. military advisor, "the Viet Cong are able to move, strike and disperse with a freedom which seems inconsistent with the imbalance of combat power."[31] Concerned scientists visiting Vietnam to assess the ecological impact of U.S. military efforts agreed. "Despite the lavish application of great wealth and superior technology," wrote one biologist in 1971, "the U.S. has made surprisingly little headway over the years against the National Liberation Front and its North Vietnamese allies."[32] The American military, it seemed, needed a new kind of "magic" to combat Vietnam's "trees and brush," and during the mid-to late 1960s no other institution worldwide seemed more magical than the National Aeronautics and Space Administration (NASA).

Seeing Vietnam from Space

The U.S. Air Force first asked NASA for technological help with the war in Vietnam on November 2, 1965. At a meeting convened at NASA Headquarters in Washington, DC, General James Ferguson and other members of the Air Force Systems Command, which oversaw the branch's weaponry needs, briefed NASA officials on "the unique technological problems arising out of operations in Vietnam," after which NASA Administrator James Webb immediately indicated "that every effort would be made to uncover those NASA solutions to problems, devices, or techniques, that might be of assistance to our forces in Southeast Asia."[33] Within a month Webb had created the NASA Limited Warfare Committee to work more closely with the Department of Defense and by the end of the following year he had secured the cooperation of administrators at each NASA center, assigned thirty-five of the space agency's engineers to research projects for the Vietnam War, and proposed seeding the program with a half million dollars through a new financial account mysteriously identified as "Special Support Projects."[34] The goal of such efforts, explained one NASA deputy administrator, was to "assure that we miss no opportunity to assist the military forces in any and all ways that are available to us."[35]

Although President Eisenhower consciously established NASA in 1958 as a civilian agency, its ties to the military were deep and often clandestine.[36] The year after its creation, for instance, on February 28, 1959, NASA launched Discoverer 1, a supposed satellite technology experiment that was, in reality, a cover for CORONA, the nation's first spy satellite. Developed cooperatively by the Department of Defense, the Central Intelligence Agency (CIA), and NASA, CORONA was a "film-return" satellite system that took photographs of Earth from polar orbit, packaged the exposed film into reentry capsules, and then jettisoned the capsules into the upper atmosphere for retrieval by aircraft. Between 1959 and 1972, when the Department of Defense abandoned the "experiment" after a Soviet submarine was detected waiting below a CORONA midair retrieval zone, the top-secret program included the launch of 144 satellites that snapped thousands of high-resolution photographs of military installations in the Soviet Union and other communist countries.[37] In 1960 NASA supplemented CORONA by launching TIROS-1 (Television Infrared Observing Satellite), an Earth observation spacecraft that forecast weather patterns for military maneuvers.[38] By the mid-1960s CORONA and TIROS had not only heated up the Cold War, but also laid the foundation for similar cooperative ventures between the U.S. military and NASA in Vietnam.

The space agency's authorization to undertake such military efforts was "fuzzy at best," as the *Washington Post* reported in 1967. While the 1958 act establishing NASA barred it from activities "primarily associated with the development of weapons systems, military operations or the defense of the United States," it permitted the space agency to make available to the Department of Defense "discoveries that have military value or significance."[39] "We are not developing anything that shoots a bullet or a missile at somebody," insisted a NASA official in 1967. Yet the space agency was developing technologies that could guide bullets and missiles, and that, explained the *Washington Post*, made NASA administrators "nervous about its military role for several reasons."[40]

On the one hand, during a year when Congress was slashing the space agency's budget by more than 10 percent to its lowest level in five years, NASA administrators such as James Webb were eager to highlight the agency's involvement in supporting the war effort in Southeast Asia.[41] Helping to end the war quickly, several internal NASA studies concluded, would also free up federal funds intended for the military and allow them to flow back to space exploration.[42] Yet Webb and his fellow administrators were equally nervous that overpromoting NASA's military research for Vietnam might cause international consternation. This was especially troubling since the space agency relied heavily on foreign countries such as Great Britain, France, and West Germany for joint space research and on others around the world such as Australia, South Africa, and Peru as hosts for ground-based tracking stations and communications outposts.[43] Webb's solution played it both ways; an internal NASA memorandum from this period explained that the space agency would admit, and even publicly promote, its technological contributions to the war effort in Southeast Asia overall, but would classify as top secret the specifics of such contributions. The public in the United States and around the world, in other words, would know that NASA was aiding the U.S. military in Vietnam but not know exactly how.[44]

Webb's approach succeeded. By the end of 1967 NASA's Limited Warfare Committee had expanded its work significantly; its budget had skyrocketed to $4 million annually and its personnel had grown to approximately one hundred NASA scientists and engineers working at the Jet Propulsion Laboratory (JPL) and the Ames Research Center in California, as well as at the Langley Research Center in Virginia, on what one NASA administrator described as "techniques and hardware that may be of direct application to the current problems of tactical warfare in Vietnam."[45] While the work of these NASA technicians was diverse, involving research on eighty-nine different projects, including an "ambush detector," "mortar counter," and a small beacon for locating downed fighter pilots that was powered by longer-lasting batteries spun off from the space agency's Surveyor, Mariner, and Ranger missions to the moon and Mars, the great majority of these efforts shared a common goal.[46] Most of the space age technologies developed by NASA's Limited Warfare Committee were intended to help the U.S. military see better into, underneath, and around the jungles of Vietnam.

One of the earliest of such research efforts by NASA involved an attempt to increase visibility for fighter pilots through the use of satellites that would forecast inclement weather over Vietnam. Members of the Limited Warfare Committee first discussed meteorology as a military aid in Southeast Asia in late 1965, and the following year assigned engineers at JPL in Southern California to research the possibility of retooling NASA's Applications Technology Satellites (ATS) to provide real-time weather reports to U.S. pilots stationed in Southeast Asia. The new technology, which built upon the CORONA and TIROS reconnaissance systems of the late 1950s, entailed the deployment into synchronous orbit over Vietnam of an ATS satellite equipped with a so-called Cloud Camera. Because the satellite would not be visible from the United States, and in order to ensure real-time data regarding Vietnam's weather, the meteorological system would necessitate a pair of ground stations in Toowoomba, Australia, and in Japan to control the satellite, and a third on the ground in Vietnam that would receive real-time reports and then forward them to fighter pilots at nearby air bases. "It would be highly desirable," explained the NASA Limited Warfare Committee, "to be able to obtain accurate and current information as to weather conditions over the target areas."[47] Internal studies throughout 1966 indicated that the retooled ATS satellite could "provide a committed meteorological cloud cover capability for U.S. operations in Vietnam" and that a prototype of the Cloud Camera could be launched as early as March of 1967 for between $100 million and $200 million.[48]

An even more elaborate scheme by NASA and the U.S. military to make the communist enemy more "legible" involved artificially illuminating the night environment of Vietnam. Known publicly by various names, including Project Moonlight, Project Reflector, and, perhaps most unfortunately, Project Moonshine, what became Project Able began in 1965 as a classified research project of the Limited Warfare Committee involving NASA and the U.S. Army and Air Force. The proposed project entailed the deployment of a giant aluminized Mylar mirror, 2,000 feet in diameter, into synchronous orbit 22,000 miles above Earth. By positioning the mirror directly over Southeast Asia at night, NASA engineers envisioned the device capturing sunlight from the bright side of the Earth and reflecting it downward, where it would illuminate a 200-mile-wide swath of Vietnamese jungle with an intensity approximately 1.7 times that of the full moon.[49] Such a satellite "could provide some level of light in Vietnam," explained George E. Mueller, Associate Administrator of NASA's Office of Manned Space Flight, to Congress soon after the project became declassified in 1966, "so as to limit enemy activity at night."[50] By the end of that year NASA had budgeted nearly a half million dollars for Project Able, assigned responsibility for it to the Manned Space Flight Center in Huntsville, Alabama, and contracted out five technical feasibility studies with space industry leaders such as Boeing, Westinghouse, and Grumman, the last of which planned to retool the technology it had developed to stabilize and control the Apollo lunar landing module to guide the giant reflector through space.[51]

Along with researching space technology that could peer around cloudy weather and turn the Vietnamese night into a dimmed-down version of day, NASA's Limited Warfare Committee also developed highly sensitive seismometers that would help the U.S. military see deep into the jungles of Southeast Asia. In the early 1960s the space agency had developed similar seismometers for its Ranger and Surveyor moon missions to assess the lunar surface for future Apollo landings, and also as part of its Apollo Lunar Surface Experiments Package (ASLEP), which each Apollo mission left behind on the moon to measure the seismic activity of the lunar crust.[52] In December 1965 NASA engineers at JPL began working with the Air Force to refashion these lunar seismometers for use in Operation Igloo White, a covert mission that involved airplanes dropping 20,000 of these updated "seismic detectors" to create an "electronic battlefield" along the Ho Chi Minh Trail.[53] One of these sensors, the "Air-Delivered Seismic Intrusion Detector," or ADSID, resembled a lawn dart whose antenna poked aboveground, disguised as a tropical plant. When the ADSIDs detected vibrations from passing military convoys they immediately transmitted "hits" to Air Force planes circling continuously overhead, which in turn relayed the data to the Infiltration Surveillance Center located on a U.S. military base in nearby Thailand. Inside the Surveillance Center, which at the time was the largest building in Southeast Asia, two IBM 360-65 computers, identical to those being used by NASA in Houston for the Apollo program, quickly analyzed the data and within five minutes conveyed bombing target coordinates to the closest available armed aircraft.[54] "We wired the Ho Chi Minh Trail like a drugstore pinball machine," bragged one Air Force officer of the U.S. military's electronic battlefield, "and plugged it in every night."[55]

Although NASA's Cloud Camera, giant space mirror, and electronic battlefield were initially top secret, the Soviet Union was well aware of the U.S. military's use of space technology to peer more clearly into the jungles of Vietnam, and it retaliated in kind. In many ways, Soviet developments in space during the height of the Cold War echoed, note for note, those of the United States. For instance, during the summer of 1962, just two years after the Air Force and NASA placed the first CORONA spy satellite into orbit, the USSR successfully launched its own version called Zenit, which, like its U.S. counterpart, took high-resolution photographs from space and returned them by means of a reentry capsule that parachuted onto Soviet territory. In 1975, less than a decade after the U.S. military researched the possibility of using NASA's ATS satellites over Vietnam to track cloudy weather, the USSR deployed its own satellites over Southeast Asia to monitor communist advances into South Vietnam and Cambodia as well as arm supplies from the United States into the region.[56] In the early 1980s the Soviets even began developing, and later successfully deployed into space, their own mammoth mirror, initially called "Star Electricity," which one American space engineer admitted "could light up a battlefield at night."[57]

The Space Race and the New Left

Although studies by NASA concluded that deploying real-time weather satellites, a giant space mirror, and jungle seismometers in Vietnam were all technologically feasible, the space agency scrapped both its Cloud Camera and Project Able during the winter of 1967. While this decision was due, in part, to opposition from American astronomers who feared that the giant mirror's brighter night would interfere with their ground-based science, it was the New Left's anti-war activism that ultimately disabled Project Able, rained on NASA's Cloud Camera, and eventually unplugged the space agency's electronic battlefield.[58]

The New Left's opposition to NASA's military efforts in Southeast Asia had its origins in the movement's open disdain for "technocracy," which many students feared was inundating university campuses during the mid-1960s.[59] The New Left student leader Mario Savio, for instance, during the December 2, 1964, protest that launched the Berkeley Free Speech Movement, compared the University of California to an "odious machine" and implored his fellow students to "put your bodies upon the gears and upon the wheels" in order to "make it stop."[60] By the end of the decade New Leftists had pushed their critique beyond analogy; they began publicly condemning universities for actively participating in the rise of what many were calling "the military-industrial-academic complex."[61] In 1967 Savio led another sit-in, not to promote free speech on campus but rather against

Navy recruitment at Berkeley, which ended with students mocking the military with a heartfelt rendition of the Beatles' "Yellow Submarine."[62] Just two years later, students at approximately thirty universities nationwide, including Cornell and Princeton on the east coast, Stanford out west, and the University of Michigan in between, were similarly employing antitechnology rhetoric to further their anti-war activism.[63]

The National Aeronautics and Space Administration was not immune to such student opposition, which was sparked, in part, by the belief that the development of NASA's space technology was siphoning federal funds away from more pressing problems on Earth. A College Poll survey conducted on more than one hundred university campuses nationwide just days after the July 20, 1969, Apollo 11 lunar landing found that "most students feel space budgets should be largely diverted to domestic problems in the future." The same survey singled out campus activists, such as those involved in the New Left, as being strongly opposed to "continuing space investments."[64] Even secondary school students had begun criticizing NASA. That same month, in a *Denver Post* article titled "Voice of Youth: Do We Really Need to Land on the Moon?," one high school senior asked readers, "What real use is there in sending a man to distant planets and having him return to an earth which is overcrowded, over-polluted, and dying from hunger and war?"[65]

Along with censuring the space agency for funneling funds from "earthly problems," much as civil rights activists had done, the New Left was equally critical of NASA for causing problems of its own, particularly when its space technology was deployed for military purposes. The movement's most prominent group, Students for a Democratic Society (SDS), had first voiced this concern in its 1962 manifesto, the *Port Huron Statement*, which criticized the Kennedy administration for making "outer space a region subject to militarization," and continued publicizing this issue through organized campus activism well into the late 1960s.[66] The entire space program, explained one SDS member from Columbia University days after Neil Armstrong set foot on the moon, was a "weapon of the military establishment which is draining our resources."[67]

This conflation during the late 1960s of space and military technology was due to a growing awareness among college students of NASA's increasing involvement in the Vietnam War. Not only were New Left activists alarmed by newspaper articles with titles such as "NASA's Role in War Grows" and "NASA to Study Military Satellites," the latter of which argued that such research "would place the agency squarely and irrevocably in the U.S. defense establishment," but these students also witnessed the expansion of NASA's military research for the Vietnam War quite literally across their campuses.[68] Whereas in 1960 the Department of Defense had contracted $1.2 billion to universities for military research, just seven years later, in 1967, the allocation had skyrocketed to more than $1.8 billion, an increase after adjusting for inflation of more than 30 percent.[69] Research and development of space technology helped drive this growth; in 1968 NASA awarded U.S. universities approximately $130 million for military-related research, much of it used to develop technologies for the war in Vietnam.[70] "Space is a great turner-offer of college people these days," reported a special July 1969 edition of *Newsweek* magazine dedicated to Apollo 11. "To many of them the astronauts, NASA and Mission Control seem part and parcel of the Pentagon, the munitions industry and the war in Vietnam."[71]

Students and faculty at the Massachusetts Institute of Technology (MIT) were just one example of "college people" who were "turned off" by NASA's involvement in Vietnam. Such concern at MIT centered around the university's famed Instrumentation Laboratory, run by Dr. Charles Draper, which in 1969 alone received more than $50 million in nearly equal parts from NASA and the Department of Defense to build, among other technologies, guidance systems that, as the *Washington Post* explained on its front page, "get missiles and spacecraft where they're headed."[72] To protest this increasingly blurry boundary between space and war research on their campus, in the spring of 1969 MIT students, with the support of the left-leaning linguistics professor Noam Chomsky, created a Science Action Coordinating Committee, modeled on the civil rights movement's Student Nonviolent Coordinating Committee, that initiated a nearly year-long campaign involving work stoppages, teach-ins, sit-ins, and campus shutdowns. Such activism culminated in a November 4 rally involving more than 800 students from MIT and other Boston-area colleges, including Harvard, Boston University, and Northeastern, who later that day surrounded Building 5 of the "I-Lab" with a picket line in order to disrupt war-related work taking place inside. The demonstration, which was covered by 118 off-campus reporters, along with similar protests forced MIT to dissociate Draper's laboratory from the university the following year.[73] "Among the New Left's numerous assaults," complained the conservative columnist Joseph Alsop in the *Washington Post* in 1969, "the most successful has been the attempted strangulation of MIT's great Instrumentation Laboratory."[74]

The New Left's success in linking NASA technology to the Vietnam War was partly responsible for the space agency's decision late in 1967 to halt research on its Cloud Camera and Project Able.[75] Such political pressure from college students also caused a major policy change at NASA regarding the space agency's military efforts in Southeast Asia. Whereas during the mid-1960s James Webb had decided to publicize NASA's cooperation with the U.S. military in Vietnam while simultaneously keeping classified the particulars of such efforts, during the fall of 1969, just months after students began protesting against NASA on college campuses, the space agency's administrators instituted a stricter public relations policy intended to eliminate all official statements concerning NASA's role in the war. "We are now requiring that *no* statements concerning our support of the Vietnam effort, even those verifying [NASA's general involvement in the conflict], be given," explained one memorandum from the space agency's Office of the Administrator.[76] Thus, while the space agency provided New Left activists with fodder for their anti-war protests during the mid-1960s, by the late 1960s the New Left's efforts against the war in Southeast Asia had begun transforming policy within NASA itself.

By the early 1970s the New Left was also beginning to influence NASA technology in Vietnam. This process began on July 6, 1970, when Wisconsin senator William Proxmire leaked to the national media his own criticism of the then-classified electronic battlefield along the Ho Chi Minh Trail.[77] As news of the top-secret military project became public, along with notification that NASA engineers at JPL were retooling lunar seismometers for the operation, students again took action on their campuses against both the Department of Defense and the space agency. In March 1971, for example, after learning from the Pentagon that the University of Michigan was at the forefront of research on the electronic battlefield, students in Ann Arbor organized rallies, marches, and fasts to pressure the school, which during the 1960s had received more NASA contract funding than any other American university, to ban all classified research from campus.[78] The following year, University of Pennsylvania students took similar action, organizing a sit-in to protest their school's investments in companies doing research and development on the electronic battlefield. Such companies included General Electric and IT&T, as well as Westinghouse, which had earlier conducted military research on NASA's Cloud Camera.[79]

Similar student activism against the space agency intensified the following year, when in April 1972 students and faculty members from more than fifteen New York metropolitan-area colleges took over the Pupin Physics Laboratory at Columbia University. Student demonstrators singled out the Pupin Lab because five faculty members conducting research inside the building belonged to the JASON Defense Advisory Group, an independent collection of scientists who in 1960 began counseling the federal government on matters involving science and technology. It was the JASONs, in 1967, who originated the idea of an electronic battlefield in Vietnam, subsequently promoting it to the U.S. Air Force and NASA.[80] During the weeklong protest, which involved the occupation of four other buildings at Columbia, students inside Pupin were joined by several veterans of the 1968 Columbia campus uprising, by organizers of the campus SDS chapter, and by the peace activist and poet Allen Ginsberg. Also participating in the demonstration was Rennie Davis, a longtime SDS leader and Chicago Seven defendant charged with inciting to riot during anti-war protests at the 1968 Democratic National Convention, who told supportive students gathered in a nearby auditorium that the "events at Columbia have been broadcast on North Vietnam radio."[81] Eventually, after ending their occupation, student protestors held a news conference announcing that they had found documents inside Pupin that directly linked Columbia "with the war machine" in Southeast Asia.[82]

Much like the New Left's opposition to the Vietnam War, which during the late 1960s gained support from the broader American public, the student movement's concerns regarding NASA's relationship with the U.S. military also began migrating off campus. Expressing such worries were several national newspapers that published political cartoons depicting in visual form the same sentiment expressed by university students in the College Poll's survey, namely that space technology, along with its large budgets, should be redirected to address issues closer to home. The *St. Louis Globe*, for instance, ran an illustration that portrayed a pint-sized civilian named "The Rest of Us" tugging desperately on a lab coat labeled "TECHNOLOGY" being worn by an overgrown technician who cradled the moon. "Could I Interest You in Some Earthly Problems?" pleaded the tiny civilian.[83] The *Washington Daily News*, perhaps taking its cue from New Left students at MIT and other universities, was even more explicit in associating NASA's space technology with that of the U.S. military. In "WHAT HAVE THEY BEEN FEEDING YOU?," another vertically challenged civilian, standing tippy-toe in a suit labeled "DOMESTIC NEEDS,"

2.1   Gene Basset, "What Have They Been Feeding You?," *Washington Daily News*, July 28, 1969.

looked up quizzically over his shoulder at an oversized, aloof astronaut with "U.S. SPACE AND MILITARY TECHNOLOGY" emblazoned across his spacesuit (see Figure 2.1).[84] As these cartoons suggest, many Americans, much like their campus counterparts, had begun to associate NASA's space technology with the U.S. military.

They also soon began questioning the role played by space technology specifically in the Vietnam War, much as New Leftists had done. The evo-

Horizons

2.2   John Fischetti, "Horizons," *Chicago Daily News*, December 26, 1968.

lution of this popular anxiety was apparent in two nationally syndicated editorial cartoons published in 1968 and 1972. In "Horizons," which appeared in the *Chicago Daily News* on December 26, 1968, just as Apollo 8 became the first manned mission to achieve lunar orbit, the Pulitzer Prize–winning cartoonist John Fischetti contrasted the technology of the space race, symbolized by a small spaceship circling the moon, with that of the war, signified by a large gun planted firmly in Vietnam's soil (see Figure 2.2).[85] The second illustration by Bruce Shanks, another Pulitzer Prize–winning cartoonist, moved beyond comparison when it appeared in the *Los Angeles Times* in early April of 1972, a moment when NASA was looking forward to two more scheduled Apollo launches but peace negotiations with the North Vietnamese had broken down yet again.[86] To capture this disparity, and to suggest that the space race was complicit in the war in Southeast Asia, Shanks juxtaposed Apollo, symbolized by "Space Progress" written into the exhaust of a successful launch in the foreground, with the stalled negotiations in Vietnam, represented in the background of the cartoon by a frowning grounded spaceship covered in cobwebs, an unused entry ladder, and a flag labeled "Peace Progress" (see Figure 2.3).[87]

2.3   Bruce Shanks, "Space Progress, Peace Progress," as printed in the
Los Angeles Times, April 9, 1972.

Americans did not merely read news coverage about NASA's growing
role in Vietnam, but also followed the New Left's lead and began to take
action. "The so-called automated battlefield, where death strikes through a
combination of sensors, computers and bombs," argued the Meriden, Connecticut, *Morning Record* in December of 1971, "is coming into focus as a
chief target for criticism in the fading Indochina war."[88] The space agency,
because it was responsible for developing this technology, became not only
a focus of these civilian critiques but also a site for popular activist opposition. Earlier that same year, in San Francisco's South Bay, an anti-war demonstration took place outside the main gate of NASA's Ames Research
Center, where engineers and scientists conducted many of the studies for

the space agency's Limited Warfare Committee.[89] Similar civic protests in 1972 involved anti-war groups, including the American Friends Service Committee, a Quaker peace and social justice organization, which produced a slide presentation depicting the horrors of the electronic battlefield and showed it to local communities across the Northeast, as well as 275 members of the American Physical Society, who, in response to Senator Proxmire's leak, proposed an amendment to the physics society's charter that would restrict military research by the group's members. Even the anti-war priest Philip Berrigan, who, along with his Jesuit priest brother, Daniel, had been sentenced to three years in prison for burning Vietnam draft cards, led eleven prisoners from the Federal Correctional Institution in Danbury, Connecticut, in a fast reminiscent of those undertaken by students at the University of Michigan. The prisoners fasted, explained Berrigan, "to protest American atrocities in Indochina—our electronic battlefields, our mining of ports and rivers, our bombing of dikes and dams."[90]

Such anti-war activism against NASA soon began influencing politics both in the United States and abroad. Just eight months before Pham Tuan's 1972 encounter with an American B-52, Wisconsin Senator Gaylord Nelson introduced a bill calling for the National Academy of Sciences to study the ecological impact of the American war effort in Southeast Asia. "A 'scorched earth' policy has been a tactic of warfare throughout history," he explained in a passionate speech before Congress, "but never before has a land been so massively altered and mutilated that vast areas can never be used again or even inhabited by man or animal."[91] In June of that same year, similar concern and condemnation arose on the international front at the United Nations Conference on the Human Environment being held in Stockholm, Sweden. At the conference Indian Prime Minister Indira Gandhi lamented the "diabolic weapons which . . . poison the land, leaving long trails of ugliness, barrenness and hopeless desolation" in Vietnam, while Swedish Premier Olof Palme, in his keynote address to all conference participants, warned that "the immense destruction brought about by indiscriminate bombing, by large-scale use of bulldozers and herbicides, is an outrage sometimes described as ecocide."[92] As a result of such widespread criticism winding its way from university campuses to the UN, later that year the U.S. military terminated its electronic battlefield program.[93]

Space Race Détente

New Left protests on college campuses against NASA's military research for the war in Southeast Asia were not only partly responsible for the cancellation of the space agency's Cloud Camera, Project Able, and electronic battlefield. They were also indicative of broader concerns, not only among America's youth but also those from West Germany, France, China, and the Soviet Union, regarding the physical dangers, and high economic costs, of the nuclear arms race. The Cuban Missile Crisis of 1962, a thirteen-day confrontation concerning the deployment of Soviet ballistic missiles on the island nation located just ninety miles off the coast of Key West, Florida, triggered such fears, which rose during the 1960s alongside the escalation of war in Vietnam. To defuse such military tensions, lower defense costs, and appease grassroots protestors on both the domestic and international fronts, in 1969 American and Soviet politicians embraced détente, a concerted easing of strained relations between the two superpowers.[94] Military treaties such as the Strategic Arms Limitation Agreement (SALT I), which when signed by the United States and the Soviet Union in 1972 placed limits on each country's nuclear weapons systems, was an example of such efforts. Terminating NASA's electronic battlefield, Cloud Camera, and Project Able, the last of which the governments of Cambodia and Laos also criticized for being a "danger to both plant and human life," were other instances of this emerging détente.[95]

Along with halting the development and deployment of space technology for military purposes, during the early 1970s the U.S. government also began using NASA technology in nonmilitary ways to more proactively promote détente with the Soviet Union. The Apollo-Soyuz space mission was a prime example. First proposed by President Richard Nixon in 1972, the Apollo-Soyuz Test Project (ASTP), as it was known, culminated on July 15, 1975, when an Apollo command module from the Kennedy Space Center in Florida rendezvoused in Earth orbit with a Soviet Soyuz spacecraft launched from the Baikonur Cosmodrome in Kazakhstan. Although the experiments performed by the ASTP crew were, as the *Wall Street Journal* argued, "a costly space circus of almost no scientific significance," the project succeeded wonderfully on several political levels.[96] This became immediately obvious at the outset of the mission; soon after the two spaceships docked, NASA's new $4 million Apollo TV camera beamed images back to Earth of the international crews shaking hands, exchanging flags and gifts, eating a meal together, and trading indigenous tree seeds that were later planted in one another's home countries.[97] "We had come a long way from flights along the Iron Curtain, secret missile tests, and the moon race," explained the Apollo astronaut and ASTP mission commander, Tom Stafford.[98] The *Saturday Evening Post* agreed, explaining to its readers that the mission's "technical objectives are secondary to a broader political aim: demonstrating to the world that the two most technologically advanced nations can work together in peace."[99]

The Apollo Soyuz Test Project not only improved political relations with the Soviet Union, but also aided the Nixon and Ford administrations politically on both the international and domestic fronts. Televised images of Americans and Russians docking sophisticated space technology in Earth's orbit, let alone hugging and sharing dinner together, raised America's political capital abroad. "Newspapers and nations in most quarters of the world Saturday hailed the Apollo-Soyuz space docking as a triumph for diplomacy," argued United Press International. Government leaders from Paris to Bangkok agreed, with a Polish politician praising the mission as indicative of "the transition from a demonstration of force and confrontation to the policy of mutually beneficial cooperation."[100] On the domestic front, the ASTP mission also supported the administration's foreign policy by diverting, even for a few days, the New Left's attention from the conflict in Southeast Asia. "The space shots, and especially the Apollo series, have the faculty of creating spectacular events," explained the Philadelphia *Bulletin*, which "become obsessions for the major mass media" and "a boost for the system as it is."[101] As a result, argued one anti-war reporter, "the Apollo program was conceived and executed to keep the people's mind off Vietnam."[102]

Yet if détente in space succeeded domestically in taking the New Left's eye off the war in Southeast Asia, and proved fruitful internationally as well by raising America's standing throughout the developing world, it nevertheless raised a central dilemma for the United States. How could America continue its global competition with the Soviet Union, as it had done in Southeast Asia, without publicly stoking Cold War animosities? Additionally, what role would space technology and earthbound nature play in this process, within the parameter of a less openly confrontational détente? In other words, how could the U.S. government exchange tree seeds with communist Russians in space but continue planting democracy firmly on the ground in developing countries like Vietnam?

NASA Technology and "Third World" Nature

One means was for the United States to redirect NASA's space technology, used during the late 1960s to make nature and communists more "legible" in Southeast Asia, toward more subtle Cold War ends in the 1970s across the developing world. Central to this effort were NASA's earth resources satellites, later known as Landsat, which the space agency first launched on July 23, 1972. Developed from both military hardware such as the CORONA spy satellite and civilian technology used clandestinely for war, including the TIROS and ATS satellites, Landsat circled 560 miles above Earth in near-polar orbit taking 13,000 square-mile "snapshots" of the planet's surface that when stitched together captured nearly the entire globe every eighteen days.[103] During the next quarter century, six additional Landsat satellites gathered data for millions of images of planet Earth. By radioing back "pictures" of Earth from space, the *New York Times* explained in mid-January of 1975, Landsat was "providing new insight into man's continuing effort to better manage earth's limited resources as well as aiding in the assessment and understanding of environmental changes."[104]

Landsat's "new insight" was based on multispectral scanners that measured from space four different wavelengths of electromagnetic radiation reflecting off objects on the surface of the Earth. Landsat satellites beamed these wavelength measurements back down to NASA receiving stations, in Fairbanks, Alaska, Goldstone, California, and at the Goddard Space Flight Center in Greenbelt, Maryland, where technicians converted the raw data into visual maps by assigning false colors to earthbound objects with different wavelengths. Landsat, in other words, made planet Earth's natural environment more readable by measuring the extremely slight temperature variations of the solar heat bouncing off rocks, trees, water, and even animals.[105] As *Science* magazine reported on the tenth anniversary of Landsat 1, the maps created from the Earth-observing satellite depicted "scarlet forests, red patchwork farms, blue city grids, brown crinkled mountains, and a delicate web of highways."[106] In July of 1982 Landsat 4 carried into Earth orbit a second-generation sensor, the Thematic Mapper, which provided higher contrast by assessing seven wavelengths and increased the technology's on-the-ground resolution from 260 feet to less than 100.[107]

Landsat's colorful maps of Earth quickly became critical tools for analyzing the state of the Earth's natural resources, and NASA immediately began promoting such capabilities through easy-to-read pamphlets and booklets with appealing titles such as "Improving Our Environment," "Ecological Surveys from Space," and "Photography from Space to Help Solve Problems on Earth." According to these publications Landsat satellites aided agriculture and forestry by making possible the inventory of different types of crops and trees, the identification of early signs of plant diseases and insects, and the assessment of soil moisture to guide future land-use practices. The space technology proved equally beneficial for Earth's hydrological and atmospheric resources; Landsat data mapped fresh and salt water around the globe, forecast drought and floods, and identified sources of both water and air pollution. It also provided geological measurements of the Earth's surface that located underground resources, including oil, natural gas, and mineral deposits, and even allowed biologists to track migratory wildlife both across the land and under the seas.[108]

By helping to manage natural resources Landsat was also helping to manage NASA's public image, which during the early 1970s was suffering from a severe case of "NASA fatigue." The civil rights movement had initiated this public questioning of the space race during the mid-1960s, and the New Left now exacerbated it. "A long mental yawn will roll over America next Sunday when Apollo 16 spits fire from its tail and streaks skyward to the moon," explained the *Los Angeles Times* in April of 1972.[109] Partly because of such apathy, between the moon landing of 1969 and the launch of Apollo-Soyuz in 1975 Congress cut the space agency's funding even more drastically then it had back in 1967, when NASA administrators had responded, in an attempt to dodge additional reductions, by agreeing to the U.S. military's request for technological aid in Vietnam. All told, during this six-year period beginning after Apollo 11, the federal government slashed NASA's budget by more than 40 percent, after accounting for inflation, to its lowest real-dollar level since 1962.[110]

In a conscious effort to reverse this trend, NASA administrators not only canceled several of the space agency's military projects but also began encouraging research and development of space technologies such as Landsat that would benefit the American public. "It is clear that if we are to move forward with a strong space program, it, too, must be useful to the people here on Earth," argued NASA Deputy Administrator George Low in 1970. "This means," he added, "that a space applications program and, specifically, an earth resources program should be the keystone for the space effort of the 1970s."[111] One result was NASA's Large Area Crop Inventory

Experiment (LACIE), a joint venture by the space agency, the Department of Agriculture, and the National Oceanic and Atmospheric Administration (NOAA) that combined crop acreage measurements obtained from Landsat with meteorological information from NOAA satellites to forecast wheat production in an effort to stabilize the commodity's price for American consumers.[112] Such efforts by NASA succeeded; not only did Congress authorize two additional Landsat satellites in 1975 and 1978, but it also increased the space agency's budget by more than 10 percent, after accounting for inflation, between 1975 and 1980.[113]

Both the federal government and the space agency quickly realized that Landsat could do for the United States internationally what it had done for NASA domestically. President Nixon early on understood Landsat's promotional potential, announcing to the United Nations' General Assembly that his country's new Earth-observing satellites would "produce information not only for the U.S., but also for the world community."[114] Space agency officials were even more explicit, focusing many of their public comments concerning productive uses of Landsat data specifically on the natural resources of poorer countries. The new space technology would "assist both the developed and developing areas of the world alike in providing maps and other important resource inventory data," explained a NASA position paper on remote sensing. In doing so, the report went on to argue, "The use of remote sensors in NASA spacecraft to aid developing countries thus represents an important way for the United States to enhance its world image."[115] By giving poor nations access to scientific data that could help them better manage their own natural resources, Landsat could raise the international standing of the United States by helping developing countries develop.

There were just two problems with such a simple scenario. First, at least initially, several developing nations openly resisted NASA's remote-sensing technology for fear that it would infringe upon their national sovereignty. While the Soviet Union was concerned that Landsat could be used for spying, countries across Latin America were more worried that developed countries would employ the technology to exploit natural resources located in the developing world; wealthy nations such as the United States could employ satellite data not only to identify within poorer countries previously undiscovered resources, such as mineral and oil deposits, but also to forecast global crop production in an effort to manipulate agricultural commodity prices.[116] To protect against such actions, in 1975 several developing nations, including Argentina, Chile, Venezuela, and Mexico, cosponsored an ultimately unsuccessful United Nations proposal that would have prohibited any remote-sensing activity relating to natural resources under a country's national jurisdiction without prior consent from the nation being remotely sensed from space.[117]

The second problem hindering the U.S. government's desire to use Landsat to raise its international standing was that scientists in developing countries were completely unprepared to receive, process, interpret, and utilize satellite data regarding their countries' natural resources. Such was the conclusion of an exasperated Verl Wilmarth, one of NASA's Earth observation managers, who during the summer of 1971 lamented the abysmal quality of proposals submitted by foreign scientists interested in participating in future Landsat experiments. The "poorly prepared proposals," he wrote, "indicate lack of knowledge of the program content and capabilities."[118] Administrators at NASA were equally concerned that even if foreign scientists did eventually understand Landsat's possibilities, they would nevertheless continue to lack the technological and scientific expertise necessary to take full advantage of the new space technology. Of particular concern was the dearth in developing countries of trained photo interpreters both to analyze the images obtained from satellites and to extract from them the types of data with economic value.[119] The space agency thus not only had to convince leaders of developing countries that Landsat did not pose a threat to their national sovereignty, but it needed as well to educate foreign scientists regarding the space technology's economic and ecological benefits.

Administrators from NASA began addressing such problems during the early 1970s by blanketing the international scientific community with press releases describing how Landsat technology worked and that also requested from foreign scientists themselves proposals that would improve natural-resource management specifically in developing countries.[120] The space agency augmented such efforts by teaming up with international institutions such as the United Nations, the World Bank, and the Inter-American Development Bank to sponsor conferences, symposiums, and workshops, some up to two weeks long, on the use of Landsat remote-sensing data.[121] Initially, the space agency invited foreign scientists, engineers, and politicians to such events held in the United States both at academic institutions such as the University of Michigan, whose faculty excelled in remote-sensing research, and also at NASA's research facilities including the

Johnson Space Center in Houston, which conducted a week-long "Earth Resources Survey Symposium" during the summer of 1975. At the Houston Landsat conference some of NASA's heavy hitters, including Apollo astronaut Rusty Schweickart, Marshall Space Flight Center director Wernher von Braun, and Johnson Space Center director Chris Kraft, addressed an audience of more than 1,200 scientists, engineers, politicians, and administrators from at least two dozen foreign countries on the practical applications of Earth-observing technology.[122]

Increasingly during the mid-1970s NASA also brought such gatherings directly to foreign scientists and government leaders within developing countries. During the summer of 1975, for instance, the space agency conducted two three-day symposiums on Earth-observing technology in West Africa in an effort both to educate scientists and government officials in the region about the capabilities of Landsat technology and to encourage them to submit scientific proposals aimed at better managing their countries' scarce natural resources. At the first conference, held in Ghana for English-speaking participants, scientists, engineers, and government leaders from nearby nations including Nigeria, Liberia, and Togo listened, along with the U.S. ambassador to Ghana and former child movie star, Shirley Temple Black, as the keynote speaker implored those present to make use of "accelerating tools" such as Landsat in order to bridge the "technological gap" between underdeveloped and developed nations. The second conference held later that summer took place in Mali, where French-speaking participants from Senegal, Chad, Zaire, Cameroon, Niger, and the Ivory Coast heard NASA scientist Bryan Erb describe how the space agency's LACIE experiments from the early 1970s could be applied to West Africa in order to lessen the severity of starvation then occurring across the drought-stricken Sahel region.[123] During the 1970s similar NASA conferences promoting the benefits of Landsat technology for developing countries took place in Asia and throughout Latin America.[124]

While NASA's conferences, workshops, and symposiums succeeded in educating participants from developing nations regarding Landsat's scientific usefulness, the space agency tried to alleviate concerns regarding the space technology's encroachment on national sovereignty by training foreign scientists to collect, analyze, and interpret Earth-observing data on their own. As with its Landsat conferences, such training took place both within the United States and abroad. In the early 1970s, for example, NASA expanded its international fellowship program to encourage foreign scientists to travel to American universities to take courses on the fundamentals of remote sensing.[125] The space agency also brought scientists from developing countries such as Brazil and Mexico to NASA centers, including the Johnson Space Center, to familiarize them with the acquisition, processing, and analysis of remote-sensing data.[126] Finally, in an effort to institutionalize such training within less developed nations, NASA, along with the U.S. government, encouraged political leaders around the world to establish their own Landsat receiving stations to collect data on their country's natural resources. In South America this process began in 1974 when Brazil built its own receiving station and continued three years later when Chile signed an agreement to build another and Venezuela formally expressed interest in doing the same. By early 1977 Egypt and Iran in the Middle East and Zaire in Africa had also established their own stations to receive and process Landsat data. Each of these host countries funded, owned, and operated their Landsat receiving stations, making their scientific experiments less dependent on the United States.[127]

Such efforts by NASA both to educate the international scientific community about Landsat and to alleviate concerns of foreign government officials regarding the technology's impact on national sovereignty proved enormously successful; by 1977 more than fifty countries worldwide were relying on Landsat data to better manage their natural resources.[128] According to *Science* magazine, the space technology's benefits were particularly significant in developing countries, which lacked other means of surveying their resources. "Nowhere is this promise clearer," *Science* reported in April of that year, "than in South America." Brazil, which early on established a Landsat receiving station, was the poster child for such efforts. Brazilian scientists were using Landsat data to map the country's soil and forecast its crop harvests; to locate underground resources including bauxite, iron, uranium, and oil; to create oceanographic charts of upwelling areas in coastal waters to enhance fishing; and, perhaps most important, to control deforestation in the country's Amazon rainforest by monitoring widespread clearing by private landowners for cattle ranching. Brazil's national geographic institute was even using Landsat measurements of the nation's landforms for "revising the official map of the country."[129] By the late 1970s NASA was similarly redrawing maps all over Latin America and providing data on natural resources to more than a dozen countries, including Bolivia, Chile, Peru, and Venezuela in the south and Costa Rica, Guatemala, Honduras, and Mexico in Central America.[130]

In Africa, which had been devastated during the early 1970s by severe drought, NASA focused most of its Landsat program on assessing the continent's food supplies. Data from the space agency's Earth-observing satellites allowed biologists from Sudan to inventory land and vegetation resources, hydrologists from Botswana to map agricultural water sources, and managers from Kenya to administer more efficiently rangeland for both domestic and wild animals.[131] The most intensive use of Landsat data in Africa, however, occurred in the Sahel region of the continent, which experienced devastating drought and famine during the early to mid-1970s. Here, NASA teamed up with local scientists and government officials as well as with several American universities to collect and assess data for western Niger in order to determine range-management techniques that could reverse the process of desertification in the region. It also used Earth-observing data to map ecological zones for more productive agricultural planning and to track, over time through repeated orbital flyovers, the deteriorating impact of sand- and dust storms on Sahelian plant communities and soil fertility.[132] The space agency conducted similar Earth-observing experiments with foreign scientists in more than a dozen African countries, including Ethiopia, Mali, Gabon, Libya, Swaziland, Egypt, and the Central African Republic. "The ability to detect, monitor, and predict the effects of drought, disease, infestation, and other calamities," concluded one National Science Foundation report on NASA's Landsat work in Africa's developing countries, "is needed by the country's planners in order to insure adequate food supplies for its people."[133]

Whereas in Africa NASA's Earth-observing satellites focused on drought, in Asia the space technology was employed for nearly the opposite reason; too much rain often destroyed annual harvests. In 1973 NASA administrators discussed redirecting the orbit of Landsat 1 to pass over regions of Pakistan affected by flooding to help that country assess agricultural damage, and in the mid-1970s the space agency teamed up with the World Bank's Agricultural and Rural Development Department to use Landsat data to map land cover in parts of India, Bangladesh, and Burma. In Burma this entailed using Landsat's multispectral scanners to delineate among two dozen categories of land types, such as wetlands versus grassland versus barren land, and also between different land uses such as agriculture and forestry. The World Bank even used Landsat data to distinguish not only among different types of agricultural crops but also among four different strains of rice grown in various regions of the country. "Since for many parts of Burma there existed no recent, large scale maps," explained a 1977 study on Earth observation in Asia, "the Landsat maps will significantly enhance that country's ability to plan its land-use and improve its food supply."[134]

By far the most surprising application of Landsat in the developing world, and undoubtedly the most politically beneficial to the United States, took place in Southeast Asia. Beginning in 1973 as the Vietnam War wound down, NASA, in cooperation with the United Nations' Mekong Committee, deployed Landsat satellites over the Lower Mekong River in an effort to help the four countries in the region—Laos, Cambodia, Thailand, and South Vietnam—more efficiently manage the natural resources in the 230,000-square-mile river basin.[135] The result was three natural-resource maps. The first two, which included a hydrological survey of basin flooding during different times of the year and a land-use map that differentiated between agricultural and forest lands as well as among different types of crops and tree species, were intended to help government officials from these developing countries better understand their current natural-resource practices. The third map, which assessed the region's "land capabilities," was essentially a soil atlas aimed at improving planning for future natural-resource management.[136] The U.S. government and NASA intended all three maps to help Southeast Asia, especially South Vietnam, transition to a more developed, and democratic, peacetime existence. "Satellite imagery," explained NASA and the Mekong Committee in an April 1976 joint report, is "urgently needed, in order to finalize a realistic post-war development program for the basin."[137]

Soviet-Style Space Détente

The Soviet Union responded to NASA's Landsat program with its own brand of space diplomacy that would similarly allow it to continue fighting the Cold War but do so within the parameters of détente. The first step in this process was the creation in 1967 of the Intercosmos Council to promote cooperation in space among socialist countries, including Bulgaria, Hungary, East Germany, Poland, Romania, and Czechoslovakia in Eastern Europe; Mongolia in Asia; and Cuba in Latin America. China, not surprisingly considering its troubled diplomatic relationship with the Soviet Union during this period, was excluded from the program.[138] Within a decade the council had launched sixteen Intercosmos satellites, five high-altitude research rockets, and dozens of weather satellites.[139] To support the council's efforts, in 1971 the Soviet Union also launched the first of nine Salyut space stations, which would serve as working laboratories in Earth orbit for experiments on space science. "A new step has been taken in space research and exploration," argued the Intercosmos Council chairman Boris Petrov in the mid-1970s. The use of the Salyut space station by the Intercosmos Council "marks the transition to an important new state in the development of international cooperation in investigating and utilizing space."[140]

Just as NASA used Landsat to publicize its efforts across the United States, so, too, did the Soviet space agency employ the Intercosmos Council and its Salyut space station to increase its own prestige among developed socialist countries. It accomplished this not only by encouraging scientists from Eastern Europe to build research equipment for Intercosmos satellites and to create experiments for the Salyut space station, but also by literally taking cosmonauts from Warsaw Pact countries up into outer space. Between the winter of 1978 and the spring of 1980 alone, the Soviets launched Soyuz rockets carrying communist cosmonauts from Czechoslovakia, Poland, East Germany, Bulgaria, and Hungary, and then promoted each mission throughout the world as evidence of the superiority of Soviet-style communism. "A Soviet space program permitting cosmonauts from socialist countries to travel with Soviets into outer space looks more like a masterly piece of public relations," reported the Associated Press in the fall of 1980. "Everyone who has gone to date has been a friend of Moscow, not Washington." Such efforts by the USSR were perfectly timed, since NASA's space shuttle program was not scheduled for liftoff until 1981. "People are proud of their cosmonauts who fly with the Russians," explained one space analyst of Eastern Europeans. "There's no question about that."[141]

The Soviets understood as well that such pride could extend deep into the developing world, and they made a concerted effort to involve poorer communist nations from beyond Eastern Europe in its Intercosmos program. The first of such efforts began in 1978, just three years after the fall of Saigon, when the Soviet space agency, in cooperation with government officials from the recently unified Socialist Republic of Vietnam, handpicked Pham Tuan to become an Intercosmos cosmonaut. The two governments immediately began publicizing Tuan's participation in the Intercosmos program as evidence of communism's benefits for the developing world.[142]

Such promotion paid off. While politicians from Laos agreed that Colonel Tuan's mission into space illustrated "fruitful cooperation based on the spirit of proletarian internationalism," and the government of Sri Lanka described it as "a source of encouragement, strength and confidence to all people struggling for peace, freedom, social progress and international cooperation," Cuban politicians held a reception at the Vietnamese embassy in Havana to celebrate the flight.[143] Even American commentators understood the public relations coup scored by Pham Tuan's journey into orbit. "The United States shared its space triumphs with third-world allies by passing out moon rocks," worried the *New York Times* two days after Soyuz 37 returned to Earth. "But in terms of national pride, taking a third world friend along for the ride may reap bigger rewards for the Soviet Union."[144] In fact, so large were the political benefits that well into the 1980s the Intercosmos Council continued launching cosmonauts from developing countries, including Cuba, Mongolia, and Afghanistan.

The political success of the Soyuz 37 mission throughout the developing world did not depend solely on Soviet technology in space; it also rested on nature back on Earth. "Long-range analysis of the earth with the help of aerospace facilities to study our planet's natural resources has become one of the most important areas" for the council, explained Petrov in the late 1970s. These "aerospace facilities" were the Soviet equivalent of Landsat; the Russian version included an "MKF-6 multizone camera," a series of transmission satellites, and data-receiving stations constructed on the ground in each of the countries participating in the Intercosmos space program. This multizone camera, which the Soviet space agency mounted on the Salyut space station, measured six different wavelengths of electromagnetic radiation bouncing off natural resources on Earth and transmitted the measurements to nearby orbiting satellites, which in turn relayed the data down to receiving stations on the ground. Such space technology, Petrov added, would "provide useful information for geology, agriculture, oceanology [*sic*] and other sectors of science and the national economy," especially the national economies of developing nations.[145]

While the United States had been using Landsat satellites for nearly a decade in Africa, Asia, and Latin America, the Soviets' first test case for this new space-based diplomacy took place high above Vietnam during Pham Tuan's Soyuz 37–Salyut 6 mission. From 200 miles straight up, the Soviet's MKF-6 multizone camera undertook soil and forest surveys to improve crop and timber cultivation; made hydrological studies of flooding, erosion, and sedimentation to enhance fish breeding; assessed the atmosphere to forecast typhoons and hurricanes; and measured geological formations in order to identify mineral deposits for future prospecting.[146] Such space-based experiments involving the "study and precise assessment of the natural resource potentials in Vietnam," explained the deputy chairman of Vietnam's National Scientific Research Center, Dr. Nguyen Van Hieu, "will serve as a basis for economic planning."[147] The Soviet space agency used its MKF-6 multizone camera to undertake similar measurements of natural resources in Cuba, Mongolia, and Afghanistan when it included cosmonauts from these countries on subsequent Intercosmos Soyuz–Salyut missions.[148]

The Soyuz 37 mission also used space technology both to assess ecological damage caused by the U.S. military during the Vietnam War and to formulate scientific plans for environmental restoration. According to the Soviet space agency, the Soyuz 37 experiments would focus especially on those areas of the country "devastated by defoliants" during the war. Two years earlier, in 1978, a joint Soviet-Vietnamese biological expedition had initiated a study, in a few locations across Vietnam, of the severe environmental consequences of chemical spraying by the United States Air Force. "Dead jungles where nothing grows still remain in various areas," explained one scientist familiar with those land-based assessments.[149] Soyuz 37 would broaden this 1978 work by measuring from space the nationwide extent of ecological damage caused by American defoliants. "From their height of 345 kilometers," explained one reporter, Tuan and Gorbatko "obtained photographs and spectral data of Vietnamese territory" that allowed scientists to study "the effects on the Vietnamese countryside, plants and forest of the enormous amounts of defoliants and fire bombs dropped during the Vietnam War." More important for the future of Vietnam, added this reporter, was that this data from space would also allow these same scientists to "develop effective methods to revive the soil."[150]

As with NASA's Landsat, the Soviet Union publicized Pham Tuan's scientific experiments from space to forward its own foreign policy goals in a new era of détente. Such goals included, on the one hand, the promotion of Soviet-style communism across the developing world through the promise of economic growth based on improved natural-resource management. Yet the Soviets also used the Soyuz 37 experiments to publicize, worldwide, the environmental atrocities committed by their superpower rival during the Vietnam conflict, as well as their own efforts to restore these damaged areas to ecological health. It was no coincidence, for instance, that Tuan's mission and its experiment to map defoliants sprayed during the war were scheduled to overlap with the 1980 Summer Olympics, which the United States boycotted on account of the Soviet invasion of Afghanistan.[151] The *Christian Science Monitor* was well aware of such tactics, reporting from the Moscow games that "the Vietnamese was preferred" for the Soyuz 37 flight because "one object of the Olympic games here is to cement Soviet influence in Asia, Africa, and Latin America."[152] Thus as they followed on radio or television the hundred-meter dash in Moscow's Lenin Stadium, citizens of the developing world were reminded, once again, of American ecocide in Vietnam.

From Seek and Destroy to Assess and Restore

At 8:15 P.M. on July 31, 1980, after one week in orbit, Victor Gorbatko and Pham Tuan undocked their Soyuz 37 space capsule from the Salyut 6 space station, reentered Earth's atmosphere, and parachuted the capsule for a soft landing in the dry desert of Kazakhstan. After being carried from the Soyuz 37 landing module, both cosmonauts followed Soviet tradition by signing their names on its side panel before Colonel Gorbatko exclaimed to the gathered reporters, "I am proud of having flown together with the first cosmonaut of Asia."[153] Colonel Tuan was proud as well, and several days later he traveled home "with pictures of Vietnam taken from the air" that not only captured the country's lush green jungles from 200 miles straight up, but which would also help Vietnamese scientists better manage their homeland's natural resources.[154] Soon thereafter the Soyuz 37 capsule also made the journey to Vietnam, as a gift from the Soviet people, where it went on a national tour through villages and cities up and down the recently reunited country. Thus even after cosmonauts Tuan and Gorbatko had returned safely to Earth, space technology and the earthbound environment continued to shape relations between their two countries.

Space hardware and nature on the ground had also influenced the Vietnam War. In a failed effort to help the U.S. military see better into the dark jungles of Southeast Asia, NASA engineers and scientists researched and developed cloud cameras, giant space mirrors, and electronic battlefields that, quite unintentionally, spurred the New Left to organize anti-war rallies, sit-ins, and fasts against the space agency. When such activism spread beyond university campuses and threatened congressional funding for space exploration, NASA administrators once again redirected their technological efforts, much as they had done in the wake of civil rights protests, by scrapping covert research for the Vietnam War and instead launching Landsat satellites to help developing nations better manage their natural resources.[155] The Soviet Union followed suit with its Salyut space station and its MKF-6 multizone camera. In both cases, space technology that had been used during the late 1960s primarily to *seek and destroy* in the jungles of Vietnam had become by the mid-1970s space technology deployed as well to *assess and restore* the environments of not only Southeast Asia but also Africa and Latin America.[156]

While this technological turnaround no doubt pleased the New Left, providing the anti-war movement with a tangible victory in an age of frustrating political failures, it was more of a mixed blessing for those across the developing world, such as Pham Tuan and his compatriots. On the one hand, Landsat measurements of natural resources from Botswana to Brazil to Burma depended on the cooperation of local scientists and politicians for success; biologists on the ground knew best which of their country's natural resources needed study from space, while native government officials had the political and economic capital to construct receiving stations and train photo interpreters. Landsat's focus on local nature, in other words, left room for local control over Landsat data.[157] The same could be said of Soviet remote-sensing efforts in developing communist countries, including Cuba, Mongolia, and Vietnam.

Yet the U.S. government, in cooperation with NASA, still fabricated and launched Landsat satellites, decided when they should be "turned on" over what geographic regions, and determined which countries could and could not participate in the program. Administrators at NASA, sometimes guided by federal bureaus such as the Department of Defense, even had the power to demand that proposals by foreign scientists for Landsat experiments be "negotiated," or revised, before being officially approved.[158] The Soviet Union retained similar control over its own remote-sensing technology. As a result, while politicians and scientists from developing countries embraced Earth-observing programs in part because they could influence them on the ground, the American and Soviet governments ultimately controlled this modernizing project from above in ways that almost always supported their own foreign policy agenda. Pham Tuan's broadcast from Earth's orbit, in other words, along with the scientific experiments he conducted from space with the Soviets' MKF-6 multizone camera, most definitely bene-fited Moscow more than Hanoi.

This foreign policy predicament for citizens of the developing world arose prior to the launch of the first Landsat satellite in July of 1972, and even before New Left activists started taking over university laboratories and shutting down college campuses to protest NASA's involvement in the Vietnam War. It began instead after World War II, when American scien-tific and government elites worked together to rebuild research and devel-opment in war-ravaged Europe, but did so in ways that benefited American political and economic interests abroad.[159] European technicians under-standably welcomed such efforts, just as scientists and government leaders from developing countries welcomed Landsat. Yet by sharing in this scien-tific diplomacy, by consenting to it and, in effect, coproducing it, postwar Europeans were less able to oppose more objectionable U.S. foreign policy initiatives. Landsat functioned similarly by enhancing America's soft power across the developing world.[160]

Such was the case regarding NASA's involvement with the Mekong Com-mittee during the Vietnam War. By enlisting Landsat in 1973 to help South Vietnam better manage its natural resources in the Lower Mekong River basin, the U.S. government and the space agency ceded some control over the project to local government and scientific officials; to verify the accu-racy of Landsat data, NASA technicians had to compare it both with aerial photographs provided by government administrators in Laos, Cambodia, Thailand, and South Vietnam as well as with field observations made by indigenous scientists from local forestry, agriculture, and other natural-resource agencies. The "short term objectives" of the Lower Mekong Landsat experiment, explained NASA's Frederick Gordon, who oversaw the project from the Goddard Space Flight Center in Greenbelt, Maryland, was "sup-ported by ground truth data and field surveys" and with "aerial photo-graphs made available by the national departments" in the basin's four riparian countries.[161] The flood, land-use, and soil maps of the basin created three years later from NASA's Landsat data were thus also coproduced, a joint effort by both the space agency and locals on the ground in Vietnam.

Yet this joint effort was not between equals. Far from it. The over-whelming ability of NASA and the U.S. government to direct Vietnam's Landsat project in ways that supported American foreign policy was quite apparent in the Mekong Committee's quarterly report from April of 1976,

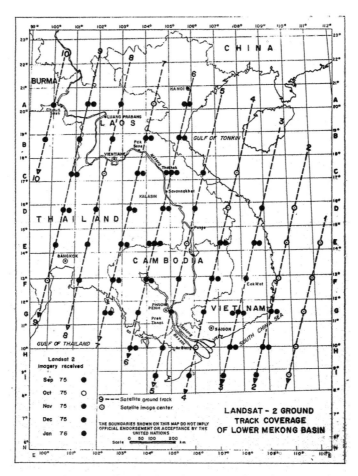

2.4   Map depicting the orbital track coverage of Landsat-2 over the
Lower Mekong Basin from September 1975 through January 1976.

which the space agency coauthored. Although NASA officials completed
the report more than six months after the fall of Saigon and the official end
of the Vietnam War, they, perhaps wishfully, referred in the text to the nation of "South Viet-Nam" even though the country no longer existed.
American interests were likewise front-and-center in the full-page Landsat
map accompanying the report (see Figure 2.4). The illustration by NASA,
which superimposed the ten orbital tracks of the satellite over a political
map of the region's national borders, refrained from identifying the recently
reunited country by its official name, the Socialist Republic of Vietnam,

and also included, quite prominently, a dotted line for the demilitarized zone that until quite recently had divided North from South Vietnam near the seventeenth parallel.[162] Additionally, while NASA did not "turn off" Landsat over Vietnam when the war ended in 1975, the U.S. government's decision to ban assistance to the victorious communist government essentially halted the Mekong Committee's remote-sensing program.[163] Thus while locals might have helped to produce Landsat maps to better manage Vietnam's water, trees and crops, and soil, it was the United States, with NASA's help, that wrote both the data's narrative and its ultimate conclusion.

The same could be said of Pham Tuan's Soyuz 37 flight. Although the world's first Asian cosmonaut no doubt helped Victor Gorbatko conduct scientific studies of Vietnam's natural environment from Earth orbit, it was politicians and scientists on the ground in the Soviet Union who no doubt approved those experiments. They also most definitely put pressure on Tuan to remind millions of viewers across Southeast Asia during his television broadcast not only that "the Vietnamese revolution has always enjoyed the great assistance of the Soviet Union," but also that "the Vietnamese people have not been able to enjoy genuine peace due to the war threats made by the U.S. imperialists" throughout the Vietnam War.[164] The United States Landsat program made similar, if less obvious, broadcasts from space. As a noted science policy analyst from the National Academy of Science explained in 1970 regarding NASA's Landsat program, "the earth resources survey satellite concept is one which, if exploited in an optimum manner, could provide an ideal opportunity for the technologically advanced nations of the world to converge their interests with the aspirations of the many developing countries in their efforts to build a just, peaceful, and economically progressive world community."[165] Stated less diplomatically, the space race and the Vietnam War allowed American and Soviet hegemony to become more subtle during the 1960s era and beyond—and therefore even more difficult to resist.

3

Thinking Globally, Acting Locally

Cape Canaveral and Whole Earth Environmentalism

In late February 1966, twenty-eight-year-old Stewart Brand placed one hundred micrograms of LSD on his tongue, climbed up to the roof of his apartment in North Beach, and took in the view. He sat there for hours, wrapped in a blanket against the winter sun, staring at the San Francisco skyline below. To the west, the Golden Gate Bridge shimmered orange above an electric bay. "The buildings were not parallel," Brand remembered later. They diverged from one another slightly "because the earth curved under them, under me, and all of us; it closed on itself." He could see that the world was not flat and endless, and he felt, on a deeper level than ever before, that it was round and finite. Brand remained on the gravel rooftop for the rest of the afternoon, tripping and trying to figure out how to share with the rest of the world his mind-altering "low-altitude view of the Earth." Before descending, from both the roof and his hallucination, he "herded his trembling thoughts together" and decided that a color photograph of the planet from space would do the trick. "There it would be for all to see," he explained, "the Earth complete, tiny, adrift, and no one would ever perceive things the same way."[1]

With such thoughts fresh in his mind, Brand rose early the next morning and ordered the printing of several hundred buttons inscribed with the question "Why Haven't We Seen a Photograph of the Whole Earth Yet?" He next visited the San Francisco Public Library, where he spent several hours searching government directories to locate politically powerful people; he wrote down the names and addresses of high-level NASA administrators, United Nations officials, Soviet scientists and diplomats, and

3.1   NASA's 22727, which later became known as both the
"blue marble" photograph and *Whole Earth*.

members of Congress along with their secretaries. When the buttons were
ready a few days later he mailed them off to those on the list. Brand then
dressed in a white jumpsuit, black top hat, and a blue Day-Glo sandwich
board that posed his question in bright red letters, and he traveled across
the bay to the University of California, Berkeley, where he began selling the
remaining buttons for twenty-five cents apiece. Students snapped them up,
faculty did as well, and an astrophysicist bought five for his friends at NASA's
Ames Research Center located nearby in what would soon be known as
Silicon Valley. After the dean's office kicked Brand off campus, sparking
coverage by the *San Francisco Chronicle*, he moved on to Stanford, where
he had majored in biology, and then crossed the country to hawk his wares
at Columbia, Harvard, and finally MIT, where his brother, Peter, was an
instructor. Within weeks his buttons began appearing on shirt collars and
jacket lapels at NASA Headquarters in Washington, DC, and a few years
later, in December 1972, the space agency finally made Brand's query moot
by releasing photograph number 22727, the first color image from space of
the entire planet, free from solar shadow.[2] (See Figure 3.1.)

So goes the origin story of the *Whole Earth* photograph. While many contemporary environmentalists remain unaware of Brand's trip-inspired advocacy back in the winter of 1966 and, for that matter, of the 1972 photograph's original numeric name, most nevertheless identify this now-ubiquitous image as having helped to jump-start the environmental movement.[3] Although NASA took the photograph, they argue, it was grassroots activists who transformed what looked like a child's "bright blue marble" floating alone in the empty darkness of space into an environmental icon by emblazing it on Earth Day flags, antipollution posters, recycling T-shirts, and "Think Globally, Act Locally" bumper stickers.[4] They also claim that *Whole Earth*, in turn, inspired skyrocketing membership in environmental organizations, including the Sierra Club; the founding of new environmental groups such as the Natural Resources Defense Council; and a barrage of "green legislation" during the 1970s, including the Clean Water Act, amendments to the Clean Air Act, and the Toxic Substances Control Act, among others. *Whole Earth* "reframed everything," explained Brand years later. "The photograph of the whole earth from space," he concluded, became a powerful cultural symbol of the "ecology movement."[5]

Less understood at the time was the scientific impact of 22727. Not only did the space agency reprint the image in the *Preliminary Science Report* for the Apollo 17 mission, but it also archived the photograph, along with similar ones of Earth from space taken during the late 1960s and early 1970s, in an effort to help scientists study the planet's air, water, and land.[6] "By taking photographs from the unmatchable vantage point of outer space," explained the *New York Times* back in October of 1968, astronauts are undertaking "scientific experiments" and "chores for science" that will "help future generations understand earth's climate of the late nineteen-sixties." Over the next several years scientists likewise analyzed photographs of the planet from space, including *Whole Earth*, to calculate continental drift, to map oceanic topography, and to trace weather patterns.[7] The space agency's 22727 was thus more than a powerful cultural symbol. *Whole Earth* and other photographs like it functioned, as well, as important scientific data sets for mapping global ecology.

This divergent reception of *Whole Earth* by grassroots activists on the one hand and earth scientists on the other neatly captures the divisions within the burgeoning environmental movement at the very moment when Apollo 17 astronauts were pointing their camera back at planet Earth.

While activists from older conservation groups, such as the Wilderness Society and the Izaak Walton League, continued to focus on the protection of open space and wildlife in rural regions, members of newer organizations, such as Stamp Out Smog in Los Angeles and Citizens for Clean Air in New York City, became involved instead in single issues that affected mostly cities. Early grassroots environmentalists thus acted mostly locally on a wide variety of concerns. To publicize their specific causes, whether to the public or politicians, most nevertheless depended on the expert knowledge and elite status of scientists, who played an equally critical role in the emerging environmental movement. While these scientific specialists did much to aid grassroots activists, they were concerned overall with a different set of issues that threatened the entire planet, such as the spread of nuclear fallout from the testing of atomic bombs, the widespread use of chemical pesticides, and overpopulation. Thus, much like the bumper sticker that adorns the cars of many environmentalists today, the fledgling environmental movement was somewhat split between scientific elites *thinking globally* and grassroots activists *acting locally*.[8]

Stewart Brand's political maneuverings at Berkeley and Stanford out west and Columbia, Harvard, and MIT back east also suggest that this emerging environmental movement, although lacking coherence, was in fact influencing the space race. For instance, Brand's "Why Haven't We Seen a Photograph of the Whole Earth Yet?" buttons not only began adorning the workday attire of NASA's top brass, but also compelled those administrators to modify Apollo 17's trajectory so that astronauts could capture an unshadowed photograph of the entire Earth.[9] Additional activism by both grassroots environmentalists in their communities and environmental scientists in their laboratories similarly pressured NASA officials to rethink their overall mission. Should the space agency focus on exploring outer space or be responsible as well for stewarding the whole Earth? Thus much like the civil rights and anti–Vietnam War movements, both of which influenced and were influenced by efforts to land a man on the moon, environmentalism and the space race also shared a common history during the 1960s era.

Many writers have described the rise of *Whole Earth* as a so-called green icon, yet few have plumbed the interrelationship between NASA and the environmental movement. For environmental historians the space agency exits the stage soon after Apollo 17 captured 22727, perhaps making another brief guest appearance more recently when NASA's scientists and technology played a starring role in identifying global warming.[10] Space historians have been even more dismissive of this courtship between astronauts, and their aerospace brethren, and environmentalists, often portraying NASA as an inept and shallow environmental organization more interested in public promotion than the ecological health of planet Earth.[11] As one space scholar has argued, "NASA's institutional blinders and groupthink failed to connect it strongly to the major science-based social movement of the late twentieth and early twenty-first centuries."[12] The space agency's environmental efforts, he concluded, were mostly for public relations purposes. While this may have been the case regarding NASA and other political movements of the 1960s era, especially with respect to civil rights, the space agency's engagement with environmentalism was more historically complex than a mere PR campaign.

Understanding this complexity requires looking back further than February of 1966, when Brand set off on his rooftop trip high above San Francisco, to the very dawn of the space age when NASA officials spurned initial overtures from environmentalists to focus instead on a very public love affair with space technology. During these years the space agency's public relations office "advertised" NASA's rockets, lunar landers, and command modules to the public through gift-shop postcards and bus tours at Cape Canaveral, in exhibits at the 1964 New York World's Fair, and in elaborately choreographed live television coverage of early Apollo launches. In the late 1960s, however, and especially after the first Earth Day in 1970, the two factions within the environmental movement became more vocal in their criticism of NASA's technology. While environmental activists condemned the hardware for both causing pollution and distracting Americans from cleaning up their own increasingly contaminated environs, environmental scientists faulted the technology, along with the engineers who designed it, for leaving little room onboard for experiments that could help solve the environmental crisis plaguing planet Earth. Officials at NASA responded to these critiques in more substantive ways than they had to those of the civil rights and antiwar movements, in this case by reframing the space agency's technology on the local level at the Cape while simultaneously redirecting it from outer space back down toward the global environment. Acting locally and thinking globally were now under one roof, and, as Stewart Brand had predicted, "no one would ever perceive things the same way."

Promoting Space Technology

In many ways it was nature that first lured NASA to the Cape. The region sat at 28.5 degrees north latitude along the eastern edge of Florida, making it closer to the equator than most other locations in the continental United States. Such geography allowed the space agency to save on rocket fuel by using the rotation of the Earth as a springboard, of sorts, when firing rockets eastward into space. The equatorial climate would also ensure year-round launches. The Cape's position jutting out into the Atlantic Ocean would be beneficial as well, providing a safe crash site for failing rockets and also facilitating easy water transportation for the enormous materials moving to and from the space center. Engineers at NASA could load the Apollo space capsule's enormous Saturn V boosters, which were developed and assembled at the Marshall Space Flight Center in Huntsville, Alabama, onto barges for a five-day float down the Tennessee, Ohio, and Mississippi rivers to New Orleans before making the journey by tug across the Gulf of Mexico, through the Florida Keys, and up the Atlantic coast via the Intracoastal Waterway to the Cape.[13] Knowledge of these environmental attributes was not new in the early 1960s; nearly one hundred years earlier, in his 1865 science-fiction classic *From the Earth to the Moon*, Jules Verne shot his space projectile from a giant cannon also located in Florida.[14]

Another natural advantage of the Cape was, according to one reporter, its "virgin" state prior to the arrival of NASA. "What is now Kennedy Space Center was virtually a semi-wilderness," wrote the correspondent in October 1968.[15] Additional adjectives used to describe the region during the early postwar period included "untamed," "unsettled," and "isolated."[16] While such descriptions ignored the long history of Native Americans in the area as well as the more recent practice of citrus growing, which began in the early nineteenth century, the natural environment of the Cape during the pre-NASA era remained predominantly undeveloped.[17] "You ask me how it was back in those days" before the space agency, joked a local resident, Jim Deese, in 1968. "It was nothing but palmettos, mosquitoes and rattlesnakes with fangs like 20-penny nails."[18] As a result, by the winter of 1964 NASA had spent just two and a half years and only $72 million to purchase more than 130 square miles of Cape Canaveral real estate, and it had negotiated with the State of Florida for the use of an additional eighty-five square miles of submerged land.[19]

Even before acquiring all of this territory, NASA began building a new technological landscape atop this "virgin" "semi-wilderness." The space agency began in October of 1962 by clearing 2.5 square kilometers of palmetto groves and orange trees and by dredging an access channel to NASA's new launch site. It then used the channel fill to plug the surrounding marshland and to build up a solid foundation for Launch Complex 39, which included a launch control center, two rocket launchpads, and the Vehicle Assembly Building (VAB), which would house the thirty-six-story Saturn V rocket and Apollo spacecraft prior to liftoff. To transport the 360-foot-tall space vehicle from the VAB, NASA paved a five-mile-long "driveway" across former swampland to its two launchpads. Additional construction projects at the Cape involved the completion of an aerospace transportation network that included landing strips, access roads, and even a deepwater port for oceangoing barges and ballistic missile submarines.[20] "A new landscape is taking shape" at the Cape, explained *Time* magazine in 1965. "Its principle feature is the tall, white, broad-hipped barn for rocket assembly."[21] General Electric engineer Ray Forbes, who worked on the Apollo program during the mid-1960s, agreed, noting in 1969 that after such construction Cape Canaveral was "not a natural environment."[22]

Such braggadocio regarding the Cape's technological makeover had deep Cold War roots. As already noted, the orbiting of Sputnik by the Soviet Union in 1957 launched not only the space race but also widespread insecurity throughout the United States regarding America's technological capabilities, or lack thereof.[23] While President Kennedy responded in 1960 by calling for a man on the moon by decade's end, American businesses began publicizing, as never before, their own technological capabilities in the hopes of alleviating such public anxiety. The 1964 World's Fair was one such public advertisement. While previous world's fairs had traditionally fixated on each nation's homegrown technologies, the corporate sponsors of New York's fair propelled such practices far into the space age.[24] The symbolic heart of the fairgrounds, for instance, was the twelve-story Unisphere, a hollow, stainless-steel globe of the Earth encircled by three satellites. Numerous rides at the fair, including Martin Marietta's "Rendezvous in Space" and General Motors' wildly popular "Futurama II," illustrated American businesses' involvement in developing space technology by taking fairgoers on simulated voyages into outer space. Administrators from NASA piggy-backed on such efforts by cosponsoring, along with the Department of Defense and the fair itself, the United States Space Park, a two-acre section of the fairgrounds that included a full-scale model of the enormous five engines comprising the first stage of the Saturn V, the Mercury capsule flown by the astronaut Scott Carpenter during the second American manned orbital flight, and full-scale mockups of the three-man Apollo command, service, and lunar landing modules.[25] The space agency invited the fair's fifty-one million visitors to walk among, pose with, and buy color souvenir postcards depicting all of these space technologies (see Figure 3.2).[26]

3.2   Postcard of NASA's U.S. Space Park at the 1964 New York World's Fair.

During the mid-1960s NASA administrators did at Cape Canaveral what they and corporate sponsors had done at the fair in Flushing, Queens; they began putting space technology on display for the visiting public. Admitting sightseers to what had been renamed the Kennedy Space Center (KSC) began in November 1964, when NASA administrators opened the launch site on Sundays for self-directed drive-through tours. When more than 30,000 automobiles arrived at the center during Christmas vacation the following year, causing massive traffic jams, NASA decided on a more formal public program involving escorted tours operated by Trans World Airlines (TWA) with Greyhound buses. The first such bus pulled out on July 22, 1966, and in 1969 alone more than one million visitors climbed aboard for similar excursions. The original tour cost $2.50 for an adult and half that for a child, covered forty miles in two-and-a-half hours, and highlighted, all along the route, NASA's large-scale space technology.[27] The trip included stops at Launch Complex 39's mission control center, a drive-by of NASA's two launchpads, and a walk-through of the enormous VAB, as well as visits to other technological attractions. "Everything you see is gigantic," explained the *Baltimore Sun* in early 1969 of this NASA bus tour. "You are awed by what man can do."[28]

At the end of the tour this technological revue continued when TWA's silver Greyhound buses deposited sightseers at the KSC's new visitor center, which housed NASA's Space Museum. Patrons entered the building, which also began operating in 1966, through two sets of glass doors that opened up onto 3,000 square feet of exhibit space, featuring, as one reporter explained, "a great collection of the hardware that has been flung aloft in the past decade or so." Inside, visitors could see up close a Nimbus weather satellite, a model of the Apollo command module with a cutaway window exposing three cramped dummy astronauts, and the actual Gemini IX capsule that had orbited Earth for three days in 1966. "You marvel at the burned and blackened heat shield," added the reporter.[29] The big draw for most tourists, however, towered beyond the museum's back wall, which was made entirely of plate glass, where NASA planted an array of technological wonders in its Rocket Garden. Out back at the Cape, as they had at the World's Fair in Queens, visitors could take photographs alongside booster rockets that had propelled astronauts into space and walk up a staircase to see inside the Apollo lunar landing module. As one Florida newspaper boasted, a "steady stream of visitors peek inside every day."[30]

After their bus tour and garden stroll, tourists often departed through the visitor center's gift shop, where the space agency continued selling, quite literally, its technological achievements to the public. "If you have not listened" during your bus tour, explained one reporter, "you can buy some space history here, in the store at the center."[31] While such purchases could involve plastic models of various NASA rockets, by far the most popular of these souvenirs were three-by-five-inch postcards depicting the technological marvels that had flitted by just moments earlier outside tour-bus windows.[32] For each of these postcards NASA supplied color photographs of rockets being assembled inside the VAB, transported from the VAB by the giant "crawler," and poised for liftoff on one of the space agency's two launchpads (see Figure 3.3).[33] Tourists could even spend five cents on mementos depicting the very building in which they stood; NASA's gift shop sold postcards of the Space Museum's interior filled to the brim with "hardware," of the visitor center's exterior with its Rocket Garden looming behind, and, in case they had missed that special Kodak moment earlier in the day, of tourists ascending steps to peer inside a mock-up of the Apollo lunar landing module (see Figure 3.4).[34]

While bus tours, space museums, and postcards, each packed with space technology, greeted tourists when they visited KSC during the 1960s, NASA sold a similar product during televised launches to those who were unable to make the trip to sunny Florida. Beginning in 1961 with Alan Shepard's suborbital Mercury flight, space agency administrators encouraged live television coverage of NASA launches by constructing a special press site at the Cape, complete with a well-lit tent filled with long writing tables, phone booths, and bleachers for reporters working on their copy. The space agency also built a platform stage for television cameras with direct views of its launchpads. Six months later, as the country prepared for John Glenn's first orbital flight, NASA again promoted its space technology to the wider world by allowing ABC, CBS, and NBC to build their own temporary television studios at the Kennedy Space Center. Each television network built "gazebos" on top of their trailers, explained *Newsweek* correspondent Edwin Diamond in 1964, "in order to 'frame' the commentators, or anchormen, properly against the backdrop of the launching pad" with its towering rocket.[35] The space agency intended such advertisements, whether they took place on launchpads, outside bus-tour windows, or inside visitor-center gift shops, to counter the anxieties sparked by Sputnik.[36]

This space age public relations blitz worked its magic on the American public. As already noted, after witnessing the technological spectacle of

3.3   Kennedy Space Center postcard of the Saturn V rocket inside the
Vehicle Assembly Building, circa mid-1960s.

Apollo 11's launch during the summer of 1969, the civil rights leader Ralph
Abernathy admitted to feeling "a great deal of joy and pride" and even
forgot, for a brief moment, why he was protesting.[37] Visitors touring KSC
during the 1960s had similar transformative experiences. "A practical man
or woman should never tour Cape Kennedy," wrote the "housewife reporter" Erma Bombeck in her nationally syndicated column "At Wit's

3.4 Kennedy Space Center postcard of visitors touring a mock lunar lander in the visitor center's Rocket Garden, circa mid-1960s.

End," after her own visit to the Cape in March of 1968. "It's not practical to create 25 square miles of research and development on scrublands with more snakes per square mile than anywhere in the country." Such impracticality nevertheless impressed this self-proclaimed Midwestern suburbanite. After touring KSC's launchpads, mission control center, and VAB, Bombeck confessed that she "began to feel the first warning pangs of optimism and hope." She then spied Apollo 6 looming in the distance. "Oh, I'll see it next week on my TV screen," she wrote, "but I will never feel the pride of seeing it gleaming in the sunlight with the U.S. flag on its side like a stamp of quality." Bombeck then ended her column with a tongue-in-cheek reference to NASA's heralded, and handsome, German rocketeer. The tour of the Kennedy Space Center, she admitted, had given her "an advanced case of Wernher von Braunitis."[38]

NASA Technology as Environmental Hazard

While the space agency's unique strain of techno-fever proved contagious for Bombeck, her readers nationwide, and many of the tens of millions who visited Cape Canaveral or watched Apollo launches on television during the mid-1960s, later in the decade this infectious enthusiasm subsided as grassroots members of the emerging environmental movement began raising concerns regarding NASA's hardware. The national media began echoing this critique in the fall of 1966 when, after an unexpectedly long extravehicular activity (EVA) in orbit during the Gemini 12 mission, the astronaut Buzz Aldrin dumped unneeded equipment out the open capsule hatch in an attempt to save fuel for his return to Earth. "Keep space clean," joked Mission Control in Houston, according to the Associated Press.[39] Such seeming irresponsibility raised additional questions concerning not only increasing amounts of "space junk" orbiting the planet, but also waste left behind on the lunar surface at each of the Apollo landing sites.[40] "Once they get to the moon, they plan to toss debris out like some Sunday driver," complained the Philadelphia *Evening Bulletin* in an article titled "Astronauts Are Neat in Their Cabin but Will Be Litterbugs on the Moon." Outer space's "first official trash dump," the paper continued, will receive a page-long list of garbage including TV cameras and cables, brackets and tools, empty bags and boxes, old urine containers, and the descent stage of the lunar landing module, all in an effort to save fuel for the return trip to Earth.[41] While one Washington reporter quipped that Apollo astronauts should take an empty beer can with them, "just for symbolism's sake," another grumbled in an article titled "Litter Is Remnant of Man's First Moon Visit" that "if little green men were on the moon, they might give astronauts Armstrong and Aldrin tickets for being litterbugs."[42]

Political cartoonists in the late 1960s seemed to agree, and they actually began drawing these little green men alongside the refuse that astronauts left behind on the lunar surface. The Philadelphia *Sunday Bulletin*, for instance, ran an illustration by Douglas Borgstedt in late July of 1969 that portrayed two extraterrestrials sporting furrowed brows and frowns while using garbage spikes and trash bags to collect litter left over from a lunar landing. "You'd think they'd at least pick up!" complained one of the pointy-headed aliens.[43] The *Minneapolis Tribune* ran a similar cartoon, which appeared a mere two days after Apollo 11 left the moon, which was even more direct in its criticism. Whereas Borgstedt's illustration scolded an unknown "they," the cartoonist Scott Long's "Mankind's Magnificent Obsession . . ." leaves little doubt regarding who, among mankind, was to blame (see Figure 3.5).[44] Not only is "United States" inscribed prominently across the lunar landing module's descent stage, but amid mission trash, including a spacesuit, two cameras, an EVA backpack, and perhaps even the

3.5   Scott Long, "Mankind's Magnificent Obsession . . . ," *Minneapolis Tribune*, July 23, 1969.

boot in which Neil Armstrong took "one giant step," is the American flag that Apollo 11 astronauts had proudly planted. The takeaway for American newspaper readers was quite obvious; NASA's technology was polluting the moon.

It was also responsible for contaminating planet Earth. After admitting that "a lot of bright individuals with highly developed technical skills" were responsible for the Apollo 11 lunar landing, the editors of *The Screw*, a left-leaning alternative newsletter published in Kansas City, Missouri, reminded readers in 1969 that "many of the industries manufacturing the components for the space shots are the very ones responsible for the chemical pollution that threatens to make the Earth as lifeless as the moon."[45] Such environmental worries were even more pronounced regarding the next generation of NASA's rockets. "One hundred tons of exhaust released sixty-two miles high could produce a cloud of smog 100 miles wide that would persist for weeks," explained *Newsweek* magazine of this environmental concern in an article titled "Sky-High Pollution." "If several countries engage in a space race," *Newsweek* warned, "man may succeed in fouling the space around him as well as his streams, his soil, and his air."[46]

These anxieties were especially pronounced regarding NASA's future space shuttle, which engineers began developing in the late 1960s, and its potential to pollute the air and water both near Cape Canaveral and beyond. Grassroots opposition to the spacecraft germinated in June 1963 when the Kennedy administration announced plans to develop the supersonic transport (SST), a commercial passenger aircraft that would travel faster than the speed of sound. While the president argued that the SST, like the Apollo program, would illustrate to the world America's technological expertise, environmentalists countered that the high-altitude jet would make too much noise with its constant sonic booms, use too much fuel during a looming energy crisis, and cause air pollution from an exhaust system that threatened atmospheric ozone. These environmentalists also took action. In the fall of 1967 the Sierra Club and the Wilderness Society both passed resolutions condemning the SST, while Friends of the Earth bankrolled the publication and distribution of a "fact book" critical of the supersonic technology. The Environmental Defense Fund, the Natural Resources Defense Council, the Conservation Foundation, and several other environmental groups eventually joined the campaign.[47] When such efforts convinced both Wisconsin Democratic Senator William Proxmire to hold hearings on the SST in May 1970 and Congress to halt funding for the technology in March of the following year, grassroots environmentalists shifted their concerns to NASA's new reusable spacecraft, which would also fly at high elevations, similarly break the sound barrier, and likewise threaten Earth's ozone with its damaging exhaust. "Another new danger would be the space shuttle," explained the *Chicago Tribune* in the early 1970s. "The exhaust of the shuttle rockets would release tons of chlorine which could then react with ozone."[48]

Along with criticizing space technology for causing pollution back on Earth, the emerging environmental movement also blamed NASA for distracting Americans from cleaning up environments already suffering from degradation. *Newsweek* magazine again expressed this widespread sentiment in early July 1969 in its cover story, "The Moon Age," which included a section titled "Good Earth?" that was illustrated with alarming photographs of garbage-strewn roadsides in New Orleans and waste being dumped into New York's Hudson River. "The same technological impulse that is carrying Apollo 11 outward to the moon," reported *Newsweek*, "is also threatening the home environment" with air pollution, ocean contamination, and the potential for global warming triggered by the industrial production of carbon dioxide. Americans' love affair with technology, the magazine concluded, not only caused such pollution but also made it more difficult to correct.[49] Such thinking reflected that of the environmental movement. "Environmentalists and other critics," reported the *Boston Globe* in the spring of 1970 just two weeks before the first Earth Day, "are still attacking what they call 'the talent and money-grabbing space program.'"[50]

That these attacks were ecological in nature is evident, once again, in syndicated political cartoons that began appearing in newspapers across the country during the Apollo 11 summer of 1969. The *Baltimore Sun* and the *Charlotte Observer*, among others, ran an illustration soon after Armstrong, Aldrin, and Collins returned to Earth that was in many respects the mirror image of the cartoons that accused Apollo astronauts of being lunar litterbugs.[51] Yet in this cartoon, instead of pointy-headed aliens it was an astronaut, outfitted in the same technology that Apollo left behind on the moon, who was himself collecting trash, including bottles, cans, and cups, from a beach back on Earth. As a guilt-ridden crowd of sunbathers looked on, one astonished bystander explained to a police officer investigating the scene, "He says he's from the moon to pick up Earth samples." A second cartoon by L. D. Warren, which appeared in several newspapers including the *Cincinnati Enquirer* and the Philadelphia *Evening Bulletin*, was perhaps even more powerful due to its sheer simplicity. In it Neil Armstrong pauses for a moment, before climbing back up the ladder for Apollo 11's return trip, to look back at a smog-shrouded Earth with the word "POLLUTION" obscuring most of North America. "Let's take a few more deep breaths, Buzz, before we leave!" Armstrong tells Aldrin. Warren reiterated the contrast between these polluted and pristine atmospheres by placing Armstrong's EVA backpack, which produced clean air for Apollo

'Let's take a few more deep breaths,
Buzz, before we leave!'

3.6   L. D. Warren, "Let's Take a Few More Deep Breaths, Buzz, before We
Leave!," *Philadelphia Evening Bulletin*, July 22, 1969.

astronauts, front and center in the cartoon (see Figure 3.6).[52] The message,
unlike the atmosphere back home, was crystal clear. In the rush to land
men on the moon, the space race in general, and NASA in particular, had
diverted the nation's attention from more pressing environmental problems on Earth.

This early indictment of NASA technology for exacerbating pollution on
Earth reflected a key concern of an increasing number of environmental

advocates in the mid- to late 1960s. Secretary of the Interior Stuart Udall, for instance, called attention to the links between the space race and America's pollution problem in his 1963 call-to-arms, *The Quiet Crisis*. "Our successes in space and our triumphs of technology hold a hidden danger," he warned, resulting in "an overall environment that is diminished daily by pollution."[53] Two years later Ralph Nader's *Unsafe at Any Speed* similarly challenged technology's "power to pollute," as did a litany of technological disasters during the late 1960s including the deadly New York City air "inversion" of 1966, the *Torrey Canyon* shipwreck and oil spill in the English Channel the following year, and in 1969 the Santa Barbara oil spill, the Cuyahoga River fire, and the declaration that Lake Erie was "dead" because of sewage and chemical dumping.[54] Such events not only helped focus grassroots environmentalists on the common concern of pollution, but also influenced the first Earth Day in 1970; the great majority of the twenty million activists participating in what became the largest demonstration in U.S. history did so by condemning sources of pollution affecting their local communities.[55] By contaminating specific places, from the Sea of Tranquility to Cape Canaveral, and by distracting Americans from cleaning up pollution in their own backyards, NASA, they argued, was one such source.

This grassroots criticism represented only one side of the environmentalist backlash against the space agency. The other emerged from environmental scientists, who first expressed discomfort with NASA's space technology during the early 1960s. In 1962, for example, Professor George Kistiakowsky, a Harvard chemist and former science advisor to President Eisenhower, complained publicly that America's space program was a "technological spectacular" more concerned with national prestige than scientific progress.[56] By the following year such lone criticism had become a "wave of protest" spilling across the pages of *Business Week*, which reported in an article titled "Sniping at NASA on Space" that "charges by scientists that the U.S. space program is losing its scientific objective show no signs of cooling down." According to the magazine, these denunciations were loud and widespread, emanating from a large number of senior scientists in a variety of research fields who had no personal stake in the space program. "They argue," *Business Week* explained of these researchers, "that true scientific purpose is being subordinated in the allocation of funds to the engineering problem of landing men on the moon."[57] The editors of *Science* agreed. "Scientific exploration," they contended during the spring of

1963, "has been accorded a secondary priority in the lunar program," evidence of which was NASA's decision not to include a scientist in the first lunar astronaut crew. "A reexamination of priorities is in order," *Science* demanded.[58]

The space agency's reluctance to outfit Apollo technology with scientific experiments stemmed from widespread concern among NASA administrators and engineers that doing so would jeopardize the overall goal of the mission, which was to land humans physically on the moon. "What were those wild-eyed scientists talking about," remembered one NASA engineer, "having astronauts diddling around on the moon setting up instruments? We weren't even sure they could walk on the surface!"[59] There was also the costly issue of weight; the heavier the scientific equipment the more fuel each mission would burn. As a result not only were NASA engineers dismissive of the space agency's life scientists, but when NASA finally established its Office of Space Science Applications (OSSA) in 1963, the division remained underfunded, disorganized, and subordinate throughout the rest of the decade.[60] In 1968, for instance, the OSSA barely beat back a proposal by engineers to eliminate all scientific experiments from Apollo 11 and later had to acquiesce, under administrative pressure, to a reduction in the number of lunar experiments from four to three. When Aldrin and Armstrong landed on the moon the following year they thus deployed a more limited version of the Apollo Lunar Surface Experiments Package, which now included only a seismometer for recording the moon's internal activity, a panel of mirrored laser reflectors to aid in measuring precisely the distance to the moon, and a solar-wind collector to capture particles and gases expelled by the sun.[61]

The decision to include such small-scale experiments on Apollo 11 did little to appease NASA's own scientists, who in the late 1960s began joining this "wave of protest" by acting with their feet. As *Science* reported in August 1969, soon after Armstrong and Aldrin returned to Earth, NASA began experiencing a scientific "brain drain," with high-level departures including Dr. Wilmont Hess, director of science at the Manned Spacecraft Center in Houston; Dr. Elbert King, curator of NASA's Lunar Receiving Laboratory; Dr. Donald Wise, chief scientist at NASA's Office of Lunar Exploration; and Dr. Eugene Shoemaker, the principal geologist for the space agency. "There is considerable disagreement over whether science is the primary justification for sending men into space," explained *Science* in response to these resignations. "The mass exit of scientists from the Apollo

Program seems to indicate that it isn't."[62] Even scientist-astronauts called it quits after realizing, rather quickly, that science, as the *Washington Post* put it, has "been forced into the back seat of the manned space program."[63] In 1967, Brian O'Leary, one of the first scientists chosen for NASA's astronaut corps, learned this lesson on his first day on the job when his boss, Deke Slayton, the director of flight crew operations and future Apollo-Soyuz Test Project astronaut, bluntly told him, "We don't need you around here." Similar indifference from NASA administrators, engineers, and astronauts followed. "It appears," O'Leary concluded in an article he wrote for *Ladies' Home Journal* regarding his resignation from the astronaut corps, "that the engineering aspect of Apollo has driven the scientists away."[64]

The wider scientific community supported their colleagues at NASA by initiating a public relations campaign during the late 1960s that not only criticized the space agency's technology for its lack of scientific applications, but which also promoted an alternative scientific mission for America's space program. The *Bulletin of the Atomic Scientists* was the forum for this public condemnation. Founded in 1945 by researchers working on the Manhattan Project to develop the atomic bomb, the activist journal originally focused on educating lay readers about the dangers posed by nuclear weapons. By the mid-1960s, however, it had begun sounding the alarm over a host of science-related issues, one of which was the space race. "To date NASA has devoted only a minor fraction of its resources to the pursuit of science," argued Allan H. Brown, a University of Pennsylvania biologist, in the *Bulletin* during the spring of 1967. To correct this situation, Brown proposed that NASA not only increase expenditures for space technology with scientific applications but also embrace what he called a "diverse goals strategy" that would reorient some of this space science away from the moon and outer space and direct it back down toward Earth. Such efforts, Brown explained, could entail the use of Earth-sensing satellites for agriculture and forestry, weather surveillance, and assessment of Earth's atmosphere.[65] Nearly a dozen *Bulletin* articles followed that advocated for space technologies that would enhance scientific understanding of our home planet's environment, as Brown argued, and also demanded congressional hearings if NASA refused to take such steps.[66]

Such forthright environmental advocacy by scientists had first emerged after World War II, when a new generation of researchers began taking concerted political action to warn the public about various environmental hazards resulting from America's headlong embrace of modern technology. Aboveground nuclear weapons testing during the late 1950s and the increasingly widespread use of chemical pesticides during the early 1960s helped spark this environmental crusade by scientists. In her 1962 bestseller, *Silent Spring*, for instance, Rachel Carson compared Strontium-90, a radioactive isotope in nuclear fallout, to "chemicals sprayed on croplands or forests or gardens" that "lie long in the soil, entering in a chain of poisoning and death."[67] Fellow biologist Barry Commoner, who perhaps more than any other scientist rallied his colleagues to educate the public about nuclear fallout, also blamed technology for causing environmental problems in his 1966 classic *Science and Survival*.[68] This new breed of activist-scientist, which also included, at opposite ends of the political spectrum, the biologist and population-control advocate Paul Ehrlich and the radical ecologist Murray Bookchin, among others, became the scientific analogue of Udall, Nader, and the other lay critics who during the late 1960s began expressing the concerns of grassroots environmentalists.[69]

The experts criticizing NASA technology in the pages of the *Bulletin of the Atomic Scientists* were participants in this scientific wing of the environmental movement, and in the early 1970s they intensified their campaign by expanding upon Dr. Allan Brown's "diverse goals strategy" for the space agency. After once again lamenting NASA's blind eye toward science in a 1971 "comment" titled "Requiem for the Scientist-Astronauts," the *Bulletin* argued that "it should be recognized" by the American public that NASA's space technology "provides the tools and the laboratory to investigate" what the journal called "the *total* environment" of planet Earth.[70] The following year the *Bulletin* reiterated this point by explaining that America's space program had become vulnerable to its critics, including scientists, precisely because too few administrators and engineers at NASA were "concerned about its contribution to understanding man's *total* environment."[71] The space agency's hardware, in other words, including the Apollo spacecraft, orbiting satellites, and NASA's Skylab program planned for later in the decade, had the potential to reinvigorate support for America's space program by providing the necessary scientific information to better understand the global environment. "It would be an American tragedy," the *Bulletin* concluded in a 1971 essay titled "The End of Apollo: The Ambiguous Epic," "if the stresses of the moment persuade us to abandon the most important program of environmental exploration any society has ever undertaken."[72]

As grassroots activists and trained scientists continued to criticize NASA technology during the late 1960s for both polluting the planet and ignoring science, the American public took notice, which in turn forced politicians in Washington, DC, to take action.[73] According to a Harris poll conducted in 1967, this backlash by environmentalists against NASA, much like that by civil rights and antiwar activists, was weakening the agency's overall popularity. "It's not worth the money," complained a Delano, Florida, housewife when interviewed by the Harris survey team. "We ought to take care of first things first, such as clearing up the air and water here on earth."[74] A subsequent Harris poll conducted two years later found that the majority of Americans would prefer to cut NASA's budget while increasing funds for federal programs aimed at controlling and eliminating air and water pollution.[75] Partly because of this growing public wariness, between 1967 and 1969 Congress decreased NASA's budget from $4.9 to $3.9 billion, a reduction of nearly 25 percent after accounting for inflation.[76] In its coverage of this particular reduction the *Boston Globe* explained that the space agency recently had been the subject of "scathing criticism." "Only part came from scientists," the *Globe* reported. Additional condemnation "came from critics ranging from Congressmen to turned-off kids" who asked "why billions for space" when there is "pollution" back on Earth.[77]

NASA Acts Locally

Beginning in 1970 NASA administrators began debating, internally, the criticisms levied by both grassroots activists and professionally trained scientists regarding the space agency's technology. Deputy Administrator George Low was the first NASA official to embrace this critique. "In the 1960s, the country was definitely looking outward, and the national priorities included the Apollo goal, because this would establish clearly in our minds and in the minds of the world technological leadership by the United States," Low wrote in a personal memo recapping discussions held during two high-level planning meetings in the summer of 1970. Beating the Russians to the moon, he explained, was the single-minded goal of NASA, and Americans supported that effort. "The situation in the beginning of the 1970s is very different," he cautioned. The United States was now an "introspective nation," concerned more with problems closer to home. "This is why anything we say about the environment or the quality of life or ecology has a great deal of appeal." The space agency's number-two administrator concluded his note by stating that if NASA wanted to move forward with a strong space program it must better illustrate the program's benefits to the environments of Earth. "This means," Low argued, "that a space applications program and, specifically, an earth resources program should be the keystone for the space effort of the 70s."[78]

During the next two years the space agency's public relations team lobbied, from within, to make Low's vision a reality by promoting a campaign that would publicize the environmental applications of NASA's technology to the outside world. The office of the Administrator for Public Affairs, explained Low, "has been pushing for quite some time that we consider, for PR purposes, the entire NASA program in terms of an environmental theme: The Study of the Earth and its Environment."[79] This public relations campaign, if implemented, would involve press releases and conferences, the production of movies and television clips, statements by astronauts and talks by NASA administrators, all of which publicized how NASA's space technology "will contribute to a better understanding of Earth's environment or the forces that affect the environment." This overarching "environmental approach," argued John Donnelly, NASA's deputy administrator for public affairs, "attaches us to a movement that ranks high on the list of public and congressional priorities and makes us 'relevant.' "[80]

Although administrative higher-ups ultimately rejected Donnelly's call to reposition NASA as an environmental agency, during the early 1970s the space agency nevertheless began taking steps to improve its standing among grassroots environmentalists. Such actions began on the local level, close to home. Although NASA had agreed to the creation of the Merritt Island National Wildlife Refuge at Cape Canaveral back in 1963, it did so reluctantly not to protect wildlife but rather to serve as a safety buffer zone between the public and its launchpads. Beginning in November 1969, however, space agency administrators identified approximately 80,000 acres of land not being actively used at KSC and enthusiastically handed it over to the Department of the Interior's Fish and Wildlife Service. In 1972, NASA again eagerly transferred additional land in order to expand the preserve to more than 140,000 acres, making it the second largest wildlife refuge in the state of Florida and among the largest in the Southeast.[81] "We are pleased to have a well-administered refuge on the Center," explained KSC director Kurt Debus in the early 1970s. "It not only preserves KSC's great natural assets but provides both employees and visitors opportunities to share in the area's historical wildlife and recreational resources."[82]

Three years later, in 1975, NASA also signed an agreement with the National Park Service for the creation of the Canaveral National Seashore along the northeastern edge of the Kennedy Space Center property. Discussions between the space agency and the Department of the Interior regarding the Cape's coastline had begun back in 1966 with a National Park Service study that determined, as Interior Secretary Stewart Udall explained the following year to NASA Administrator James Webb, that the Kennedy Space Center shoreline was "one of the outstanding seashore areas on the Atlantic Gulf Coasts" and thus the perfect site for a national park.[83] Although Webb initially opposed Udall's plans, for fear of relinquishing land that might be necessary for future launches involving larger rocket systems, during the late 1960s environmentalism's increasing popularity convinced him to reconsider.[84] "The broad appeal of conservation and preservation of 'wild' areas," explained Webb's associate administrator, George Mueller, regarding the national seashore proposal, "indicates that execution of the agreement at the Administrator's level, with appropriate ceremony, should be considered."[85] After several years of painstaking negotiations, in the mid-1970s the space agency finally granted the National Park Service the use of more than 40,000 acres of KSC land, including twelve miles of coastline, for the Canaveral National Seashore.[86] "When we first started talking with them" in 1966, explained one state conservation official who was involved in the negotiations with NASA, "their attitude was 'leave us alone, we don't want to talk to you.'" Since that time, the official added, "NASA's attitude has completely changed."[87]

The space agency also became actively involved in the environmental stewardship of both of these Department of the Interior sites. Kennedy Space Center personnel and funding helped the wildlife refuge during the early 1970s to hold annual bird counts and to band ducks to track their migratory patterns, to transport brown pelicans from the Cape for reintroduction in Louisiana where they had become nearly extinct, to relocate alligators from heavily visited areas of KSC to more wild locations on the refuge, and to monitor, in partnership with the local Audubon Society, the resurgence on the Cape of bald eagle populations, which the space center's newspaper noted had been decimated by "the cumulative effects of pesticides, such as the long-lived DDT."[88] The space agency undertook similar actions along Canaveral National Seashore by helping to initiate a tag-and-release program for endangered baby sea turtles in 1973, by sponsoring an ecological study of local fish species in 1974, and by cooperating with a radar-monitoring program for manatees in 1976.[89] Funds from NASA and plans by its engineers also helped halt dune-barrier erosion, caused by increasing numbers of visitors, along the Cape Canaveral shoreline during the early 1970s.[90] "It is a unique co-existence of Saturns and natural sciences, of engineering and ecology," concluded NASA of its cooperative arrangement with neighboring national preserves.[91]

As it became increasingly involved in the ecological stewardship of Cape Canaveral, NASA, along with the Department of the Interior, began promoting such local environmental activism to visiting tourists. Such publicity entailed dozens of press releases, the construction on KSC land of a new wildlife refuge visitor center complete with auditorium, exhibit rooms, and a classroom, the creation of nature trails for hikers, and the clearing of a campground inside the refuge for "conservation-oriented" youth groups that could purchase a "package deal" that also included a tour of the Kennedy Space Center. Administrators at NASA were equally involved in the development of an overall public education program that entailed wildlife classes and nature study groups, "open houses" and ecology field trips for secondary school classes, and the screening of environmentally themed films with titles such as *NASA's Wildlife Neighbors*.[92] In May 1976 the space agency and the Fish and Wildlife Service even opened a "wildlife drive," a self-guided automobile tour that led motorists "through various habitats, including river, shore, savannah and marshland."[93] Such promotion worked; in 1975 alone more than 325,000 tourists visited the Merritt Island National Wildlife Refuge and the Cape Canaveral National Seashore to see for themselves how America's space program was preserving, rather than polluting, the lands of Cape Canaveral.[94]

Visitors to the Cape's new wildlife refuge and national seashore often stopped off at the adjacent Kennedy Space Center, where in the early 1970s they could still hop aboard a Greyhound bus tour, stroll through the center's Space Museum and Rocket Garden, and then exit through the gift shop. Yet whereas during the mid-1960s these tourists could spend a nickel on postcards that highlighted NASA's technological milestones, a mere five years later such souvenirs had changed dramatically. While towering rockets, sophisticated lunar landers, and the enormous VAB still populated NASA's postcard collection, the technology was now surrounded by Cape nature. This shift in the visual culture on sale at the KSC gift shop began in 1968 with Apollo 8 and intensified throughout the early 1970s with the launches of Apollo 14 through 17. For each of the postcards documenting

3.7   NASA gift shop postcard of Apollo 11 framed by palm trees, circa summer of 1969.

these missions NASA's public relations office supplied color photographs that framed its space technology, not as it had in the mid-1960s when television networks had unobstructed camera shots of Mercury, Gemini, and Apollo rockets poised on their launchpads, but rather so the space agency's hardware appeared surrounded by the wild flora and fauna of Cape Canaveral (see Figure 3.7).[95] In case tourists missed this visual juxtaposition, NASA often repeated it in written form on the address side of these three-by-five-inch keepsakes. "In this contrast between nature and the space age," explained the caption for the postcard of Apollo 11, perfectly bordered by Cape Canaveral palm trees, "the transporter carries the 363-foot high Apollo 11 Saturn V space vehicle along the crawlerway to Launch Complex 39-A."[96]

Criticism by grassroots environmentalists in the late 1960s thus transformed NASA during the early 1970s. Public condemnation regarding littering in outer space and dumping on the moon; protestations that the production and launching of Apollo rockets fouled nearby water, air, and land; and complaints that the space race as a whole distracted the nation from correcting pollution problems back on Earth: all compelled NASA to become an environmental steward on the local level. As a result, what had been a safety buffer zone originally created to protect NASA hardware had become, by the mid-1970s, an environmental showcase presenting wild life and wild lands to the visiting public. Environmentalists took note. "We are impressed with the substantive measures KSC is taking to protect the environment as it proceeds with its mission to explore space," announced the president of Save Our Waterways, a local Cape Canaveral environmental group, in a speech to NASA administrators, the public, and "scores of conservationists" gathered in June of 1972 to celebrate the expansion of the Merritt Island National Wildlife Refuge. "We commend you for your wisdom."[97]

Helping Scientists Think Globally

While grassroots environmentalists compelled NASA to act locally at the Cape, complaints by scientists were also forcing the space agency to think more globally about our home planet's atmospheric and aquatic resources. Such contemplation within NASA began officially in August 1971, when the Langley Research Center convened approximately fifty earth scientists from both within and outside the space agency for a workshop on the possibilities of using satellite-sensing technologies to detect gaseous, liquid, and particulate contaminants in the Earth's air and water. The conference's final report, titled *Remote Measurement of Pollution*, not only outlined the space technologies that could be used for measuring an array of pollutants—including carbon monoxide, sulfur dioxide, and fluorocarbons in the atmosphere as well as oil, thermal effluents, and chemical and toxic wastes in water—but also specified the accuracy requirements necessary for effective monitoring from space. "Problems of the environment," explained the authors in their preface, "have been matters of serious concern and active study during recent years." Two hundred pages later, the report concluded that satellites in space, if developed properly, could provide essential data regarding pollution "that can not be obtained by any other means."[98] For the first time in its history, NASA was considering developing space technology specifically to benefit scientific research on the Earth's ecological well-being.[99]

Back in the 1960s NASA administrators had been extremely reluctant to develop nonmilitary satellites that could remotely sense environmental problems on Earth. Instead, engineers built several generations of hardware, including the Television Infrared Observing Satellite (TIROS), the early Nimbus satellites, and the Advanced Technology Satellite (ATS), to forecast the weather by photographing cloud cover and transmitting the images back to Earth.[100] This hesitancy was also evident in a June 1966 NASA report titled "Space Science and Applications: Where We Stand Today," which failed even to mention remote sensing of the Earth environment.[101] Frustrated by such inaction, three months later Secretary of the Interior Stewart Udall announced with great public fanfare, and to the surprise of many NASA officials, that his department planned to move ahead with its own Earth Resources Observation Satellite (EROS) to help fulfill its mission of protecting and preserving the country's natural resources.[102] "The time is now right and urgent to apply space technology toward the solution of many pressing natural resources problems," Udall proclaimed at a packed press conference.[103] It was a brilliant political ploy; while the secretary knew Congress would never fund a Department of the Interior satellite, he believed, quite correctly, that his announcement would unleash a torrent of negative media coverage for NASA regarding its technological foot-dragging and along with it waves of support from the very earth scientists who had criticized, and later quit, the space agency. This outcry also lit a fire under NASA administrators, who were forced to accelerate the development of their own Earth-observing satellite in order to maintain programmatic control over the new technology.[104] The result was Landsat, NASA's first piece of hardware designed specifically to study the Earth's environment.

While during the Vietnam War the space agency used Landsat to help developing countries monitor their own natural resources, as the conflict wound down NASA scientists also began deploying the satellite to monitor global pollution. "The application with the broadest current user interest is detecting elements of water pollution, tracing them to their source, and in measuring the dispersion and concentration of the pollutants," announced the space agency after testing Landsat sensors over Massachusetts Bay and Boston Harbor.[105] The experiment, undertaken less than a year after the Santa Barbara oil spill of 1969, produced images that clearly identified discharge flowing from storm sewers, power plants, and wharf complexes leaking petroleum. Landsat could not only "'see' through clouds," the space agency argued, but also "act as a sentinel to detect and track spreading oil spills" such as that which washed up on Santa Barbara's pristine beaches.[106] To support such research, in the mid-1970s NASA began developing Seasat, the first Earth-orbiting satellite designed specifically for remote sensing of the Earth's oceans. Although it short-circuited in June 1978, only 106 days after launch, Seasat collected more scientific information about the Earth's water than had been gathered in the previous hundred years of shipboard research, and it also spawned numerous subsequent Earth remote-sensing satellites and instruments that tracked changes throughout the planet's waters.[107]

The space agency's remote-sensing conference at Langley back in 1971 also signaled NASA's newfound commitment to developing satellites that could monitor air pollution. "Only within the past few years have advances in optical instruments permitted these gasses to be traced remotely," *Remote Measurement of Pollution* explained. The report went on to argue, forcefully, that satellites orbiting Earth offer "extraordinary new opportunities for measuring the distributions of these gasses."[108] Such conclusions by the fifty scientists attending the conference, many of whom hailed from within the space agency, compelled NASA administrators to redesign the final satellite in their Nimbus series of orbiting weather forecasters as a rechristened "pollution-patrol" satellite.[109] While the technology would retain the meteorological capabilities of previous Nimbus satellites, NASA now planned to incorporate into the hardware several scientific sensors devoted to detecting various atmospheric pollutants. This new technology, argued NASA officials, "is one of the tools which can be utilized in expanding man's understanding of his environment."[110]

Administrators at NASA were especially interested in developing the new Nimbus-7 satellite for ozone monitoring. Although earlier in 1971 grassroots environmentalists had successfully grounded the SST jet by pressuring Congress to halt its funding, scientists at that time had begun raising similar concerns regarding the space shuttle's impact on the ozone layer. The physicist Robert Hudson from the Johnson Space Center, the geophysicist I. G. Poppoff of the Ames Research Center, and the chemist James King of NASA's Jet Propulsion Laboratory, all of whom feared that chlorine emitted from the space shuttle's exhaust system might damage the ozone layer, almost certainly raised such worries from within the space agency.[111] "The concern on the environmental impact of the solid rocket booster exhaust during shuttle launch operations has once again been the subject of a number of NASA Headquarters' meetings," explained a 1974 internal memorandum from the space agency's shuttle operations office.[112] The following year, when a required Environmental Impact Statement for the new technology confirmed these fears, scientists beyond NASA began voicing similar concerns. "A recent environmental impact study of the space shuttle," explained two physicists from Princeton's Center for Environmental Studies in the *Bulletin of the Atomic Scientists*, showed that NASA's new vehicle "could have harsh environmental impacts" on the stratospheric ozone layer.[113] As such news went public, NASA administrators became understandably nervous that just as grassroots environmentalists had successfully grounded the SST over the ozone issue, so, too, would the science community scrub the space shuttle.

Administrator James Fletcher took decisive action during the early to mid-1970s both to insulate NASA from such criticism by scientists and to position the space agency as the key player in the scientific study of ozone depletion. He began in 1971, soon after President Nixon asked him to leave his post as president of the University of Utah in order to lead NASA, by establishing an ad hoc group to brainstorm how the space agency might address the possibility of shuttle-induced ozone depletion. The following year, in anticipation of congressional hearings on the ozone crisis, Fletcher institutionalized this informal study group by pulling funds from several other NASA operations in order to create the Stratosphere Research Program. He also began specifically promoting NASA's Earth-observing satellites as crucial technologies in understanding the ozone problem through his Office of Space Science, connections at the National Science Foundation, and communication with President Gerald Ford's science advisor.[114]

Fletcher then pitched Congress. "NASA is called the space agency, but in a broader sense, we could be called an environmental agency," he explained to the Senate's Committee on Aeronautical and Space Sciences in March of 1973. "Virtually everything we do, manned or unmanned, science or applications, helps in some practical way to improve the environment of our planet and helps us understand the forces that affect it." The head of NASA then ruminated, "Perhaps that is our essential task, to study and understand the Earth and its environment."[115] When it came to ozone, Congress agreed, and in 1975 it passed legislation that awarded NASA $7.5 million to conduct a comprehensive research program to monitor the upper atmosphere and more than $115 million for satellite development. Two years later Congress also instructed NASA, through an amendment to the Clean Air Act, to prepare a biennial report to Congress regarding ozone depletion.[116] Then, on October 24, 1978, less than a year and a half after he left NASA, Fletcher sat back and watched as the space agency launched Nimbus 7, its pollution-patrol satellite.[117]

Nimbus 7 carried eight new instruments, four of which scientifically measured atmospheric pollution. While the Stratospheric and Mesospheric Sounder (SAMS) monitored atmospheric gas concentrations of water, methane, carbon monoxide, and nitric oxide, the Stratospheric Aerosol Measurement (SAMS-II) sensor determined the distribution of stratospheric particulates. Complementing this pair of air-pollution technologies were two instruments developed specifically to measure atmospheric ozone. Engineers and scientists at NASA designed the Limb Irradiance Monitor of the Stratosphere (LIMS) to calculate vertical gas concentrations of water vapor, ozone, nitric acid, and nitrogen dioxide in an attempt to determine whether the latter two caused ozone depletion, while the Total Ozone Mapping Spectrometer (TOMS) measured the amount of ozone in a column of air 30 to 125 miles wide projecting from the Earth's surface to the top of the atmosphere.[118] As one NASA publication explained, Nimbus 7 was considered "the single most significant source of experimental data from Earth's orbit" relating to atmospheric processes.[119]

Just as NASA administrators promoted their environmental stewardship at Cape Canaveral through public nature programs, the opening of wildlife automobile drives, and sales of postcards that framed space technology within local nature, so, too, did they publicize their new Earth-observing hardware to scientists who had previously criticized the agency's space technology on environmental grounds. The oceanographer and popular underwater filmmaker Jacques Cousteau proved instrumental in this promotional campaign. Deputy Administrator George Low had begun wooing Cousteau in December 1974 during a five-day Caribbean cruise aboard *Calypso*, the filmmaker's well-known research ship. The courtship continued the following year when Low invited Cousteau to spend nearly a week touring four NASA labs, where the Frenchman examined scientific data from Landsat and learned of future plans involving ocean research for Nimbus 7, Seasat, Skylab, and the space shuttle, the last of which Low emphasized would have a large "ocean studies" component. "I think this is the beginning of a very rewarding relationship," wrote Low in his personal notes regarding the visit, "not only for Cousteau but especially for NASA."[120]

It was indeed. While later that year NASA awarded Cousteau a contract to perform experiments on *Calypso* off the Bahamas that would help determine whether Landsat could measure ocean depths, in 1977 Cousteau returned the favor by using footage of his crew undertaking the experiment, along with narration explaining NASA's Earth-sensing efforts, in *Troubled Waters*, a segment in one of his famous underwater film series. Titled *Oasis in Space*, the series departed from Cousteau's usual underwater fare depicting brightly colored fish darting among coral reefs; instead he examined the growing problem of oceanic pollution.[121] Low and other NASA administrators hoped that such publicity by Cousteau, who himself believed that "the space program will contribute a great deal to oceanography," would help promote similar sentiment among scientists.[122]

The space agency's efforts to convince earth scientists that Nimbus 7 was useful for tracking air pollution were even more direct. In order to collect the most useful data, NASA established a Nimbus Experiment Team (NET), consisting of between three and eight scientific experts from outside the space agency, for each of the satellite's eight instruments. These NET scientists decided on what data was needed, developed the methods for acquiring this data, and evaluated the data in order to validate these methods. To manage these extremely large data sets, NASA also created the Nimbus Observation Processing System (NOPS), which similarly engaged the broader scientific community by dispersing the data-processing function to outside facilities, including the National Oceanic and Atmospheric Administration, the National Center for Atmospheric Research, and several academic institutions, including Oxford and the universities of Colorado and Florida. These NOPS facilities then distributed the processed data to NASA's NET scientists as well as to the National Science Satellite Data Center archive at the Goddard Space Flight Center in Greenbelt, Maryland, which passed it on to the wider science community. "The program goal," explained NASA's operations manager for the Nimbus 7 satellite, "was to broaden the participation of the science community."[123] For NASA administrators, achieving this goal also meant quieting the concerns of scientists who had criticized the space agency's technology for its scientific myopia.

Opposition by trained scientists, much like the criticism of grassroots environmentalists, thus also transformed NASA. The "brain drain" of researchers quitting the space agency in protest, a journalistic campaign in the pages of the *Bulletin of the Atomic Scientists*, and demands for a new NASA research program charged with scientifically monitoring the "total environment" of planet Earth all raised public awareness that put political pressure on NASA administrators to radically reform the agency's space technology. As a result, while NASA engineers during the 1960s had focused on developing rockets that could transport humans to the lunar surface, during the following decade they also worked closely with scientists to develop satellites that monitored the planet's atmosphere and water. Landsat and Nimbus 7 were just two of the most important examples of this technological turnabout.[124] "The move marks the transition to a new phase of the space age," announced NASA in a December 1976 news release, "from an era of exploring and probing the secrets of space to a time of emphasis on exploiting near-Earth space for the benefit of this planet's inhabitants."[125] What NASA administrators failed to realize, however, was that by taking this new "environmental approach," the space agency and its new technology were also transforming environmentalism.

Earth Photographs and Global Data

Stewart Brand was not the only environmentalist to claim that NASA's 22727 photograph jump-started the environmental movement. David Brower, the Sierra Club's first executive director and one of the most prominent environmentalists of the post–World War II era, also paid homage to what became known as *Whole Earth*, which he, too, believed gave birth to the movement by symbolically depicting an imperiled planet in dire need of ecological stewardship. For years Brower traveled the country giving what he called "The Sermon," a public talk at universities, local meeting halls, and once in an actual cathedral that outlined his environmentalist "religion." After covering the six days of Genesis and evolution, Brower focused his homily on the recent emergence of environmentalism by holding up NASA's *Whole Earth* photograph. "This is the sudden insight from Apollo," he told his audiences. "We see through the eyes of the astronauts how fragile our life is."[126] Brower, like Brand, thus also gave *Whole Earth* credit for initially enlightening environmentalism.[127]

Brand, at least, should have known better. He was well aware that photographic images of the entire planet already existed; back in 1968 he had placed one such image on both the front and back covers of the inaugural edition of his *Whole Earth Catalog*, a counterculture publication, similar to a Sears Roebuck catalog, that listed and reviewed small-scale and alternative technologies. That earlier photo, which Brand explained to his readers was the "first full-Earth picture," originated from a high-resolution color TV film shot in November 1967 from NASA's ATS satellite.[128] Photographs of portions of Earth from space had an even deeper history, beginning in 1962 when John Glenn walked into a Cocoa Beach, Florida, drug-store, purchased a Minolta 35mm camera off the shelf, and used it to take pictures of Earth through the small window of his Mercury spaceship. "Friendship 7 was going to be the first spacecraft that had a window over the pilot's head where you could take a picture," Glenn recounted years later, "and I thought we should take a camera."[129] Since then every astronaut journeying into space has carried one, including Apollo 8's William Anders, who on Christmas Eve in 1968 captured *Earthrise*, another iconic photo-graph of our home planet.[130]

These early Earth photographs did not gather dust on an archive shelf far from public view. Instead, the public relations team at NASA immedi-ately circulated the global snapshots during the late 1960s to local and na-tional newspapers, mainstream magazines such as *Life* and *Newsweek*, and television networks, and also offered them up at a dollar a piece to the gen-eral public in eleven-by-fourteen-inch lithographs.[131] In late December 1968, for instance, the *Washington Post* and the *Spartanburg Herald* both ran photographs of Earth taken by Apollo 8 on their front pages, while the local newspaper in Spokane, Washington, buried the image on page 16 along-side photographs of fifty-five debutantes attending a cotillion.[132] Corpora-tions likewise capitalized on the images; beginning as early as 1969, companies including McDonnell Douglas, AT&T, IBM, and others, most of which worked on the Apollo program, peppered the pages of *Fortune*, *Forbes*, and *Personal Business* magazines with images of Earth taken from space in an attempt to publicize their technological wares.[133] Officials at NASA even included astronaut photographs of Earth from space, taken years be-fore 22727, in reels for the popular View-Master 3-D stereoscope.[134] During the late 1960s it was thus nearly impossible for most Americans to avoid encountering photographs of Earth from space, and it was quite difficult for them to distinguish these earlier images from the nearly identical *Whole Earth* photograph captured by Apollo 17 in 1972 (see Figures 3.8 and 3.9).[135]

These pre–*Whole Earth* photographs rarely sparked concern for a pol-luted planet among environmentalists during the late 1960s.[136] Instead, a different idea dominated early public discussions of these images of Earth from space. On Christmas Day in 1968, for instance, just hours after Apollo 8 astronauts beamed back live images of Earth onto television sets world-wide, the *New York Times* published a short essay titled "Riders on the Earth" written by the Pulitzer Prize–winning poet Archibald MacLeish.

3.8   Photograph of Earth from space, taken by NASA's
ATS-3 satellite on November 10, 1967.

3.9   Photograph of Earth from space, captured from a live
transmission from the crew of Apollo 8 in late December of
1968.

"Men's conception of themselves and of each other has always depended on their notion of the earth," MacLeish began. He then argued that peering back at Earth during Apollo 8's mission had altered this conception once again. "To see the earth as it truly is, small and blue and beautiful in that eternal silence where it floats," he explained, "is to see ourselves as riders on the earth together, brothers on that bright loveliness in the eternal cold—brothers who know now they are truly brothers."[137] Readers across the world connected deeply with the essay, and the images of Earth that sparked it, and the American press responded in the days, weeks, and months that followed by reproducing it often, almost always accompanied by a photograph of the planet from space.[138] Soon MacLeish's notion of worldwide harmony dominated the cultural meaning of Apollo 8's Earth images and those that followed.[139] Looking back at Earth from space in the late 1960s meant global unity, not planetary environmental concern.

This cultural disconnect between early photographs of Earth from space and environmentalism continued for the next twenty years, as is evident in the movement's visual culture. Earth Day promotional materials are a case in point. To publicize the first Earth Day, held on April 22, 1970, organizers and activists relied on a handful of symbols that they plastered across posters, flyers, pamphlets, T-shirts, and even across their own bodies. Dying trees, traffic jams under polluted skies, and a garbage-strewn environment all appeared prominently.[140] Yet it was the gas mask—worn by mothers pushing strollers through parks in the pages of *Life* magazine, by young men trying desperately to sniff spring flowers during one of the day's many protests, and by an Earth depicted by a classroom globe appearing on an announcement for the event—that emerged as the undisputed poster child for Earth Day 1970.[141] That an Earth Day poster for the event used a gas-masked Earth, represented by a schoolroom globe, is significant; in only a small fraction of these promotional materials as well as photographic coverage of the first Earth Day did an image of the Earth appear, and none included a photograph of the planet from space.[142]

This dearth of Earth photographs adorning Earth Day's visual culture continued throughout the 1970s and 1980s. During the celebration's tenth anniversary, which rallied far fewer participants than its 1970 counterpart, images referencing the OPEC oil crisis, the Three Mile Island nuclear accident, and endangered species, especially whales, dominated Earth Day's visual culture.[143] It was not until the 1990 Earth Day celebration, almost two decades after Apollo 17 captured *Whole Earth*, that the Earth photographed from space became a popular Earth Day feature, appearing on enormous flags unfurled at rallies in the nation's capital, on pin-up posters announcing the celebration in small towns across "Middle America," and even in a public service announcement television spot depicting the Muppet Kermit the Frog singing "It's Not Easy Being Green" from inside a whole Earth stage set floating in outer space.[144] Even the official logo of Earth Day 1990 included a depiction of Earth from the vantage point of space.[145] Thus as this early history illustrates, Stewart Brand's origin story for *Whole Earth*'s role in launching the environmental movement is in dire need of an edit.

While grassroots environmentalists ignored *Whole Earth* well into the 1980s, environmental scientists had been trying to study the whole Earth since the early post–World War II period. The first step in this direction occurred back in 1957, when thousands of earth scientists from sixty-five nations joined together for the International Geophysical Year (IGY), a United Nations–sponsored program of cooperative experiments aimed at studying the Earth as a "single physical system." To do this, IGY scientists collected data from more than 4,000 research stations worldwide as well as from the world's first three orbiting satellites, Sputnik 1 and 2 and the United States' Explorer 1, all three of which were launched specifically for the IGY program. Although the IGY was unable to capture a truly global environmental data set—the Southern Hemisphere, for instance, was sparsely covered by the program's collection efforts—its unprecedented compilation of regional scientific information illustrated to researchers worldwide the necessity of both orbiting satellites to collect, and computer models to process, global data.[146]

During the 1970s NASA was instrumental in developing both of these technologies. Although the space agency had begun collecting *regional* meteorological data through its Earth-orbiting satellites, including its ATS program initiated in 1967, the TIROS satellite launched just two years later, and Landsat in 1972, such efforts took a decidedly *global* turn with Nimbus 7, the space agency's pollution-patrol satellite. While the eight sensors on Nimbus 7, which circled the Earth every six days, were able to collect data that was truly global in scope, it, like all satellite data, was also problematic; it was overwhelmingly large, often inaccurate, poorly calibrated, and incomplete when clouds made it difficult to "see" through to the Earth's atmosphere. To make this data *function* as global, during the late 1960s the space agency also began developing computer models to "smooth out" these inconsistencies.[147]

At the forefront of such efforts was NASA's Goddard Institute for Space Studies, a subdivision of the Goddard Space Flight Center, which since its creation near Columbia University in New York City in 1961 served as the space agency's theoretical modeling and data analysis center. Beginning in 1969 and lasting through the early 1970s, the institute took a leading role in the Global Atmospheric Research Program (GARP), an international effort to create global data sets that could be used by scientists to assess pollution of the Earth's atmosphere. While NASA's Nimbus series of satellites gathered such data, computer scientists at the institute involved in GARP generated mathematical models to assess the data's accuracy and to determine if and how it could be smoothed into global data sets. In early 1979 NASA tested such models through GARP's Global Weather Experiment, which involved seven satellites gathering data continuously for two sixty-day periods. The result, explained one space historian, was the world's first "global, quality-controlled, extensive meteorological dataset."[148]

During the mid- to late 1980s earth scientists combined NASA's satellite data and computer models to transform two local scientific discoveries into the most important global environmental issues of the postwar era. The detection of both the Antarctic ozone hole and global warming each began on the local level with regional data. While scientists from the British Antarctic Survey relied on ozone data gathered in the early 1980s from two research stations within the Antarctic region, Charles Keeling of the Scripps Institution of Oceanography used atmospheric readings of carbon dioxide taken between 1957 and 1971 at the Mauna Loa Observatory in Hawaii.[149] When both groups of researchers published their findings in the late 1970s and early 1980s, few beyond the scientific community took note.[150] Earth scientists within NASA, however, were not only alarmed by such findings but also took steps to deepen the space agency's involvement in this important scientific research.[151] Less than a year after the British team published its ozone research in *Nature* magazine, for example, NASA used data collected by sensors on Nimbus 7 to corroborate such research. "Satellite observations have confirmed a progressive deterioration in the earth's protective ozone layer above Antarctica," explained the *New York Times* in November of 1985. The Goddard Institute's director, Jim Hansen, took similar steps regarding global warming by directing his team in New York to develop

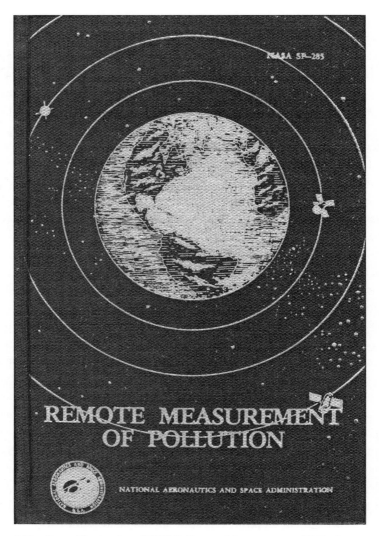

3.10   Cover illustration of NASA's *Remote Measurement of Pollution*, 1971.

highly sophisticated computer models that not only confirmed Keeling's observations in Hawaii but also broadened such research to cover the entire planet. The global warming projected for the next century, warned Hansen in several pathbreaking papers published in the mid-1980s, "is of almost unprecedented magnitude."[152]

Researchers at NASA not only corroborated these local scientific research efforts through the collection and modeling of worldwide data, but they also helped scientists transform these data sets into signifiers of global environmental crises. Earth scientists accomplished this by embracing the visual culture that environmentalists had been overlooking since the late 1960s. Beginning in the early 1970s, for instance, NASA scientists began by "covering" their data detailing global environmental degradation with images of Earth from space. The space agency initiated this practice on the cover of *Remote Measurement of Pollution*, the report from its groundbreaking 1971 conference on the uses of satellites for environmental monitoring, which depicted an image of the full Earth circled by three orbiting satellites (see Figure 3.10).[153] Other scientists followed NASA's lead. The important 1971 *Study of Man's Impact on Climate*, which was a report written by more than thirty scientists attending an MIT-sponsored conference in preparation for the 1972 United Nations Conference on the Human Environment in Stockholm, along with the Club of Rome's 1972 classic, *The Limits to Growth*, as well as numerous articles and books written during the 1970s and early 1980s by James Lovelock and Lynn Margulis on their Gaia hypothesis, all sported photographs of the Earth from space on their covers.[154]

As the 1980s wore on, scientists not only covered their global data with images of Earth from space, but also began "covering" images reminiscent of *Whole Earth* with global data. Such efforts were most pronounced regarding the ozone crisis. The British Survey scientists who first published their findings in *Nature* in 1985 illustrated their data through a graph revealing a local decline in Antarctic ozone (see Figure 3.11).[155] When later that year scientists from NASA's Ozone Processing Team at the Goddard Space Flight Center in Maryland confirmed such findings with the aid of Nimbus satellites and computer models, they instead illustrated their global data with *Whole Earth* in mind. First, NASA's computer modelers "smoothed" the enormous variation in ozone satellite data by assigning a handful of false colors to specific value ranges within the data set; light and dark blue represented the lower ranges of ozone depletion while red and yellow signified the most intense. Space agency scientists further simplified this data by connecting points of equal value with contour lines to give the illusion of continuous measurement across geographic space. When NASA used computers to combine these false colors and contours, the result was a *Whole Earth*–like image layered with global data that was more

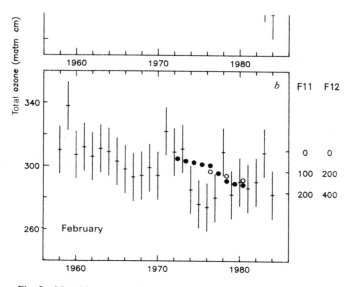

Fig. 2 Monthly means of total O_3 at Halley Bay, and Southern Hemisphere measurements of F-11 (●, p.p.t.v. (parts per thousand by volume) $CFCl_3$) and F-12 (○, p.p.t.v. CF_2Cl_2). *a*, October, 1957–84. *b*, February, 1958–84. Note that F-11 and F-12 amounts increase down the figure.

3.11   British Antarctic Survey's ozone-depletion graph, May 1985.

readable by the lay public (see Figure 3.12).[156] The result was also a very different environmental crisis. Whereas few media outlets reported on the British Survey's findings regarding local ozone "depletion" in the Antarctic, six months later the *New York Times* immediately ran a story on NASA's research that for the first time reframed the issue as a global crisis involving an ozone "hole."[157]

The immediate, widespread, and alarming reaction to NASA's ozone-hole image within both scientific and lay circles seems to have persuaded the space agency in 1988 to institutionalize this process of converting raw data into readable images by establishing the Scientific Visualization Studio at the Goddard Space Flight Center. Much like the informal efforts of the Ozone Processing Team two years earlier, the studio took numerical data, usually collected by satellites such as Nimbus 7 orbiting the Earth, and combined it with computer models and animation software to make the scientific information more understandable for nonscientists.[158] "Visualization is that link between the flood of data coming down from space and the ability of the human mind to interpret it," explained a NASA oceano-

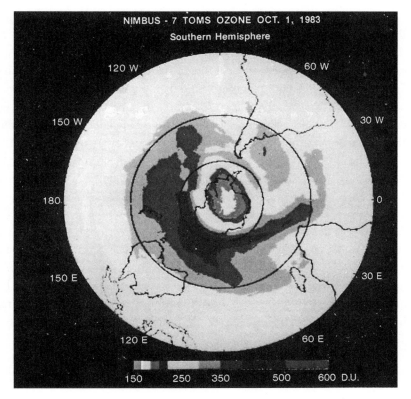

3.12   NASA's ozone-hole illustration, August 1985.

grapher who worked closely with the studio. "They are able to take this fire hose of data coming down and turn it into images," he continued, "that then allows the general public to see this data in ways their brains can interpret and study."[159] The studio also quite consciously turned these images into stories by becoming a key player within the Goddard Center's Earth Science Storytelling Project. This collaborative effort involving scientists, science writers, animators, and editors ensured that the studio's scientific visualizations told interesting scientific tales.[160] "Like their Hollywood counterparts," explained the *Washington Post* of NASA's scientific visualization efforts, the studio and its partners "refer to their finished products as releases, but the scripts are composed of data and the script writers are some of the world's most brilliant scientists."[161] The space agency's depiction of an "ozone hole" in 1986 was one such story.

A second visual narrative helped NASA publicize global warming. While Charles Keeling's upward-sloping graph tracked temperature increases at Mauna Loa throughout the 1970s without public fanfare, from the mid-1980s onward NASA began confirming such findings by publishing *Whole Earth*–like images created from satellite data and computer models that caused an environmental firestorm. The first step in this process involved scientists at Goddard's New York Institute collecting data from NASA satellites regarding average global temperatures. This team then analyzed the data for anomalies by comparing it to previously collected data in order to identify statistically significant trends. It was then that the Scientific Visualization Studio down at Goddard's headquarters in Maryland stepped in to transform these data anomalies into visual images, or moving images, that began to tell the scientific story of global warming.[162] An important component of that story, as it was during the ozone crisis, was *Whole Earth*, or the hint of it underneath NASA's climate-change visualizations. In a more contemporary example of such work, NASA's Scientific Visualization Studio created an image in 2016 depicting the planet's rising temperatures that could easily have been mistaken for *Whole Earth*, except that the "big blue marble" was burning bright red.[163]

Making Whole Earth Environmentalism

Stewart Brand's acid trip atop his San Francisco rooftop during the winter of 1966 is thus only one chapter in the *Whole Earth* origin story, and an incomplete chapter at that. Chapter two took place a few years later when Brand met, quite by accident, an officer in the military who handled investigations for NASA on the West Coast. "You're the guy!" the officer exclaimed after Brand mentioned, in passing, his "Why Haven't We Seen a Photograph of the Whole Earth Yet?" buttons. It turned out the officer had received a request from the U.S. government to investigate whether or not "some character" distributing literature on college campuses about photographs of the planet from space represented a threat to the U.S. government. Being a military man, the officer diligently researched the "character's" background and interviewed numerous unnamed sources familiar with Brand's bizarre button behavior. He then wrote back to his superiors in Washington, DC: "This is California," the officer explained. "People here take very strange enthusiasms, and quite often they do it by themselves." Then, after stating that Brand represented no threat whatsoever, he concluded his memo with the question: "P.S. By the way, why *haven't* we seen a photograph of the whole Earth yet?"[164]

As this chance encounter suggests, while Brand may have acted by himself back in 1966, he was not alone in his belief that environmental concern could alter America's space program. Grassroots activists and elite scientists involved in the early environmental movement felt similarly. While the former criticized NASA's space technology for littering the moon, fouling the air and water at Cape Canaveral, and distracting Americans from increasingly severe pollution problems in their local communities, earth scientists complained instead that the agency's hardware excluded experiments that would help scientists analyze the global environment. Administrators at NASA became acutely aware of both sets of criticisms and responded in kind. At the Cape they acted locally to protect and preserve the wildlife and wild lands of the Kennedy Space Center, all the while publicizing such efforts to the visiting public. At the same time, engineers and scientists at NASA also began to think more globally by developing new space technologies, such as the Nimbus 7 satellite, that could monitor air and water pollution across the entire planet.

Yet as it responded to these environmental critiques, NASA was in turn altering environmentalism, just not as Brand had envisioned. The space agency did produce stunning images of Earth from space, but these cultural artifacts were not initially environmentally important; grassroots activists in the late 1960s worried mostly about pollution, and the photograph of the "bright blue marble" from space simply failed to capture, visually and emotionally, this concern. Beginning in the 1970s NASA began remedying such environmental anxieties not only by both redirecting its satellites back down toward Earth to collect global environmental data and building powerful computer models that processed and smoothed out such information, but also by superimposing this newly acquired global data onto scientific images that hinted at older, more culturally familiar photographs of Earth from space. While such images during the 1960s had symbolized global unity, NASA's updated scientific versions instead communicated environmental concern regarding global problems including the ozone crisis and climate change. Such association went both ways; the space agency's photographs of planet Earth acquired similar environmental traits. Thus it was not *Whole Earth* that jump-started environmentalism, as Brand had argued, but rather NASA and its technology that transformed 22727 into *Whole Earth*.[165]

In merging cultural images and scientific data, NASA also helped to unify the burgeoning environmental movement. It accomplished this by telling stories, through scientific visualizations, that spoke to activists and scientists alike. The ozone-hole image created by NASA technicians in the mid-1980s, for instance, joined together the one-world story of 22727 told first by Archibald MacLeish with the alarming account of ozone "depletion" above the Antarctic narrated more recently by British scientists. The result of this storytelling amalgamation was both *Whole Earth* and the ozone "hole." In many respects, such communication and transference between the cultural and the scientific were not new. *Silent Spring* became a worldwide best-seller not because the science in it was novel—it wasn't—but rather because Rachel Carson skillfully packaged data about DDT and other chemical pesticides within accessible and beautifully written prose that made the science both understandable and emotionally significant to nonscientists.[166] The same could be said of NASA's ozone-hole image, as well as the space agency's subsequent climate-change visualizations. All of these cultural productions, whether verbal or optical, helped grassroots activists and earth scientists speak a common language, and understand shared stories, about the environment. The result, which left room for both culture and science as well as for activists acting locally and scientists thinking globally, was a more mature movement that one might call *Whole Earth* environmentalism.

Stewart Brand hinted at the need for this broader, more inclusive movement soon after NASA released 22727. Although he admitted to phrasing his question "Why Haven't We Seen a Photograph of the Whole Earth Yet?" in order to tap into what he believed was an American obsession with conspiracy, Brand chose the word "we" for quite different reasons. "It was we," he explained, who paid for Apollo with tax dollars, and "we" who watched as the program landed men on the moon. It must, therefore, also be "we," all of us from grassroots activists to earth scientists to the general public, who both demand a picture from space of our planet and become more invested in the stewardship of Earth. "With all the photographic apparatus in the world, we hadn't turned the cameras that 180 degrees to look back," he concluded philosophically. "We had designed beautiful cameras but no mirrors."[167]

4

Heavenly Bodies

"Manned Spaceflight" and the Women's Movement

To avoid a storm gathering over Florida's Cape Canaveral, at 6:49 A.M. on June 25, 1983, the space shuttle *Challenger* crossed over the California coast high above Ventura, emitted twin sonic booms that "rattled the ground like cannon shots," and quickly descended into clear blue skies before touching down on a dusty, dry lake bed in the middle of the Mojave Desert. Minutes later STS-7's five crew members climbed down from *Challenger* and waved to the crowd of test pilots and several hundred Edwards Air Force Base employees who had arrived early for work to watch the landing. The *Challenger* crew waved again to reporters and fans later that day, after flying to a homecoming ceremony at the Johnson Space Center in Houston, but stopped abruptly when a NASA protocol officer walked forward with a red and white bouquet of roses and carnations. Sally Ride looked him squarely in the eyes and "shook her head." When that failed, she "turned her back" on the space agency official, who stood alone, awkwardly, with an armful of flowers. Other NASA administrators then rushed forward and distributed single red roses to the wives of each of the *Challenger*'s male crewmates but had nothing for the astronaut Steve Hawley, Sally Ride's husband.[1]

Although interpreted by some at the time as impolite, the refusal of a bouquet of flowers by the first American woman in space upon her return to Earth should have caught nobody at NASA by surprise. From the moment she and five other women broke NASA's ultimate glass ceiling in 1978 by gaining admittance to the space agency's all-male astronaut corps, Ride had insisted on being treated as equal to her male colleagues. Such sentiment only intensified on April 19, 1982, when NASA announced that Ride would be the first woman assigned to a space shuttle mission. In a news conference just prior to her historic flight, NASA public affairs official Hugh Harris reminded the press of Ride's repeated efforts to minimize the attention she received as the first American woman astronaut and her deep desire to be seen instead as a mission scientist who just happened to be a woman.[2] Sally Ride reiterated such beliefs every chance she got. "There's been absolutely no difference because I'm on the crew as far as our training and all the things associated with the training," she explained days before liftoff. "I'm just one of the guys."[3] Ride stated this once again, without uttering a single word, when after returning to Earth she publicly rejected an armful of flowers.

While NASA officials wanted nothing more than to quickly forget this awkward social moment, women across the country went out of their way to remember Ride and her landmark flight. In one seven-day period two weeks after landing in the desert, Ride received a seventeen-foot scroll signed by 714 female government workers, served as guest of honor for a luncheon organized by a "high ranking" group known as Executive Women in Government, participated in a conference honoring outstanding women in the space field, accepted praise from the Secretary of Labor for showing "thousands of American women that their contribution to the achievements of this nation need not be confined to traditional roles," and appeared before the National Press Club to discuss her newfound responsibilities as a heroine for millions of young girls both in the United States and abroad.[4] "When I was growing up I didn't have the opportunity to have an astronaut as a role model," explained Ride to the gathered journalists, and because of that "it honestly never occurred to me that I could be an astronaut." Today, she added, the situation is quite different, with many more women from a wide variety of professions serving as exemplars. "I thank the women's movement for that," Ride stated at the conclusion of her talk. "And I think I owe them a real debt and it's a debt I intend to repay."[5]

By the time Sally Ride became America's first female astronaut in 1983, many women in the United States felt equally indebted to second-wave feminism. Most had read *The Feminist Mystique*, Betty Friedan's 1963 best-selling book that critiqued suburban gender norms and which many women credited with sparking the feminist movement. They had cheered early legislative victories—such as the Equal Pay Act of 1963 and Title VII of the 1964 Civil Rights Act, which prohibited gender discrimination in the workplace—and then enthusiastically joined, as did Sally Ride, the National Organization for Women (NOW) soon after it was formed in 1966 in part to ensure that such laws were enforced. During the early 1970s as NOW's political influence grew, along with that of the liberal feminist movement, tens of thousands of women also started reading *Ms.* magazine, first published by Gloria Steinem in 1971, while millions took to the streets, and the halls of Congress, to address a wide range of issues regarding women's sexuality, domestic violence, maternity leave and child care, and, perhaps most important, passage of the Equal Rights Amendment to the U.S. Constitution. Yet in the late 1970s as state ratification for the amendment faltered, and definitely by the time Sally Ride thanked the women's movement for helping to propel her into space, many of these same women regretfully acknowledge that second-wave feminism, weakened by both conservative opponents and internal divisions, had begun to ebb as a movement.[6]

While Ride reminded the National Press Club journalists of the connections between space exploration and second-wave feminism, her actions at the Johnson Space Center ceremony highlighted a history that is more complex than the women's movement of the late 1960s and early 1970s simply paving the way for women astronauts in the early 1980s. For instance, Ride could claim she was "just one of the guys," in part, because of her scientific and technological expertise. Not only was she a Stanford University–trained physicist, but during her five years of astronaut training in Houston she also had become extremely skilled at manipulating the shuttle's robotic arm, which would be used during the STS-7 mission to deploy two communications satellites and to deploy and then retrieve a package of orbiting scientific experiments. "You get people who sit in the lab and think like Einstein, but they can't *do* anything with it," explained shuttle commander Robert Crippen, who had personally chosen Ride for the mission because of such talents. "Sally is flying because she's good," Crippen told reporters, "and she's not being treated differently from any other astronaut."[7] President Ronald Reagan agreed. "Your handling of that long arm," he explained to Ride in a congratulatory telephone call just hours after *Challenger* touched down, "indicate[s] that you were there for one reason: because you were the best *person* for the job."[8] Then before hanging up Reagan joked, somewhat inappropriately, that sometimes "the best man for a job was a woman."[9] Similar to other females struggling within historically all-male professions, Ride used her scientific and technological expertise to bolster her claims for equality, and in this instance, to claim a seat on the space shuttle.

Ride's refusal to be feted with flowers at the Johnson Space Center also suggests that nature played a role in her fight for equality. The carnations and roses that she rejected have for centuries symbolized distinction and beauty, especially with regard to women.[10] In Houston, however, the flowers' meaning was as dependent on Sally Ride's physical body. By shaking her head from side to side, and when that failed, by turning her back on the bouquet, Ride was rebuffing such gendered associations and instead reinforcing her desire to be considered "just one of the guys." She did likewise by refusing to be televised from space doing what would have been considered women's housework, such as preparing food and cleaning the shuttle's "bathroom," even though the entire crew shared these tasks.[11] Many second-wave feminists—from those advancing women's health to others promoting the birth control pill—similarly turned to the female body in their fight for women's rights.[12] Yet Ride's body politics proved even more problematic when situated within the free-floating environment of outer space. Not only did NASA engineers install a candy-striped curtain around the shuttle's toilet to give Ride a semblance of privacy, but the American media obsessed in the months leading up to the flight about whether or not the vacuum-powered commode would function correctly for the nation's first female astronaut.[13] While Ride avoided such potty talk in public, her crewmates sometimes felt compelled to speak out on her behalf. "There's not much privacy in the shuttle," explained Commander Crippen in a veiled reference to the spaceship's exposed toilet. "But we'll work around that," he stated, much as Ride would have. "I think Sally will be just one of the guys."[14]

Many National Press Club journalists at the time reported on Ride's acknowledged debt to the women's movement, yet few writers since have covered the deeper connections between the space race and second-wave feminism. Space historians and feminist scholars, for instance, while noting Ride's support from, and participation in, the movement, have explored neither how activists paved the way for Ride's historic flight nor what impact that flight had on feminist politics.[15] Even fewer commentators have analyzed the role of space technology in shaping the experiences of Ride and her fellow female astronauts, even though historians of technology have written extensively on the gendered implications of both technology and technological systems.[16] Additionally, while environmental historians

interested in understanding our relationship to the natural world have recently turned to the human body, whether it be through an examination
of bodily health, physical labor, or outdoor recreation, here, too, such
analysis has rarely engaged feminism and has never made it into Earth's
orbit.[17] Thus, much like the unexplored criticism and collaboration between
the space race and the civil rights, antiwar, and environmental movements,
the fight for women's equality during the 1960s era also had an uneasy
partnership with what NASA officials constantly referred to as "manned
spaceflight."

That partnership began unintentionally back in the early 1960s, when
Sally Ride was about ten years old and had no American female astronauts
to look up to. During this time NASA publicized the health and fitness of
its all-male astronaut corps, much like it had advertised its space technology at the Cape, as signifying a strong, virile United States. While the
unknown hazards of outer space posed a threat to these spacemen, the
space agency again turned to technology for a resolution; Apollo spacesuits
would protect astronaut bodies by transporting a small portion of the
Earth's environment into space, while mechanical simulators would train
these bodies to function within and across foreign environments in space and
on the lunar surface. Women's bodies, officials at NASA argued forcefully,
repeatedly, and publicly during the late 1960s and early 1970s, were simply
unfit for both of these technologies. Liberal feminists rejected such excuses and under NOW's guidance opposed such discrimination by picketing outside NASA Headquarters, organizing letter-writing campaigns
against sexist Apollo astronauts, and even by holding mock beauty pageants, with NASA administrators as fake contestants, inside Mission Control in Houston. When such discrimination splashed across the pages of
Ms. magazine and then spilled beyond the women's movement proper,
NASA was again forced to react, much as it had in response to protests by
civil rights, antiwar, and environmental activists. The result was female astronauts such as Sally Ride as well as a debate within second-wave feminism regarding women's bodies that both energized yet further divided the
movement.

The Comely Cosmonaut

Back in 1963, just three weeks after Sally Ride's thirteenth birthday, her
chances for a seat on the space shuttle should have increased dramatically

when on June 16th Valentina Tereshkova became the first woman in space. During her mission aboard Vostok 6 the twenty-six-year-old Soviet traveled more than one million miles, orbited the Earth forty-eight times, and spent almost three days in space, which amounted to more miles, orbits, and hours traveled beyond Earth than the combined total logged by all American astronauts up to that time. Yet the *New York Times* began its article announcing this feat with a description of Tereshkova's good looks. "The world's first spacewoman is a pleasant-looking, gray-eyed, athletic young woman with wavy, dark blond hair," explained the newspaper.[18] *Life* magazine chimed in similarly in an article subtitled "A Blue-Eyed Blonde with a New Hairdo Stars in a Russian Space Spectacular," which included a photograph of Tereshkova at the hairdresser "primping for orbit." "Man's exclusive role in space," *Life* griped, "was at an end."[19]

In Moscow the Soviets did not focus on Tereshkova's beauty or lament that space was no longer a men's-only club. Instead, the country's space scientists celebrated her body for its biological fitness. Since part of the research agenda of the Vostok 6 mission was to determine the resiliency of female biology in outer space, Soviet scientists tracked Tereshkova's heart rate and respiration through telemetric data obtained by ground control stations, and they determined that "the organism of Lieutenant Tereshkova was in a physiological normal state during the flight."[20] The Russian state media translated such scientific techno-speak to the Soviet public by praising Tereshkova's "sheer physical stamina in cosmonaut school," by noting that her "physical toughness and courage astounded her male colleagues and at times made them almost envious," and by illustrating Tereshkova's training regimen with photographs of her rowing, skating, skiing, and pedaling on a racing bike.[21] Yet while strong, the world's first woman in space was by no means exceptional. The Soviet media instead portrayed Tereshkova as the consummate Soviet woman; not only was she physically hearty, explained official press biographies, but her "father was a tractor driver and her mother a worker in a textile enterprise."[22]

Soviet politicians also used Tereshkova's physical fitness to demonstrate the equality of the sexes under socialism. In his speech in Red Square upon her return to Earth, Soviet Premier Nikita Khrushchev handed Tereshkova an enormous bouquet of flowers and then explained to 60,000 Muscovites that she had spent more time in space than all the American astronauts combined. "That's the 'weaker sex' for you," he then shouted, sarcastically, to thunderous applause. "She has demonstrated once again that women raised under socialism walk alongside men in all the people's concerns."[23] Other Soviet officials parroted the premier, focusing in particular on Tereshkova's physical capabilities. "Like the men astronauts, women are chosen for the space program with special attention to good health, strong physical endurance, and great courage," explained one Soviet diplomat stationed in Washington, DC. "She has proved that women can serve in all fields of human endeavor as men can."[24] Women in Moscow certainly agreed; news reports described them as dancing in the streets after Tereshkova's flight to celebrate "the emancipation of the Communist woman."[25]

American feminists greeted news of Tereshkova's historic flight not with fancy footwork in the streets but rather by arguing, publicly, for equality of the sexes in the United States. "Soviet Russians put a woman into space because Communism preaches and, since the Revolution of 1917, has tried to practice the inherent equality of women," wrote Clare Boothe Luce, a former ambassador to Italy and outspoken feminist, in *Life* magazine, which at the time was owned by her husband, Henry. "It symbolizes to Russian women that they actively share (not passively bask, like American women) in the glory of conquering space." Although Luce cited dozens of statistics illustrating Soviet women's progress in the scientific, engineering, and medical fields, the "active sharing" she had in mind depended more on the equal distribution of physical labor; women in Moscow, she explained, mixed cement, drove buses, pitched hay, and swept streets. "It is against this background of the participation of Russian women in every effort, from sweeping the stables to combing the stars," Luce concluded, "that we must view the flight of the first woman cosmonaut."[26] Other feminists, including *The Feminine Mystique* author and future NOW president Betty Friedan, concurred.[27] Women were equal to men in the Soviet Union, and they should be considered similarly in the United States.

Astronaut Bodies as Cultural Symbols

Administrators at NASA begged to differ and, from the outset of the space race, publicly promoted the necessity of an all-male astronaut corps. This process began on April 9, 1959, when NASA Administrator T. Keith Glennan introduced the nation's first astronauts, hailed as "the Mercury Seven," to reporters and photographers who had packed a makeshift auditorium in the space agency's Washington, DC, headquarters. "They are trim-figured young men, lean jawed, tanned," explained the *Los Angeles*

Times the following day. "The seven," the newspaper continued, were "chosen from a rigorous selection program, testing both physique and personality."[28] The point person for this selection process was Dr. Randolph Lovelace, chairman of NASA's Life Sciences Committee, which comprised medical professionals responsible for choosing the Mercury astronauts. In order to whittle down the original list of 110 military test pilots, Lovelace ran each candidate through an intensive seven-day trial at his clinic in Albuquerque, New Mexico, that involved not only cardiological, neurological, hematological, and pulmonological examinations but also what Lovelace called "physical competency tests" to determine, and then rank, each candidate's general bodily condition.[29] According to the *Baltimore Sun*, the Lovelace clinic relied on the physical tests used for submariners, frogmen, and special assignments in the Army and "extended those tests to the ultimate."[30] "Our twentieth-century Mercury," Glennan explained at another publicity event held at New York's posh Astor Hotel, will have "outstanding" physical qualifications.[31] "What we are looking for," explained another NASA spokesperson, is "supermen."[32]

Immediately after selecting the Mercury Seven, NASA administrators strove tirelessly to maintain these men's overall bodily health. Such efforts initially involved on-the-ground physical training undertaken by each individual astronaut. While some astronauts enjoyed swimming and others played handball to keep in shape, John Glenn ran two miles every day, in under fifteen minutes, no matter where he was stationed.[33] When this hands-off approach failed, and especially if astronauts began gaining weight that might jeopardize future missions, NASA instituted a more routinized workout regimen overseen by "training specialists" in a gymnasium built specifically for the Mercury Seven at the space center in Houston.[34] Such monitoring of astronauts' overall bodily health also extended from the gym on the ground in Houston to the Mercury, Gemini, and Apollo space capsules in outer space. For each of these programs NASA developed sensors, including special electrocardiograms, that attached to astronauts' bodies to track a host of vital signs from blood pressure to respiration to body temperature to heart rate.[35] When John Glenn became the first American to orbit Earth on February 20, 1962, for instance, NASA doctors located in Kano, Nigeria, tracked his physical health for the seven minutes he was directly overhead during each of his three orbits around Earth.[36] Other tracking stations evenly dispersed along Glenn's route monitored his body in similar seven-minute increments. "We were literally wired up to give them medical information," explained a fellow Mercury astronaut, Scott Carpenter, in 1960.[37]

The space agency went out of its way to publicize its strong, healthy astronauts as representing the epitome of masculinity. Such efforts began just a few months after the Mercury Seven announcement when NASA signed an exclusive deal with *Life* magazine, the best-selling weekly in the country, for the "personal stories" of the astronauts. According to the magazine's main astronaut correspondent, Loudon Wainwright, the space agency needed to maintain public support in order to ensure congressional funding and "*Life* was a superb vehicle for that."[38] So were virile astronaut bodies, so much so that NASA, which according to the original contract had the right to review *Life's* coverage prior to publication, pushed stories that praised its spacemen's physiques. The magazine, for instance, ran articles on John Glenn's diet and exercise program, the latter of which included running on the beach at Cape Canaveral, and photographs of Gemini astronaut Ed White doing backyard pushups on a horizontal bar alongside his son and daughter. *Life* even included descriptions of the Apollo 7 astronaut Donn Eisele undertaking "short sprints down the block with his 94-pound wife on his back."[39] Other news media followed *Life's* lead. Newspapers across the country ran articles with titles such as "Astronaut: The Magnificent Male" and "Brains as Well as Bodies Have to Rate with the Best."[40] The space agency, concluded Norman Mailer in *Life* magazine in 1969, "was vending space," in part, by deploying strong astronaut bodies as billboards for a strong space program.[41]

Strong, masculine, physically active astronauts appealed to Americans because they symbolized a nation of strong, masculine, physically active men. "Each of the astronaut crews," explained the *Washington Daily News*, "has represented the ideal of American manhood."[42] Yet this masculine ideal proved particularly powerful because NASA and the national media situated astronaut bodies not only along Kennedy's New Frontier in space but also, by analogy, along America's old frontier beyond the Mississippi. During the early 1960s, for instance, *Life* magazine constantly compared virile astronauts journeying beyond Earth to rugged individuals settling the American West. The Mercury Seven were "splendidly conditioned" "pioneers" with a "destiny" similar to those who explored the "U.S. West," the magazine argued.[43] Their adventure in space, *Life* added, "seems a natural undertaking for the American people, who are a venturesome lot."[44] Natural, that is, as long as you were male and white. *Newsweek* was perhaps the most forthright in exposing the particulars of the astronauts' gender, nationality, and race. Neil Armstrong and Buzz Aldrin, explained the magazine in its special *Moon Age* issue in July of 1969, were "two earthlings representing both sexes (though they are men), all races (though they are pinkish-white beneath their white spacesuits) and all nations (though they are from the United States, as you might infer from the patches on their sleeves)."[45]

Such tight cultural associations between strong astronaut bodies and a strong American nation became even more important on April 12, 1961, when Soviet cosmonaut Yuri Gagarin beat out the Mercury Seven to become the first human to orbit the Earth. Similar to his American counterparts, Gagarin had also undergone extensive medical tests, in his case with approximately 200 other Soviet air force pilots at the Central Scientific Research Aviation Hospital in Moscow, to assess his physical capabilities for space travel. When these examinations reduced the candidate pool to twelve, they, too, like the Mercury Seven, began an intensive on-the-ground training regimen that included jogging, two hours of daily calisthenics, and, after the Soviet space agency built its own cosmonaut gymnasium, swimming. "I felt as if I could enter the Olympics," exclaimed fellow cosmonaut German Titov, who followed Gagarin into Earth orbit in August of 1961.[46] Soviet space doctors, like their counterparts in Houston, also used bodily sensors to measure cosmonauts' respiration and heart rate.[47] Yet while NASA and *Life* magazine promoted America's astronauts as extraordinary physical specimens, the Soviet propaganda machine in Moscow instead depicted Gagarin and his fellow cosmonauts, much like they had portrayed Tereshkova, as examples of everyday Russian citizens. "He was an ordinary man," explained TASS, the Soviet news agency, of Gagarin, who himself added: "Just imagine an ordinary man like me being entrusted with such a great task of national importance."[48] Thus while Americans promoted their own virile astronauts as models of American exceptionalism, the Soviets publicized strong cosmonaut bodies as representing Russia's proletarian everymen.[49]

This Cold War battle over muscles, rather than missiles, proved even more culturally relevant during the late 1960s when middle-class Americans became increasingly anxious regarding the physical state of the nation's men. President-elect John F. Kennedy had sounded this alarm in a December 1960 article for *Sports Illustrated* that decried the flabby shape of the American male. "Our struggles against aggressors throughout our history have been won on the playgrounds and corner lots and fields of America," he warned. "Our growing softness, our increasing lack of physical fitness, is a menace to our security."[50] *Life* magazine had voiced similar concerns, complaining a year earlier in an article titled "The Sixties: Decade of Man in Space" that the "energies that drove America to true greatness, lately seem diluted."[51] The space agency's macho astronauts were for Kennedy and many middle-class Americans the antidote to this perceived masculinity crisis.[52] America's spacemen, argued the Washington, DC, bureau chief of the *Columbus Dispatch* in an editorial coinciding with the 1969 Apollo 11 launch, are the sons of men and women who had the "pioneer spirit" and "vigor" that made America great.[53]

Outer Space as Extreme Environment

Flabby male physiques on the home front and cosmonaut everymen throughout Russia were not the only threats to NASA's able-bodied astronauts. So were potatoes. At least that was the theory of Frank A. Brown, a biologist at Northwestern University who in 1965 proposed launching a potato into orbit to determine if the space environment disrupted "biological clocks," which at the time were thought to regulate organic rhythms in all living things. "If the potato dies," argued Professor Brown in the *New York Times*, the space program "better be checked and rechecked before a man is sent out there." Although comic today, "*Spud*nik-1," as the *Times* nicknamed it, was in the mid-1960s quite understandable since the new field of aerospace medicine remained ignorant regarding both the natural environment of outer space and the effects it might have on human bodies.[54] This was precisely why NASA funded Brown's experiment with $75,000 for the construction of a "potato space cabin" scheduled for orbit in late 1965. Although Professor Brown's potatoes never made it off the launchpad, NASA did spend the better part of a decade wiring up, and firing off, other biological entities, including fruit flies, mice, and monkeys.[55]

Such animal experiments were intended to help medical experts back on Earth better understand the physical characteristics of the unfamiliar, alien space environment. "For all our calculations and planning," explained Neil Armstrong in September 1964, "it will be a voyage into the unknown."[56] The *Los Angeles Times* put it less tactfully, complaining that "when it comes to the ways a long stay in space may affect biological systems, of which man is among the most complex, NASA doesn't know what in hell it is doing."[57]

Due to such scientific unknowns the popular press tended toward the sensational. Outer space, argued the *New York Times* in 1965, is "an environment as hostile to a human being as is the atmosphere to a deep sea fish."[58] *Life* was even more melodramatic, explaining that same year that the "the principal hazards of outer space are fire and ice, asphyxiation and explosive decompression, all bundled up into an irradiated package labeled 'Instant Death.'" In case readers missed the obvious, the article, subtitled "Freezing, Frying, Bubbling to Death," concluded that "it's easy to read the label on the human package that is to wander within this murderous environment: that label reads PERISHABLE, in big capital letters."[59]

Aerospace scientists working with NASA spent the better part of the 1960s trying desperately to parse fact from fiction regarding the environmental hazards of outer space.[60] Dr. Charles Berry, who was at the forefront of such efforts as the space agency's astronaut physician and director of medical research, admitted in the *Journal of the American Medical Association* to "a large number of expected problems involving man and his hardware or vehicle in the space environment." The lack of oxygen, high levels of radiation, and zero gravity in space, Berry warned, might cause a host of ailments, including decreased white blood cell counts, the loss of bone calcium, and increases in muscle atrophy, respiratory difficulties, and heart "deconditioning" or arrhythmias.[61] Small meteorites the size of pebbles traveling at bulletlike speeds, as well as wild lunar temperature swings from a balmy 250 degrees Fahrenheit at noon to minus 200 degrees by midnight, added yet other environmental risks.[62] Because of these possible complications, explained *Reader's Digest* in the mid-1960s, "hundreds of scientists—at NASA research centers and at scores of university and industrial laboratories—are involved in an increasingly pressing race against time to devise the means and the strategies of safeguarding our astronauts against lunar hazards."[63]

Launching Earth Environments into Space: Spacesuits

The cornerstone of such efforts was the Apollo spacesuit, which would protect astronaut bodies by physically transporting pieces of the Earth's environment with them into outer space.[64] During the early 1960s NASA accomplished this by updating high-altitude pressurized flight suits from World War II for use during the low-Earth orbital flights of the Mercury and Gemini missions. By mid-decade, in preparation for Apollo's lunar

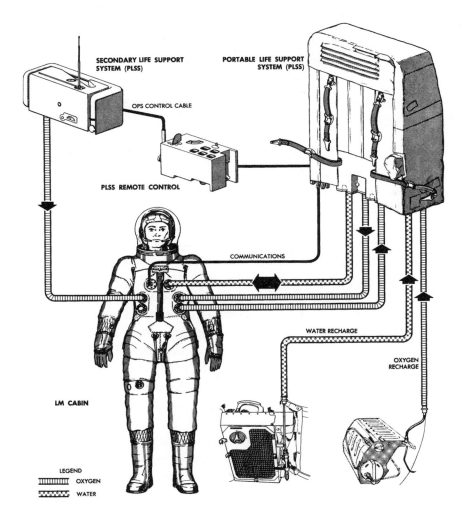

4.1   Diagram of NASA's Portable Life Support System.

landing, engineers and scientists at the space center in Houston oversaw
the overhaul of the suit once again. The result was the A7L spacesuit, which
NASA publicized through illustrated flow diagrams appearing in its Apollo
press releases (see Figure 4.1).[65] The twenty-one-layer wearable technology
could either be hooked up to Apollo's Environmental Control Subsystem
through oxygen, water, and electrical "umbilical" tubes, seen in the diagram
emanating from the astronaut's stomach and chest, or operate independently

of the capsule in order to permit astronauts to move their bodies into space and across the lunar surface during extravehicular activities (EVAs).[66] Because of this technology, explained one commentator, "a real space suit is a little envelope of Earth conditions."[67]

Circulating within that little envelope was the life-sustaining air that astronauts had left behind 240,000 miles back down on Earth, or at least a healthy approximation of it. To offset dangerously low concentrations of oxygen in astronauts' lungs and blood, which resulted from the low pressure of outer space, all Apollo moonwalkers donned a backpack-sized Portable Life Support System that pumped out pure oxygen rather than a more Earth-like mixture of 20 percent oxygen and 80 percent nitrogen. This personal "environmental control system," as NASA called it, even reprocessed exhaled carbon dioxide and other contaminants by passing them through a canister of lithium hydroxide.[68] The A7L also replicated Earth's atmospheric pressure through its Pressure Garment Assembly, a five-layer bodysuit surrounding a neoprene-coated nylon bladder that filled with oxygen to offset the lack of air pressure on the moon.[69] The space agency's $300,000 Apollo 11 suits, explained the Associated Press during that mission, "are really an attempt to bring their earth atmosphere with them."[70]

To protect astronaut bodies from freezing, frying, and bubbling to death on the lunar surface, the A7L also mimicked the Earth's climate.[71] To keep astronaut bodies from overheating during physically taxing EVAs, the innermost layer of the spacesuit consisted of a Liquid Cooling Garment that resembled a pair of long johns interlaced with capillary-like tubing. Chilled water circulated through this network to cool the astronaut's body, by direct conduction, to a comfortable seventy degrees.[72] Engineers fashioned the spacesuit's outermost covering, the sixteen-layer Integrated Thermal Micrometeoroid Garment, to offset a quite different range of temperatures. "It combines the basic functions of the thermos bottle," explained the *New York Times*, protecting the astronaut's body against temperature extremes involving the blazing 250-degree heat of the moon's sunlit areas and the frigid minus-250-degree chill of lunar shade.[73] The A7L similarly protected astronauts from severe lunar weather; while the thermal outer garment served as an umbrella, of sorts, against a deadly lunar rain composed of micrometeorites the size of dust particles traveling at speeds of 64,000 miles an hour, the suit's gold-coated Lunar Extravehicular Visor Assembly deflected harmful infrared radiation and ultraviolet rays, which were more intense on the moon because of the lack of air, from crewmembers' eyes.[74]

"We earthlings use sunglasses," admitted the *Los Angeles Herald-Examiner*. When astronauts joined NASA, the newspaper concluded, they traded in their test-pilot Ray-Bans for Apollo's gold-tinted visor.[75]

Although the A7L protected all astronauts equally from the physical hazards of space, each suit functioned slightly differently in the lunar environment because engineers designed them to fit individual crewmember's bodies. "You cannot, of course, walk in and get a pressure suit off the rack," wrote astronaut Walter Schirra in a 1960 *Life* magazine article titled "A Suit Tailor-Made for Space." "To fit it properly requires more alterations than a bridal gown."[76] The International Latex Corporation of Frederica, Delaware, more popularly known as Playtex, secured the contract for the Apollo suits and undertook no fewer than sixty-four body measurements, in different poses and postures, and continually updated them through follow-up "fit checks" that measured each astronaut's body against an idealized masculine norm (see Figure 4.2).[77] Only then did Playtex's highly skilled seamstresses, who had spent most of their careers making women's

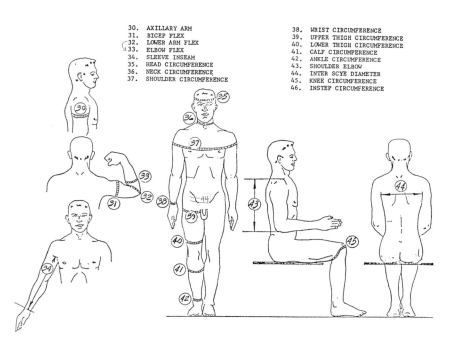

30.	AXILLARY ARM	38.	WRIST CIRCUMFERENCE
31.	BICEP FLEX	39.	UPPER THIGH CIRCUMFERENCE
32.	LOWER ARM FLEX	40.	LOWER THIGH CIRCUMFERENCE
33.	ELBOW FLEX	41.	CALF CIRCUMFERENCE
34.	SLEEVE INSEAM	42.	ANKLE CIRCUMFERENCE
35.	HEAD CIRCUMFERENCE	43.	SHOULDER ELBOW
36.	NECK CIRCUMFERENCE	44.	INTER SCYE DIAMETER
37.	SHOULDER CIRCUMFERENCE	45.	KNEE CIRCUMFERENCE
		46.	INSTEP CIRCUMFERENCE

4.2   International Latex Corporation's anthropometry astronaut body chart, mid-1960s.

girdles and brassieres, begin sewing each of the suit's twenty-one layers by hand to within one sixty-fourth of an inch from each seam.[78] Because the A7L spacesuit was custom-tailored for each astronaut's body, crewmen took a special interest in "their" suits. "To the Girls of Frederica—Thank you for sewing straight and careful," wrote Apollo 13 commander Jim Lovell to the Playtex seamstresses. "I would hate to have a tear in my pants on the moon."[79]

As Lovell's quaint note suggests, the cultural significance of these space-suits, much like their production, was highly gendered. Soon after Playtex won the contract to fabricate the Apollo spacesuit in 1965, the company spun off its aerospace division from the newly christened "International Playtex Corporation" in an effort to distance its spacesuit brand from the production of women's underwear. Two years later when this contract came up for renewal, the corporation's aerospace division also pitched NASA by associating their product with the most manly of sports; it dressed one of its engineers in a pressurized A7L spacesuit and filmed him, for several hours to account for the length of an Apollo EVA, playing football at a nearby high school field.[80] It then included the film with its renewal application. Administrators at NASA likewise promoted Apollo spacesuits to the American public as symbols of American manhood. In a 1964 press release, for instance, the space agency compared its spacesuit to medieval armor, even including side-by-side photographs of each above a caption reading "Nothing New under Ye Olde Sun?"[81] The national media passed such symbolism on to readers. An Apollo astronaut on the moon, reported United Press International in 1964, will be the equivalent of "a medieval knight in a suit of armor."[82]

Building Space Environments on Earth: Simulators

The A7L spacesuit alone could not ensure the success of Apollo. Along with protecting their bodies, astronauts also had to learn how to maneuver them through foreign environments in space and on the moon. As Neil Armstrong explained in September 1964 regarding his future lunar landing, astronauts "master each separate step with the help of an assortment of strange and sophisticated machines called simulators."[83] While the majority of these technologies duplicated space hardware in order to teach astronauts how to use it, others replicated back down on Earth the various natural environments encountered in space.[84] Such simulations had historical antecedents; increasingly after World War II military agencies had begun building laboratories within the United States that mimicked various global climatic conditions, especially those of extreme environments in the arctic and tropics, in order both to learn how to protect armed forces deployed in these physically demanding regions as well as to teach soldiers the best means of maneuvering across these hazardous landscapes.[85] The space agency did similarly, constructing a sprawling network of simulators stretching from the flatlands of Houston to the woodlands of Virginia to the volcanic mountaintops of Hawaii.[86] "The best we can do," concluded Armstrong, is build mechanisms that "break the mission down into its components—launch, rendezvous, lunar landing, lunar liftoff, and re-entry."[87] For each component, NASA engineers developed different simulated environments.

To imitate the increased intensity of Earth's gravity during launch, NASA used enormous machines on the ground that spun astronauts round and round. For the Mercury Seven the space agency borrowed a centrifuge at the Naval Air Development Center in Johnsville, Pennsylvania that had been used to test fighter pilots.[88] With the opening in Houston of the Manned Spacecraft Center in 1963, NASA began constructing its own centrifuge, which *Time* magazine called a "gruesome merry-go-round."[89] Spanning more than 150 feet in diameter and rising four stories tall, the Flight Acceleration Facility could spin three astronauts at the end of a sixty-foot arm at speeds high enough to generate a maximum force of 30 Gs, far more intense than the approximately 4 Gs experienced by Apollo astronauts during launch. Such simulations taught astronauts how to counteract the debilitating physical side effects of liftoff and reentry into the Earth's atmosphere by using breathing techniques and muscle contractions. "When you try to inhale, it's impossible to reinflate your lungs," explained astronaut Michael Collins of his experiences whirling around in the Johnsville centrifuge. Such practice runs, he argued, helped to "develop an entirely new method, keeping the lungs almost fully inflated at all times and giving rapid little pants 'off the top,'" a technique that he and his fellow astronauts used during liftoff.[90]

After launch, astronauts faced a four-day journey to the moon, and here again NASA developed simulators back on Earth, in this case to replicate as much as possible the weightlessness of space. During the Mercury era of the early 1960s the space agency flew airplanes, with their seats removed and their walls padded, in parabolic arcs twenty seconds long to re-create for astronaut passengers the sensation of zero gravity. To extend the duration of such experiments, in the mid-1960s Mercury astronaut Scott Carpenter suggested looking to the ocean, rather than the air, as an equivalent for weightlessness. Carpenter made his recommendation after taking a leave of absence from NASA in 1963 to work on the Navy's SEALAB program. "We had a lot of tasks to perform in the water outside SEALAB," explained Carpenter of his experiences with the experimental underwater habitat, "and the problem in the water is you don't have traction."[91] The same was true of the space environment, Carpenter argued, and soon NASA began training its astronauts underwater. Such exercises began in 1967 when two dozen astronauts traveled to Key West, Florida, for scuba training in an outdoor pool that *Life* magazine called "the nearest thing to the weightlessness of space that is attainable on earth."[92] Soon after, NASA brought such training in-house by constructing the Water Immersion Facility at the Manned Spacecraft Center in Houston. Training at the facility, which was basically a large indoor pool, involved placing astronauts in specially designed water-filled suits that countered their own buoyancy to emulate the weightlessness of space and having them maneuver their bodies through a variety of underwater obstacles in order to simulate movements within the tight confines of the Apollo space capsule.[93]

Because it was the most complex and dangerous maneuver of the mission, NASA built several environmental simulators that mimicked the Apollo spacecraft's descent from lunar orbit to the Moon's surface. The most sophisticated of such technologies was the Lunar Module Mission Simulator, which included a full-scale mockup of the lander's interior surrounded by walls of display screens that projected into the spacecraft's four windows a visual simulation of the lunar crust from distances of approximately 15,000 feet to touchdown. To accomplish this illusion, NASA teamed up with optical specialists to create what it called the "Infinity" visual display. The highly complex system relied on numerous visual cues, including artist renderings of the moon's surface based on data gathered from lunar satellites, actual lunar satellite photographs that were "photomosaicked" into filmstrips, and photographs of low-altitude lunar surface scenes taken during earlier Apollo missions.[94] Engineers at NASA then used reflective optics techniques involving dozens of mirrors and lenses to interleave the projection of these images at different distances to achieve the illusion of a continuously approaching three-dimensional lunar landscape.[95] To visually simulate touchdown, NASA synchronized the lunar landing module's descent controls with a closed-circuit television camera in a nearby room that moved across a sixteen-foot-wide lunar terrain model, made from plaster and painted to account for the sun's shadows, of specific landing sites that had been chosen for each Apollo mission.[96] As a result, even before Apollo 14 commander Alan Shepard landed in the Fra Mauro region of the moon on February 5, 1971, he had seen exact replicas of this landscape, as well as specific features from it, hundreds of times.[97]

After landing their spacecraft safely on the moon, NASA's astronauts relied on bodily practices learned within environmental simulators back on Earth to take their first giant steps for mankind across the lunar landscape. Such training began in April of 1963 when NASA asked the U.S. Geological Survey (USGS) to scour the country, and to look beyond its borders as well, for geographies comparable to the moon's. Geologists then took Apollo crewmen to these locations to train them in rock identification and sampling techniques, to familiarize them with landscapes similar to possible Apollo landing sites already chosen by NASA, and, perhaps most important, to provide astronauts undertaking moonwalks with the experience of moving their bodies across lunarlike terrain. By 1967 NASA required each Apollo astronaut to participate in ten of these USGS field trips to locations including Bend, Oregon; Medicine Lake, California; and the Sonora Desert in Mexico, among others.[98] "We went to the San Juan Mountains in Colorado, where we got a feel for the Apennine Mountains of the moon," explained Apollo 15 Commander David Scott. "The Rio Grande Gorge near Taos, New Mexico," he added, "served as our familiarization course for Hadley Rille," more than three miles of which Scott and his crewmate James Irwin explored on August 2, 1971, with Apollo's lunar rover.[99]

Apollo astronauts on USGS field trips did not merely take a hike across these moonlike surfaces. They specifically practiced moving their bodies, sometimes fully space-suited, along routes with topography and distances that mimicked, as closely as possible, the extravehicular missions they would undertake on the moon. When the Apollo 17 crew practiced on the crest of Mauna Kea on the island of Hawaii, for instance, USGS scientists not only plotted the astronauts' route along a series of traverses around the volcano's summit that approximated those the crew would follow up the Taurus-Littrow valley on the moon, but also designated rock-sampling sites at intervals that replicated the timeline that had already been carefully plotted by NASA for the actual moonwalk. As one engineer working on the

Mauna Kea training explained, "This Hawaii simulation was about as good as we could get in obtaining a high fidelity rehearsal before the real mission was under way."[100]

When nature on Earth was just not quite right, NASA, with the help of the USGS, also "improved" on it. Sunset Crater, just northeast of Flagstaff, Arizona, was the most important practice site for the Apollo astronauts because it was, as a NASA engineer involved in these field trips explained, "the closest copy of a moonscape that existed anywhere on Earth." Yet to make this landscape even more authentic, USGS scientists first analyzed a photograph of the moon's surface taken by an unmanned lunar satellite. They then carefully measured the diameter and depth of each crater in the photo, and also determined the crater's relative age based on the ordering of overlapping layers of crust material ejected from each depression when it was formed by meteor impact. The farther below the ejected material, the older the crater. Geologists then selected a ten-acre site just south of Sunset Crater and laid out a grid of explosives, arranged according to the force they would generate, in order to create properly sized craters in the correct locations. Scientists even timed these explosions to go off in the sequence that would provide the correct ejection layers, as observed on the real lunar surface (see Figure 4.3).[101] When USGS and NASA scientists finally detonated the explosives, the lunar landscape that had been captured in the satellite photograph was re-created for Apollo astronauts in the high desert of Arizona.

Finally, when NASA administrators wanted to keep their astronauts closer to home just prior to liftoff, they trained them to moonwalk not on natural replicas thousands of miles away but rather on simulated lunar surfaces built in the space agency's very own neighborhood. At the Kennedy Space Center in Cape Canaveral, Florida, NASA constructed an imitation lunar surface from gravel, sand, and rock so that Apollo astronauts could review specific extravehicular tasks that were either modified or added to the mission at the last minute before launch. Unfortunately, the center's fake moonscape was often unusable, because its "craters" often filled with water during high tide.[102] Across the space crescent in Houston, NASA also began fabricating a much more elaborate moonscape in the mid-1960s; engineers covered the football field–sized simulation with tons of slag from a nearby blast furnace and large chunks of lava rock, and they excavated several craters fifty feet wide by fifteen feet deep. Just as USGS geologists had studied photographs of the lunar surface in order to construct a copy in

4.3   NASA and the USGS detonating craters in the Arizona desert, 1968.

Arizona, NASA engineers at the Manned Spacecraft Center similarly analyzed the best available observation data, in this case of a possible Apollo 11
landing site near the Kepler Crater in the Oceanus Procellarum region of
the moon, and then rebuilt a scaled-down version of it in Houston. To
make the landscape even more true to life, engineers included a mockup of
the lunar lander not only so astronauts could practice moving their bodies
into and out of their spaceship but also so NASA could make time-and-
motion studies of the crewmen moving in their spacesuits.[103] "At our headquarters in Houston, NASA geologists are building a simulated lunarscape,"
explained Neil Armstrong in the mid-1960s, "which we hope is so like the
real thing that when we step out onto the moon for the first time it will be
almost as familiar as our backyards."[104]

According to Apollo astronauts, NASA's simulators succeeded in making
some lunar landscapes as recognizable as their own backyards, perhaps even
more so because NASA's intense training regimen left little time for these
spacemen to relax at poolside barbecues.[105] "When we actually descended

the ladder," explained Armstrong in a postmission press conference during the late summer of 1969, "it was found to be very much like the gravity simulations we had performed here on Earth."[106] *Popular Science* concurred, noting that simulations conducted by the USGS involving astronauts moving across lunarlike landscapes back on Earth were "rehearsing procedures so they would become habitual."[107] As a result of such bodily practices, the space environment for Apollo astronauts had become second nature, one mediated by technology but no less natural. "I knew exactly what to do, after more than a year of training," concluded James Irwin of his exploration of Hadley Rille near his Apollo 15 landing site. "I felt right at home."[108]

Women's Bodies and the Space Environment

After landing in the Mojave Desert during the summer of 1983 and stepping from the space shuttle *Challenger* with her four male crewmates, Sally Ride also felt "right at home," especially since she had been raised most of her life in nearby Los Angeles. Yet for Ride and women like her, this had not always been the case regarding NASA's astronaut corps. "I didn't day dream about being an astronaut when I was in high school," admitted Ride a few months before her *Challenger* flight. She couldn't; back then in the 1960s there were no female astronauts in the United States. There were likewise few female scientists or women engineers working at NASA or for its thousands of contractors. Instead, the only women popularly associated with the space race were the wives of the Mercury Seven, who appeared on the September 21, 1959 cover of *Life* in smart dresses, pearl earrings, and red lipstick, and inside the magazine's pages as dutiful wives, caring mothers, and happy homemakers.[109] For Ride, who was eight years old at the time, such cover stories served little purpose. "I suppose I just took it for granted that it was pretty much a closed club," she explained in reference to NASA's astronaut program. "I just assumed there would never be a place for women."[110]

Officially, NASA did not prohibit women from becoming astronauts. Yet it accomplished this indirectly by making one of the requirements at least 1,500 hours of flying time as a qualified jet pilot, the thinking being that test pilots not only had the necessary skills and bravery for space exploration but also had already volunteered to risk their lives for their country. Never mind that flying a spaceship and piloting a jet had little in common, or, for that matter, that NASA originally planned to "fly" its spaceships by computer from the ground in Houston with astronauts simply along for the ride. Racking up such jet time nevertheless proved impossible for female pilots because the Women's Armed Service Integration Act of 1948 banned them from combat cockpits, including those of fighter jets.[111] "Under conditions as they exist in aviation at the present time," explained NASA Administrator James Webb, who was quoted in a 1962 *Saturday Evening Post* editorial titled "No Ladies in Orbit," "these qualifications are more readily met by men than by women."[112] The *New York Times* put it more bluntly, explaining just days after Webb's remarks that "the criteria effectively ruled out women, since there apparently are none, in this country, who have tested high-performance jets."[113]

Unofficially within the space agency, a more blatant form of sexism underlay these astronaut criteria. "The men go off and fight the wars and fly the airplanes," explained John Glenn in 1962 regarding the question of women astronauts. "The fact that women are not in this field is a fact of our social order."[114] That same year another high-ranking NASA official added that women astronauts would be a "waste in space, a luxury the United States space effort cannot afford."[115] Such thinking inside the space agency merely mirrored widespread beliefs beyond Cape Canaveral and Houston. The female astronaut "would probably use her insidious influence to get softer cushions in the rocket ship, more room on the inside, curtains over the portholes, antimacassars, throw rugs and pastel walls," complained the *Los Angeles Times* in a September 1960 editorial. The only sensible solution, according to the paper, was to "leave women at home until men had time to prepare the raw ground, as they did at the frontiers of earth."[116]

Editorial cartoons from the late 1960s repeated this sexist mantra in visual form. The *Christian Science Monitor*, for instance, ran an illustration during the Apollo 11 mission that depicted a wife and mother of two struggling to gain the attention of her distracted husband who stares intently at a large moon, lit up by a living-room spotlight, lodged between his legs. "Okay dear, come on . . . children are hungry . . . something's wrong with the water . . . ," she complains to no avail (see Figure 4.4).[117] The Pulitzer Prize–winning cartoonist Bill Mauldin was even more explicit in relegating women to the home front while their husbands played astronaut. In a cartoon also published during Apollo 11 an exasperated homemaker, dressed in a cooking apron and skirt, stands alongside a broken oven with its unhinged door held together with string. ". . . And I can't get the car fixed, and the dishwasher broke down again, and the TV quit just as you were landing on the moon," she complains to her husband, who wears an astronaut

'Okay, dear, come on . . . children are hungry . . . something's wrong with the water . . .'

4.4   Guernsey LePelley, "Okay, Dear, Come On . . . Children Are Hungry . . . Something's Wrong with the Water . . . ," *Christian Science Monitor*, July 26, 1969.

spacesuit and a confounded look on his face while carrying a simple toolbox labeled "U.S. TECHNOLOGY."[118] With such coverage appearing almost weekly in local newspapers, it's no wonder that a young Sally Ride was unable to daydream about a future in space.

Sexist opponents of women astronauts within NASA attached their most vitriolic rhetoric to one site in particular: the female body. It was along its curves and within its internal organs that they made their most ardent defense of the space agency's all-male astronaut corps. "All I can say is the male astronauts are all for it," joked NASA rocketeer Wernher von Braun in response to a question regarding women astronauts posed after a speech he gave at Mississippi State College in 1962. To allow for this possibility, he added with a smirk, the space agency was "reserving 110 pounds of payload for *recreational* equipment."[119] Similar rhetoric circulated beyond the space agency. Female companionship, like "an occasional cocktail," argued the American Psychological Association in 1958, could not only help alleviate the isolation and boredom of long-distance spaceflights but also jump-start a new generation of "space children."[120] A decade later the *Washington Post* echoed this sentiment. "The question of man's sexual needs on flights," the newspaper reported in an article titled "Women's Place in Outer Space," "has to be considered."[121] As such remarks suggest, during the early 1960s aspiring female astronauts were given two options for space travel: homemaker or whore.

Many opponents of women astronauts both within and beyond NASA couched their sexist rhetoric within paternalistic attitudes regarding the physical hazards of outer space, which many believed were even more threatening for female bodies.[122] Newspapers such as the *New York Times* constantly worried about the health of potential female astronauts during the intensified gravitational force of liftoff and the extreme weightlessness of spaceflight.[123] Of most concern, however, were the dangers of space travel for women's reproductive health. "What would be the effect of low gravity, for instance, on the extraordinarily complicated reproductive mechanism of the female body?" wondered science fiction writer Isaac Asimov in a 1971 *Ladies' Home Journal* article titled "No Space for Women?"[124] *Science Digest* likewise noted concerns by obstetricians and gynecologists that the female menstrual cycle might cause weight gain, vision problems, and mental and emotional shifts that would jeopardize space missions.[125] Thinking within the space engineering community was similar, if a bit less scientific. "Over the course of a month most girls vary from deep depression to great happiness," explained one NASA contractor in a 1972 article for *Space World* magazine. "The males on-board might well want to have a calendar on each girl." The author went on to argue that tracking each female astronaut's menstrual cycle would allow the men on board to know "who wakes up in a cheerful mood and who is a grouchy bear."[126]

Administrators defended such chauvinism by noting that NASA did not currently have spacesuits that could protect female bodies from the physical threats posed by the space environment. Developing such suits, the space agency argued publicly, would not only be extremely costly and thus imperil America's efforts to beat the Russians to the moon, but would also be an engineering nightmare. "There is the problem of designing and fitting a spacesuit to accommodate their particular biological needs and functions," explained one NASA official during the fall of 1960.[127] The Apollo spacesuit, added another spokesperson more than a decade later, "would be damaging to the soft structures of the feminine body."[128] There was also the issue of bodily waste. By the mid-1960s the space agency had already spent millions of dollars developing a urinary collection device that slid over each crewman's penis, but the female anatomy, NASA administrators claimed, presented additional engineering difficulties in the weightlessness of space. "There was no way to manage women's waste," argued NASA's Director of Life Sciences, David Winter. "If you can't handle a basic physiological need like that, you can't go anywhere."[129] The national media became obsessed with this particular issue, publicizing NASA administrators' concerns to the broader American public. "Before a woman could be catapulted into space," explained *Parade* magazine in 1967, "NASA would have to design 'his' and 'her' spacesuits."[130]

By claiming that Apollo spacesuits could not accommodate female physiology, NASA was also suggesting that women were culturally unsuited for space exploration. A portion of the American public seemed to agree. "Space suits will have to be made more attractive if there are to be many feminine volunteers," argued an editor of the *Los Angeles Times* during the summer of 1960. "You might talk a shapely young thing into wearing one of those multiple-layer, bubble-headed sacks," the newspaper added, "but, out there in space, who would zip her?"[131] Such thinking was reinforced in editorial cartoons, such as that appearing in the *Daily Oklahoman* in 1962 depicting "his" and "her" spaceships that suggested the need for separate male and female toilets (see Figure 4.5), as well as in popular films such as *Barbarella*, which opened in 1968 and starred Jane Fonda as a comedic bikini-clad space traveler.[132] Unlike the muscular bodies of male astronauts, which the space agency guarded with space age medieval armor, women's curves and protrusions were just not fit for travel beyond Earth.

Critics also defended their opposition to women in space by noting that the female body had not yet been fully vetted for space exploration in NASA's environmental simulators. "In order to determine adequately women's physiological and psychological resistance to the stresses of a space environment," explained Brigadier General Don Flickinger, who served as a

Two-Rocket Pads?

4.5   Jim Lange, "Women in Space, Too?," *Daily Oklahoman*, July 19, 1962.

bioastronautics consultant for NASA during the early 1960s, "biomedical studies would have to be made on hundreds of the fairer sex." The numerous simulators developed by NASA to assess the bodies of its male astronaut candidates, Flickinger argued, would have to be radically retailored to accommodate women. Doing so, added Houston's space center director, George Low, in a 1962 article titled "Space Women Expensive," "will slow down the national space program and might prevent the U.S. from placing a man on the moon by 1970."[133] Defenders of NASA's all-male astronaut program thus argued not only that the A7L spacesuit was unable to protect women's bodies from the environmental hazards of outer space, but also that the space agency did not have the resources to develop environmental simulators to test these bodies back down on Earth. As a result, Flickinger concluded, "The rule of 'ladies first' does not apply to outer space."[134]

Feminist Opposition to "Manned" Spaceflight

Many women during the early to mid-1960s did not buy the argument that their gender was not fit for space travel because they did not fit, both biologically and culturally, inside the A7L spacesuit. "I'm hard pressed to believe," complained Mrs. Emily Kozakoff in a 1967 *Parade* magazine article titled "Would-Be Astronauts: Legion of Angry Women," "that NASA can send a rocket to the moon but doesn't know what to do about the problem of feminine hygiene." Nor did American women believe the alibi that there were no environmental simulators with which to assess how female bodies would react to the physical hazards of outer space. Instead, they began blaming NASA. "It's a kind of discrimination that is not spelled out," argued Mrs. Gladys Philpott of Sunnyvale, California, in the same *Parade* article. "NASA thinks the American ideal is for women to marry, have kids and stay home."[135] Yet rather than adhering to this ideal, American women began leaving their homes in droves to protest the space agency's discriminatory astronaut-selection process.

During the early 1960s such activism entailed walking the halls of the federal government in order to pressure NASA from within to accept women astronauts. Two female pilots spearheaded this lobbying effort. Jerri Cobb, at the time one of the most well-known female aviators in the country, along with fellow pilot Jane Hart, who was the wife of Michigan Senator Phillip Hart, secured a meeting with Vice President Lyndon Johnson in the spring of 1962 through Hart's husband's Senate office. Cobb and Hart chose Johnson specifically; as head of the National Aeronautics and Space Council he had the ability to compel NASA to reconsider its astronaut requirements. During the half-hour meeting the conversation quickly turned to the female body. "There's no need of muscle in space— anymore than in driving a car or in flying a plane," explained Hart to the press in reference to her talk with the vice president. "Why must we handicap ourselves with the idea that every woman's place is in the kitchen despite what her talents and capabilities might be?"[136] While Johnson listened politely, afterward he publicly deferred to NASA regarding its astronaut-selection process while privately criticizing Cobb's and Hart's efforts.[137] The conversation with the vice president "was a miserable meeting," remembered Hart years later. "He kept saying there wasn't much that he could do about it."[138]

Cobb and Hart were particularly qualified for such a high-level meeting because the year before they had been two of thirteen women who had put their bodies on the line and passed an exhaustive set of physical examinations that were identical to those initially performed on the Mercury Seven. Dr. Randolph Lovelace, who oversaw the original clinical trial for the Mercury astronauts, also conducted this second set of tests independently of NASA because he believed that women would be needed in future missions as laboratory technicians and communications officers, the space age equivalents of pink-collar lab secretaries and telephone operators. Cobb, Hart, and the other female pilots underwent all of the first-round tests performed on Glenn, Shepard, and the rest of the Mercury Seven, and they also had to endure a gynecological exam. Lovelace, who privately funded the tests, called them the Women in Space Program.

Yet when it came time for a second round of examinations, which like the Mercury tests would assess each subject's physiology within spacelike environments, Lovelace was unable to obtain the necessary aeromedical equipment; both the U.S. military and NASA refused him access to pressure suits and simulators, such as the centrifuges and weightlessness machines located in Johnsville, Pennsylvania, and Houston.[139] Moreover, when Lovelace petitioned the U.S. military to use such technologies to determine, as his clinic phrased it, "the fundamental differences between male and female astronauts," the Pentagon replied with yet another veiled reference to the female body, writing, "If you don't know the difference already, we refuse to put money into the project."[140]

Less than a year after their failed meeting with Vice President Johnson, Cobb and Hart continued their campaign by successfully pressuring the House Committee on Science and Astronautics to convene a special subcommittee to investigate women's place within the U.S. space program.[141] In their own testimony before the subcommittee, both women turned the American frontier analogy, which NASA administrators had used to support their all-male astronaut corps, on its head. "There were women on the *Mayflower* and on the first wagon trains west, working alongside the men to forge new trails and new vistas," explained Cobb in her opening statement. "We ask that opportunity in the pioneering of space." When it was NASA's turn to testify, its spokespeople instead enlisted the female body to oppose female astronauts. In reference to Lovelace's Women in Space Program, John Glenn argued, by way of analogy, that his "mother could probably pass the physical exam they give preseason for the Redskins, but I doubt if she could play too many games for them." Glenn then concluded with a more subtle, and more insulting, reference to the female body, stating that if NASA could find qualified women "we would welcome them with open arms, so to speak." The nation's most famous astronaut then smiled as the politicians in the room erupted in laughter. The subcommittee's final report, not surprisingly, supported NASA's all-male astronaut corps.[142]

Jane Hart then shifted gears, from working within the halls of government to pressuring NASA from beyond Capitol Hill through a network of Washington, DC, women's clubs. Such efforts began in March 1962 when she invited James Webb to speak on the future of the U.S. space program at a Woman's National Democratic Club (WNDC) luncheon. Although Hart introduced Webb to the audience without mentioning her own campaign for female astronauts, a member of the audience, undoubtedly with Hart's blessing, asked the NASA chief during the question-and-answer session why there were no women in space. "Mrs. Hart promised she would not ask that question," Webb complained before explaining that it was not that NASA was "not fond of the ladies," but rather that the space agency was "simply looking for the best qualified people."[143] The following year when Webb refused an invitation to another WNDC luncheon, Hart shrewdly invited his wife, as well as Lady Bird Johnson, to hear an address by New Mexico Senator Clinton Anderson, who also happened to be both chairman of the Senate Aeronautical and Space Sciences Committee as well as an outspoken advocate of female astronauts. "I'd trust Jane Hart and Jerri Cobb with the job," Anderson proclaimed as the luncheon guests, perhaps with the exception of Lady Bird and Mrs. Webb, applauded.[144] Such informal lobbying quickly spread to other women's clubs. At its annual convention in 1963, more than 1,500 delegates from the National Federation of Business and Professional Women's Clubs adopted a resolution demanding that NASA give the thirteen women who had passed Lovelace's initial set of physical exams, including Hart and Cobb, equal opportunity during the astronaut-selection process.[145]

Political activists from the burgeoning feminist movement took note of Cobb's and Hart's actions in support of women astronauts. Early in 1966, for instance, Betty Friedan telephoned Hart to praise her 1962 congressional testimony and also to invite her to become a founding board member of a new women's group that Friedan was forming, called the National Organization for Women. Hart agreed and went on to found local NOW chapters in both Washington, DC, and her home state of Michigan.[146] During the next few years, as the feminist group gained national recognition, Hart, Friedan, and other NOW leaders spearheaded a grassroots campaign to open up NASA's all-male astronaut corps to women. Such formal protests began in early December 1969, when NOW's board members voted to file a formal legal complaint charging NASA with discrimination against women in its astronaut training program. As Friedan exclaimed, quite simply, when announcing the vote to the press, "President Nixon should send a woman to the moon."[147]

Such activism intensified a few years later when NOW members became alarmed by sexist comments made in public by NASA's astronauts. Here, again, the female body took center stage. The first such incident occurred in September 1973, when an Apollo 12 astronaut, Pete Conrad, replied to what he felt were annoying questions about the possibility of female astronauts by asking the women in the audience if they were "sexually deprived." Leaders at NOW responded by initiating a letter-writing campaign that inundated both NASA Headquarters and the U.S. Senate with expressions of outrage.[148] The following month NOW supplemented such actions by inviting Apollo 13 astronaut Jim Lovell to a mock award ceremony. "The annual Barefoot and Pregnant Ball is being held especially in your honor," wrote the president of the Tampa, Florida, NOW chapter in an open letter to Lovell, "as you have uncontestedly merited one of our five annually announced Barefoot and Pregnant awards for the chauvinistic attitudes you have displayed towards women and their struggle for recognized equality."[149] Not surprisingly Lovell, who during a NASA publicity tour of Brazil earlier that year had portrayed women as sex objects by stating publicly that "we will fly women into space and use them the same way we use them on earth—for the same purposes," failed to show up and receive his award.[150]

NOW targeted NASA administrators as well, less for their sexist comments than for their poor record of hiring females. Such activism also peaked in 1973, when NOW's president, Wilma Scott Heide, demanded a meeting with NASA Administrator James Fletcher to discuss the space agency's "woman problem."[151] When Heide and several NOW leaders finally met with Fletcher on November 29 at NASA Headquarters in Washington, DC, the first three items on NOW's agenda included discussion of women's exclusion from the astronaut program, sexist comments about women made by male astronauts, and the failure of the space agency's Equal Employment Opportunity (EEO) Program that had been created in

1971.[152] The NOW leaders were especially concerned about NASA's poor record in hiring women scientists, engineers, and, of course, astronauts; from 1958 to 1970 only between 2 and 3 percent of the technicians and engineers at the space agency were women.[153] According to an official statement issued by NOW immediately following the November 29 meeting, Fletcher responded to such concerns by resorting to vague excuses regarding "goals and timetables" for NASA's EEO Program, refusing to publicly criticize male astronauts for making sexist comments, and standing by the space agency's astronaut-selection criteria that essentially banned women from space.[154]

Leaders at NOW countered by publicly calling for Fletcher's resignation. The women's group took this step at its seventh annual national conference, held in Houston, Texas. In a shrewdly written news release, NOW reminded the press not only that the meeting was taking place in "Space City U.S.A," but also that Houston's spacecraft center had recently been renamed the Johnson Space Center to honor the president who promoted gender equality in the workplace by securing passage of Title VII of the 1964 Civil Rights Act.[155] "Any impression of equal opportunity" at NASA, the press release concluded, is "an insulting illusion."[156] National newspapers quoted liberally from NOW's release, with the *Washington Post* adding that the women's group objected to "astronauts who went around the country making sexist statements that women will be needed on space flights to relieve the sexual tensions of men."[157] To ensure that both politicians and the public understood this criticism, NOW's "National Coordinator to Launch NASA into Feminist Space" petitioned New York Representative Bella Abzug to submit the group's demands for Fletcher's ouster to the *Congressional Record*. Such a submission, NOW explained to Abzug, "would further facilitate public awareness."[158]

This grassroots campaign by NOW against NASA also caught the attention of Gloria Steinem, cofounder of *Ms.* magazine, which after its initial publication in 1971 quickly became the feminist movement's megaphone. Steinem publicized the campaign against NASA in September 1973 by publishing a cover story titled "Yes We Do Have Women Astronauts." Similar to NOW's protests, the magazine's detailed exposé focused on the bodily experiences of women, in this case the thirteen female pilots who participated in the Lovelace Clinic's Women in Space Program. "Jerri Cobb," explained *Ms.* to its tens of thousands of readers, "swallowed three feet of rubber tubing, drank a pint of radioactive water, and had a syringeful of ice water injected into her inner ear, so that the resulting disorientation known as vertigo could be observed." "In short," the article added, "no single, solitary inch of her five-foot-seven-and-a-half-inch frame escaped analysis." The cover story then announced that NASA administrators had initially attempted to keep these tests secret and then abruptly canceled them when news of the physical examinations was leaked to the public. "It is inconceivable to me," argued Jane Hart in the *Ms.* magazine article, "that outer space should be restricted to men only, like some sort of stag club."[159]

The combination of NOW's heels-on-the-ground activism and *Ms.* magazine's feminist bully pulpit, along with the media coverage each garnered, encouraged other feminists to take action on their own against NASA in support of women astronauts. In September 1973, for instance, the same month that Steinem published the female astronaut exposé, a women's rights group at the University of South Dakota organized what they called a "quiet protest" when the Apollo 17 crew presented the school's museum with a moon rock. During the official presentation the female students silently held up three signs, one of which read "Where Are the 13 Women Astronauts Now?," and later confronted astronaut Pete Conrad at a public forum with a question about why there were no women in space.[160] Female students took similar action on both coasts. At Mills College in Oakland, California, the all-female student body organized a teach-in symposium titled "Women's Place in Space Science," while those at Harvard's Radcliffe College held a conference in 1975 titled "Earth in the Cosmos: Space for Women," which included talks by female space science researchers as well as discussions regarding the possibility of female astronauts.[161]

Such grassroots feminist opposition to NASA's all-male astronaut program also circled back to the space agency itself. During the early summer of 1973, a group of women working at the Johnson Space Center became concerned when the Employment Activity Association in charge of planning social functions at the center decided to hold an employee election to name a "Queen and Five Girls in the Court" for Houston's upcoming Lunar Landing Festival. According to the ballot, which asked voters to choose from a list of forty-eight female employees, the winners would be "required to attend the public coronation ball, social teas, and private parties connected with this social function." To protest what was basically a beauty contest, the group of concerned women, who had already complained privately about NASA's poor female hiring record, snuck into the space center at night and secretly replaced the original ballots with fake ones listing

forty-five male employees, including center director Chris Kraft and astronauts Rusty Schweickart, Pete Conrad, and Deke Slayton, as possible "Queen
and her Court" nominees. "Selection should not be based on merit," explained the substitute ballots. "Vote for a pretty face and a good bod." After
the fake election materials were distributed to employees through the center's interoffice mail, and an embarrassed Kraft launched an unsuccessful
internal investigation to uncover the culprits, the perpetrators snuck back
into the center and taped a list of "winners" from their alternative election to
every bulletin board and elevator bay in the complex. The results listed
Gloria Steinem as write-in favorite for Festival King, and it included a note
that ridiculed NASA's efforts at equal employment for women.[162]

By holding this mock beauty pageant within the inner sanctum of NASA's
Mission Control, the Johnson Space Center feminists placed women's
bodies squarely within the space agency's internal debate over female astronauts. The main vehicle for this was a set of cartoons drawn by the protestors that accompanied the fake election results. In "He'll Give Her Tha
Moon," a coquettish "Miss Moon Maid" standing on a "Height of Equality"
pedestal suggested not only that women at NASA were judged according
to their good looks and bodily figures but also that the space agency
considered them only "¾" equal to men. The cartoon's caption, which read
"Just About as High as She Can Go," reaffirmed the perception of this glass
ceiling at NASA (see Figure 4.6). In stark contrast was the cartoon "I'm the
Winner of Course," which depicted a hairy-legged, muscular king holding
a rocket aimed at the moon. To ensure that viewers fully understood this
imagery, before posting the cartoon throughout the space center one of
the protesters hand-wrote "White male supremacy based on possession of
penis, signified by size of rocket" in the corner of the image (see Figure 4.7).
The fake ballot included other illustrations as well, such as the sarcastically titled "One Giant Step for Mankind . . . ," which implied through its
depiction of a large astronaut boot stepping on the back of the festival's
"queen" that if women did, indeed, make it to the moon they would do so
as doormats for NASA's astronauts (see Figure 4.8).[163]

In an effort to better coordinate these and other local protests, during
the summer of 1974 NOW initiated its "ZAP action" campaign involving
theatrical demonstrations that "Zeroed in for Action and Progress." Demonstrations highlighting gender discrimination within NASA were central
to this synchronized effort scheduled to take place on August 26, which the
feminist movement began celebrating in 1970 as Women's Equality Day to

4.6   Anonymous, "He'll Give Her Tha Moon," summer 1973.

commemorate the anniversary of the Nineteenth Amendment giving women the vote. During one of these ZAPs, NOW members picketed outside NASA Headquarters with five enormous placards, shaped like feet with toenails painted bright red, which, when lined up in a row, read "ONE" "GIANT" "LEAP" "FOR HUMAN" "KIND." Spokespeople from the women's group explained to the gathered press that NASA should make that "giant leap" by hiring women astronauts. The largest ZAP of the day, however, involved several local NOW chapters from the Washington, DC, area distributing 1,000 posters listing an "FBI's Ten Most Wanted Men for Offenses Against Women." The pin-up posters, which explained that NOW's "FBI"

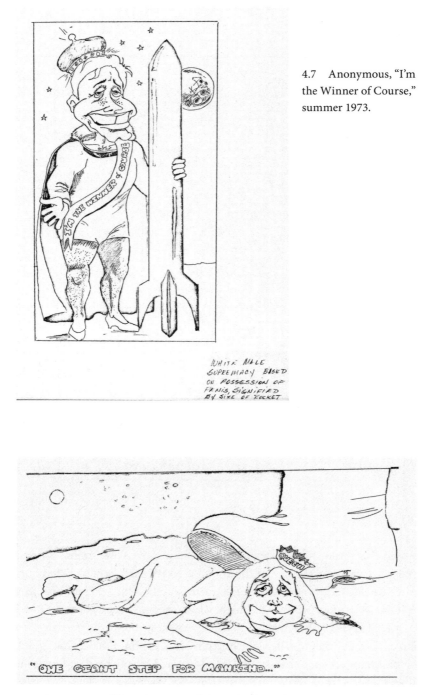

4.7   Anonymous, "I'm the Winner of Course," summer 1973.

4.8   Anonymous, "'One Giant Step for Mankind . . . ,'" summer 1973.

stood for its "Feminist Bureau of Investigation," appeared suddenly on telephone poles, street-corner garbage cans, and on the sides of buildings throughout the capital with James Fletcher's name listed as one of America's most wanted chauvinists.[164]

Shattering NASA's Glass Ceiling

Lobbying by Jane Hart and Jerri Cobb, nagging questions during luncheons hosted by the Woman's National Democratic Club, muckraking journalism published by *Ms.* magazine, and a national grassroots campaign orchestrated by NOW all nourished a growing dissatisfaction among American women during the late 1960s and early 1970s for NASA and the space race. According to a Harris Survey conducted in February 1969, while 46 percent of men polled favored landing a man on the moon, only 32 percent of women did. The same survey found that a majority of female respondents, accounting for 54 percent, opposed the Apollo 11 mission.[165] Feminist activism, much like that by civil rights leaders, New Leftists, and both grassroots environmentalists and environmental scientists, was thus aggravating NASA fatigue, which had peaked in the late 1960s and was now persisting into the mid-1970s. While supporters of these other political movements increasingly opposed the space agency for ignoring the plight of minority neighborhoods in cities, participating militarily in Vietnam, and polluting the environment while failing to scientifically study it, women began to turn on NASA because the space agency had a long history of turning against women.

While disapproval by these other movements had put pressure on politicians to cut funding for the space agency, feminists succeeded instead in compelling Congress to hold hearings regarding NASA's seemingly discriminatory hiring practices.[166] The first of such hearings, which took place in early January 1973, concluded that because of the space agency's "extremely poor record" it would now be required to report quarterly to Congress regarding its hiring of female employees. The Senate's Committee on Aeronautical and Space Sciences called a second hearing toward the end of the month specifically to review NASA's EEO Program. As NOW's Vice President for Legislation, Ann Scott, and its National Compliance Task Force chairperson, Gayla Salinas, argued in a written statement submitted during the hearing, "the failure of NASA's EEO Program is of grave concern to the National Organization of Women." The space agency, NOW demanded, "must be made answerable to women."[167] As a result of this and other testimony, the Aeronautical and Space Sciences committee supported the recommendations made during the first hearing and recommended as well that women be placed on NASA promotion and selection boards.[168]

Feminist activism also succeeded in pressuring the space agency to conduct medical examinations of women to determine, once and for all, whether the space environment threatened female physiology. The five-week study, which NASA announced to the public on September 21, 1973, a mere three weeks after publication of the *Ms.* cover story on Lovelace's Women in Space Program, took place at the space agency's Ames Research Center in Northern California and involved twelve U.S. Air Force flight nurses.[169] During the study's initial two weeks the nurses, who ranged in age from twenty-three to thirty-four, not only underwent physical tests similar to those endured by male astronaut candidates, but also were subject to several new exams developed specifically to assess the female cardiovascular and endocrine system, as well as female body chemistry, potassium levels, and body temperature. The nurses, explained *Time* magazine, "patiently endured such triumphs of technology as electronic heart monitors tucked into their bras and pill-sized sensors that transmitted temperature data from the vagina by way of a 'bio-belt' worn around the waist."[170] Each test subject also ingested small capsules that transmitted exact internal temperature readings to laboratory computers.[171] According to NASA's Director of Life Sciences, Dr. David Winter, who oversaw the tests, these medical procedures were developed specifically because space agency officials were considering "expanding our population" of potential astronauts.[172]

After evaluating each nurse's baseline health, aeromedical scientists at Ames spent the remaining three weeks of the study placing the women into a variety of simulators to determine how the female body might react to the space environment. To replicate the physical deconditioning that occurs from weightlessness, scientists put eight of the twelve nurses on bed rest for seventeen days, during which time they were unable to exhibit excessive muscular movement and only allowed to raise themselves up on one elbow during mealtimes. The other four nurses served as a control group. Boredom was relieved by watching television, listening to the stereo, reading books through prismatic glasses that allowed the women to remain horizontal, and, according to a NASA press release, by "a lot of needlework."[173] After this bed rest, doctors wired up the women's bodies and spun them in the center's centrifuge at three Gs to simulate reentry from space. The Ames medical team also re-created the effects of gravity on astronauts returning to Earth by placing the nurses from the waist down in a partial vacuum that concentrated blood in the lower part of their body.[174] The Ames tests, stated one NASA spokesperson in October of 1973, "will prove or disprove" women's physical qualifications for spaceflight.[175]

Three years later Dr. Winter undertook another medical trial that began to address NASA administrators' second concern regarding a lack of women's spacesuits. The "Shuttle Re-Entry Acceleration Tolerance in Male and Female Subjects Before and After Bedrest" study, which involved twenty-seven female civilian volunteers and an equal number of male control subjects, was unofficially known as the "housewives in space tests" and also occurred at the Ames Research Center. The study included even more simulations, such as a weightlessness test that plunged female subjects into a giant water tank, similar to the Water Immersion Facility in Houston used to train Apollo astronauts, and a "performance simulator" that mimicked the turbulence of space capsule reentry to assess each woman's hand-eye coordination. Yet because this study incorporated a more demanding G-force centrifuge test, NASA provided the twenty-seven female subjects with the same pressure suits, albeit in smaller sizes, worn by male astronaut candidates. "When you put it on," explained Dale Graves, one of the female test subjects, "you couldn't help feeling a little closer to blastoff."[176] Soon thereafter the space agency also announced that engineers at Hamilton Standard, the contractor then responsible for NASA's spacesuit production, were redesigning parts of the space shuttle suit to "accommodate extra-small sizes for women."[177]

Both of these Ames Center studies concluded that women's bodies were capable of withstanding the physiological rigors of space travel. The 1973 bed-rest tests, for example, found that when compared with men women experienced similar muscular deconditioning and comparable subsequent recovery after simulated weightlessness.[178] The "housewives in space" study reinforced such findings.[179] Yet both studies also noted differences. According to the 1976 examinations the female subjects experienced greater cardiovascular changes during the partial vacuum gravity simulation, but were also more radiant resistant, recovered faster from centrifuge simulations, and may be better at withstanding the combined stresses of space travel than members of the male control group.[180] The women, admitted

Dr. Harold Sandler, who, as chief of the Ames Biomedical Research Division, oversaw the "housewives in space" tests with Dr. Winter, "had done a superior job, compared to the men, of adapting to the physical and psychological challenges of the series of tests." As a result of such differences, predicted Sandler, "in space, women are going to beat men."[181]

These medical findings, along with the nearly fifteen-year grassroots campaign by feminists that compelled such tests, finally convinced NASA administrators during the summer of 1976 to open up the space agency's all-male astronaut corps to women. Officials at NASA announced the decision in a press release that described the space shuttle's need for nonpilot mission scientists and then openly encouraged women to apply for the positions.[182] In public comments regarding the announcement NASA spokespeople also went out of their way to alleviate widespread concerns regarding the perceived problems associated with transporting female bodies into space. "The space shuttle has been designed to accommodate women astronauts," explained one administrator to the New York Times soon after the announcement. "The waste management system is the only problem really and that has been designed for both males and females."[183] Officials at NASA also took concerted steps to ensure that women did, indeed, apply. They hired Nichelle Nichols, who played Lieutenant Uhura on television's Star Trek, to produce a recruitment film titled Space Is for Everyone. They organized several weeklong symposiums at NASA centers, similar to those held at Mills and Radcliffe, which highlighted the future of women in space, and targeted as well women's magazines with NASA's affirmative action press release. The space agency's EEO Program even contacted women's organizations on college campuses across the country to explain the space agency's new policy as well as its need for female scientist astronauts.[184]

It was during the spring of 1977 that one of these women's groups, the Center for Research on Women at Stanford, publicized NASA's call for female astronauts in the university's newspaper. "I was on my way out of the room to apply while I was still reading the notice in the paper," remembered Sally Ride.[185] Less than a year later, in mid-January 1978, NASA accepted Ride and five other women into its astronaut training program. When asked at a news conference why it had taken so long, Johnson Space Center Director Chris Kraft initially fell back on NASA's pat answer, explaining that in the past there had simply been no acceptable women. When pushed by reporters, however, Kraft admitted that "in the last few years, because of the women's movement frankly, more women have been qualified."[186] In other words, rather than admit that feminists had forced NASA to change its policy, Kraft implied that second-wave feminism had somehow magically improved women's qualifications for the astronaut corps. America's first woman in space knew better. As *Ms.* magazine explained in a cover story detailing her historic *Challenger* flight, "Ride warmly credits the Women's Movement with changes in NASA since the late 1960s."[187]

Feeding a Feminist Debate

While feminists such as Clare Booth Luce and Betty Friedan used Valentina Tereshkova's 1963 orbital flight to champion the equality of the sexes, later in the decade members of the women's movement began emphasizing a competing feminist ideology. Pilot and NOW activist Jane Hart expressed such alternative thinking in January 1969 after Apollo 8 astronaut Frank Borman admitted to reporters, after smiling wryly, that while he did not think women needed to go into space he did find them "very pretty." "Colonel Borman's reply was frivolous," Hart told reporters. "It sets the whole thing on a biological level." Hart then turned such thinking to her advantage, explaining that several medical studies, including the one she and Jerri Cobb had undergone at the Lovelace Clinic back in 1961, had actually determined that women were physically more adaptable to space travel than men. "A woman's cardiovascular system is much less vulnerable," she explained. "She requires far less life support equipment, breathes less oxygen, consumes less food, excretes less."[188] The space agency's medical examinations undertaken in the mid-1970s confirmed Hart's conclusions, and they also uncovered additional biological differences that suggested women might in fact be better astronaut candidates than their male counterparts.

These discrete arguments made by Luce and Hart, in this case regarding the appropriateness of women's bodies for space travel, had a much deeper history within the liberal feminist movement. Known variously as a debate between "minimizers" and "maximizers," radical and cultural feminists, and essentialists and social constructionists, this divide pitted feminists calling for outright equality between the sexes against others who argued instead that biological differences between women and men mattered both theoretically and politically. The historical split between these "equality feminists" and "difference feminists" arose most intensely during the late 1960s and raised the ultimate question that would bedevil second-wave feminism during the last quarter of the twentieth century. Should the liberal

feminist movement promote women as equal to men, which would necessitate the downplaying or wholesale renunciation of women's own unique
biology? Or should they instead acknowledge and even celebrate the physiological differences between women and men, knowing full well that doing
so meant admitting that true equality between the sexes is unattainable?
To grossly oversimplify, to best fight gender discrimination should women
be more like men or admit they are different from them?[189]

At stake in this debate was nothing less than the role of the female body
within second-wave feminism. For equality feminists, such as Luce, human
physiology mattered less as a marker of inequality because they believed
that as the feminist movement gained ground, biology would wane as a
basis for social organization, as it seemed to have in the Soviet Union.[190] This
was also the hope of Sally Ride, who argued again and again during the
months leading up to her own historic flight that her astronaut training was
no different from that of her male crewmembers because she was "just one
of the guys." Difference feminists, on the other hand, were historically less
apt to dismiss the importance of the female body. For some in this camp,
biology made women separable from men and imbued them with particular
traits, such as the capacity for empathy and morality, which should be nurtured by the women's movement rather than snuffed out in the name of
equality with men.[191] For others, acknowledging women's particular physiology was central not only to understanding sexual discrimination against
women but also to formulating political strategies against it.[192] Such was the
thinking of Jane Hart, who in January 1969 countered Borman's sexist smile
with the simple caveat that "medical studies show that women are more
adaptable to life in outer space than men."[193]

This embrace of women's biological difference among liberal feminists
fighting for female astronauts did not occur in isolation. While NOW was
picketing NASA Headquarters with enormous red-painted toenails and
space agency employees were plastering the Johnson Space Center with cartoons of bathing-suited beauties being stomped on by big-booted astronauts,
other feminists were similarly turning to the female body in their fight
against gender discrimination. In the spring of 1969, for example, just a few
months after Hart deployed women's biology to promote female astronauts,
twelve women attending a "Women and Their Bodies" workshop at a women's
liberation conference in Boston, Massachusetts, decided to embrace female
physiology to promote women's health. After realizing that they had each

experienced frustration with male doctors who, like NASA administrators, were ignorant of the female body, the twelve responded, as Hart had, by turning to female biology; within two years they had published *Our Bodies, Ourselves*, a handbook written "by and for women" about female health, and in doing so launched the women's health movement.[194]

Ms. magazine helped broadcast these contending feminist discourses to the wider feminist movement in its 1973 exposé on the Women in Space Program. Similar to *Life*, which after Tereshkova's 1963 flight published Luce's call for equality between the sexes, the *Ms.* article initially also lamented the inequality of women and men within the American space program. "Women are still not taking part," explained *Ms.*, "unless you want to consider the appearance of the commercial breaks during launch and splashdown, when they can be seen pitching Tang breakfast drink" with the tagline "if it's good enough for the astronauts, it's good enough for my family." Later in the article, however, *Ms.* pushed beyond Luce's equality ideology by noting the benefits of female biology for space travel, as Jane Hart had done back in 1969. "In the opinion of scientists evaluating the test results," the magazine explained, "women were *as* capable and *as* suitable as men for spaceflight; in some ways, *more* suitable."[195] The article concluded by describing other scientific studies that also showed women were less prone to heart attacks, and more able to endure loneliness, heat, cold, pain, and noise, all of which astronauts encountered in space. Thus in part because of the space race, the feminist movement in the early 1970s began relying on the biological *difference* of women as a political strategy to promote the *equality* of the sexes.

Covering Sally Ride's Back

Soon after Sally Ride "turned her back" on an armful of flowers during the Johnson Space Center homecoming ceremony commemorating her flight, NASA sent America's first female astronaut on a whirlwind international publicity tour. It kicked off in Washington, DC, when Ride, dressed in a "pale pink formal gown," attended a state dinner at the White House where she was seated next to the guest of honor, the emir Isa bin Salman Khalifa of Bahrain. During the dinner, NASA had Ride present the Middle Eastern leader with a photograph of Bahrain taken from the shuttle. Sixteen events followed in that week alone, including a reception at the Smithsonian

FLIGHT OF THE SPACE SHUTTLE CHALLENGER AS SEEN BY THE MEDIA

4.9   Ken Alexander, "Flight of the Space Shuttle *Challenger* as Seen by the Media," *San Francisco Examiner*, June 20, 1983.

Institution's National Air and Space Museum, congressional appearances, and breakfasts, luncheons, and additional dinners. The public relations office at NASA then packed Ride's bags for a six-nation European tour. "The space agency can't be blamed for spotlighting one astronaut," argued the Associated Press. "Miss Ride's flight," it added, "came along just about the time when interest in the program had declined to the point when even the locals didn't turn out any more to see a launch."[196] Such promotion worked, at least according to "Flight of the Space Shuttle Challenger as Seen by the Media," a syndicated editorial cartoon first published by the *San Francisco Examiner* that depicted Ride flying high above the Earth, arms outstretched like wings, beneath NASA's STS-7 spacecraft (see Figure 4.9).[197] As the cartoon suggested, the space agency that had turned its back on women during the 1960s, was in 1983 resting America's space program squarely on Sally Ride's.

Second-wave feminism caused this dramatic shift at NASA. Just as feminists reading the *Examiner*'s cartoon no doubt condemned its stereotyping of Ride as a happy homemaker, with frying pan in hand, so, too, had they criticized statements like John Glenn's congressional comments about welcoming women into the astronaut corps "with open arms" and Wernher von Braun's equally snide aside that in preparation for the possibility of female astronauts NASA was "reserving 110 pounds of payload for *recreational* equipment." Space agency officials literally laughed off such criticism by women in the early 1960s. Yet as the decade wore on, and feminists including Jane Hart promoted women astronauts to the vice president, female students held "quiet protests" during moon-rock ceremonies at their universities, and grassroots organizations such as NOW launched a nationwide campaign that included legal complaints, letter-writing drives, and street protests, all of which highlighted gender discrimination within the American space program, NASA officials realized that the space agency's "woman problem" was no laughing matter. The result was the launch of Sally Ride, immediately after which her mother, Joyce, exclaimed "God bless Gloria Steinem."[198]

Such feminist activism, much like the *Examiner* cartoon, also highlighted the important role played by both technology and nature in NASA's painstakingly slow embrace of female astronauts. The cartoon frying pan and spatula in Ride's hands, not to mention the pair of eggs orbiting sunny side up, suggested that the media viewed the nation's first female astronaut as a woman first and foremost. So did many NASA administrators. Yet Ride countered such sexist thinking by promoting herself as a scientist who wielded not the tools of the kitchen but rather those of the technician, in this case an extendable robotic arm that was essential to the success of the mission. Feminists during the late 1960s and 1970s supported such efforts by turning as well to nature; they successfully pressured the space agency to design environmental simulators that would accommodate female bodies and to develop spacesuits that could protect those bodies from the environmental hazards of space. These body politics played out most publicly in the pages of *Ms.* magazine's cover story illustrating that women were physiologically as capable as men, and perhaps more so, of undertaking spaceflight. Sally Ride's body in the *Examiner* cartoon, which comfortably supports the space shuttle *Challenger*, illustrates the press's willful ignorance of such struggles.

By transforming NASA, feminists also altered their own movement, albeit in ways that were less obvious than decorating the nation's first female astronaut with a pink evening gown and sitting her next to a sheik. On a more subtle level, the protests and pickets and pretend beauty pageants influenced feminist thinking regarding women's equality with, and difference from, men. This shift in political discourse among second-wave feminists was neither a linear transition from equality to difference, nor an abandoning of one concept for the other. The 1960s and 1970s were a messier and more fluid historical moment than that. Rather, grassroots activism against NASA during this period demonstrated a feminist movement grappling with each of these seemingly contradictory ideas in an attempt to forge a political ideology that made room for both. The historian Joan Scott has theorized about this equality-versus-difference debate, arguing that the resolution of what she calls "the difference dilemma" must not rest on an equality that implies sameness, but rather on one that depends on differences.[199] Second-wave feminists sought such a resolution by picketing NASA with signs depicting women's feet, toenails painted bright red; by holding mock beauty pageants inside the Johnson Space Center that asked participants to vote for "a pretty face and a good bod"; and, most important, by demanding that NASA undertake a new set of medical tests to determine, once and for all, whether women's bodies could physically withstand the environmental hazards of spaceflight.

More obvious was the way such protests helped to galvanize the emerging women's movement. Although liberal feminism faltered as a political crusade soon after Sally Ride "turned her back" on NASA's flowers, back in the late 1960s and throughout the following decade feminist opposition to the space agency brought together leaders such as Betty Friedan and Gloria Steinem with on-the-ground activists from local women's clubs, female colleges, and from within NASA itself. NOW not only drove much of this activism, but, just as important, also rode it to national political prominence in the early 1970s. This shared opposition gave the movement a new set of heroes, as well, including Jerrie Cobb, Jane Hart, Sally Ride, and even Valentina Tereshkova. Due to the efforts of all of these women, noted Gloria Steinem after Sally Ride's *Challenger* launch, which she attended as one of NASA's VIP guests, "millions and millions of little girls are going to sit by their television sets and see that they can be astronauts, heroes and explorers, and scientists." This milestone, Steinem concluded, should convince all of us "to conquer more inner space along with outer space."[200]

5

The New Right's Stuff

The Hippie Counterculture and the Rise of the Conservative Crescent

At 4:00 A.M. on December 21, 1968, four hours before the launch of Apollo 8, the underground press reporter James Kunen ate breakfast in the lobby of his Cocoa Beach, Florida, hotel and once again felt happily out of place. "The seat of this futuristic enterprise is ten years behind the time," he explained of his fellow hotel guests in *US: A Paperback Magazine*, an alternative publication whose motto claimed "all the news that's fit to eat." "The boys wear stovepipe chinos 6 inches above the ankle; the women wear skirts below the knee; Alley Cat is on the juke." Kunen's physical appearance, not to mention his musical tastes, was radically different from those gathering at the Cape, so different, in fact, that he quickly became part of the Apollo 8 news coverage. "There's a hippy," announced a *Women's Wear Daily* reporter soon after Kunen arrived at the Kennedy Space Center's press viewing site. After asking him several questions regarding the counterculture's take on Apollo, the photographer in tow snapped pictures of Kunen's long hair and attire, including his blue jeans, open loose-fitting shirt, and small beads dangling around his neck. The young journalist seemed to revel in his outsider status. After mocking the elite "reportorial class" that had gathered in NASA's press bleachers, he "confronted" conservative columnist and *National Review* founder William F. Buckley Jr., who was also covering the launch, most probably in a jacket and tie that would have seemed all too familiar to those eating two eggs, toast, coffee, juice, and grits for 65 cents at Kunen's Cocoa Beach hotel.[1]

Kunen's and Buckley's contrasting tastes in clothing correlated to their quite different take on the Apollo 8 launch. For Buckley, the technological wonder of NASA's Saturn rocket rising slowly across an inlet just three and one half miles away represented "the nearest thing to total beauty that science ever created."[2] Kunen thought otherwise, noting that the thirty-six-story Saturn was a "big clunky thing" whose seven-and-a-half-million-pound thrust symbolized "the ultimate phallic symbol." "I felt," he added sarcastically, "that we should troop up to it, genuflect, and leave baskets of liquid oxygen at its base." For Kunen, it was Cape nature that stole the show. He described the sunrise at "T minus 70" minutes before the early-morning liftoff as "an event of considerably greater magnitude than the launching," and he noted that a half hour later a lone seagull was gliding silently and "thinking very little of the flight." Just sixteen minutes before Apollo 8 roared from the Earth, Kunen was still playing Thoreau by documenting the local environment. "The jungle, pushed back from the launch site, patiently waits to move back in," he explained.[3]

The chance encounter of *US* reporter James Kunen and *National Review* columnist William F. Buckley Jr. represented in microcosm a clash of cultures playing out during the late 1960s. Members of the counterculture such as Kunen, like the activists from the civil rights, anti-war, environmental, and feminist movements, also condemned Apollo but were less specific in their criticism.[4] The space program, they complained, was part of what they called "the Establishment," a derogatory term for the entrenched powers that be.[5] Additionally, while many in the counterculture participated in these other movements' marches, demonstrations, and sit-ins, their own attempts at social change appeared less overtly political because they emphasized instead a variety of alternative lifestyles involving, for instance, nontraditional clothing, rock and roll music, relaxed sexual practices, and unconventional, often communal, living arrangements. For many in the counterculture, in other words, Kunen's long hair at Apollo 8 was no less political than Abernathy's Poor People's Campaign protest during Apollo 11.[6] Even the counterculture's nomenclature proved difficult to pin down; while participants often embraced "hippie" to describe themselves, their critics often slung the term back as an insult for what they believed to be a self-absorbed mob of unruly youths rather than a true social movement.[7]

William F. Buckley was one such critic, as were the conservatives reading his magazine and his nationally syndicated column, "On the Right." During the early to mid-1960s these so-called Middle Americans stood by quietly as activists on the left took to the streets against racial and gender injustice, the Vietnam War, and an increasingly polluted environment. They also sat idly by as hippies on both coasts dropped out and dropped acid. Until, that is, Republican presidential candidate Richard Nixon called them out in a late September 1968 campaign speech in St. Petersburg, Florida. The upcoming November election, Nixon shouted over the roar of 10,000 enthusiastic supporters, would be "a day of protest for the forgotten American," for those "who obey the law, pay their taxes, go to church, send their children to school, love their country."[8] After these conservatives propelled him into office, President Nixon continued courting what he began referring to as the "silent majority," so named because he felt their political voices had been drowned out by the more vocal minorities demonstrating in the streets.[9] Buckley's *National Review* agreed and christened this quiet crowd the "New Right," an emerging grassroots movement in its own right composed primarily of white, middle-class, religiously conservative residents of small towns and suburbs scattered throughout America's heartland but especially across the Sunbelt South.[10] Although perhaps silent during the early 1960s, by the time of the Apollo 8 launch these conservatives had found their voice and were cheering passionately for NASA's space program not only as a means of winning the Cold War and promoting economic prosperity, but also to counter the growing popularity of the counterculture.

Although Kunen and Buckley wrote extensively about Apollo 8, few since the launch have explored the role of the space race in what has been dubbed "the culture wars" of the 1960s era. Writers chronicling the American space program, for example, have traced the impact of conservative ideology on NASA, primarily through the presidential administrations of Eisenhower, Nixon, and, later, Reagan, yet have refrained from introducing liberal thinking into such accounts.[11] Cultural critics have explored this relationship from the opposite vantage point, following instead the influence of Apollo during the 1960s and 1970s on avant-garde art, alternative music, underground literature, and novel fashion trends. Yet while some of these critics invite conservative thinkers into their analysis, none connect this exchange between the space race and American culture to the grassroots political movements of the period.[12] Nor do they make room for space technology and earthbound nature, as Kunen and Buckley did during their coverage of Apollo 8. Writers have thus failed to acknowledge what both of these reporters understood only too well: that conservative culture and the counterculture not only shared a relationship to the space race, but also used Apollo in their battle against one another for political authority.

That battle began during the mid- to late 1960s, when NASA administrators started promoting Apollo technology as a "triumph of the squares."
Conservative writers, thinkers, and politicians, including Buckley, Ayn
Rand, and Barry Goldwater, nurtured this idea by arguing publicly that the
free-market production of such technology, along with the hardware's
ability to control wild nature both on Earth and in outer space, signified
the superiority of American-style democracy over Soviet communism.
During the late 1960s, however, the counterculture refuted such thinking
by criticizing NASA's administrators, engineers, and especially its astronauts for being hopelessly un-hip, and the agency's rockets, command
modules, and lunar landers as more symbolic of an extraterrestrial land
grab along the New Frontier. To express their disdain, hippies neither
marched on Cape Canaveral nor picketed outside NASA Headquarters, but
instead wrote poems, created art, attended music festivals, and migrated
in droves to rural communes where they embraced what they felt was more
appropriate technology in order to live, as they put it, more in tune with
nature. When such opposition further weakened popular support for space
exploration in the early 1970s, NASA appealed to Buckley's New Right.
The result was a failed attempt to plant suburban-like colonies in outer space
and the very real spread of conservative sprawl across the space crescent.

Promoting NASA as Square

Soon after Kunen and Buckley bumped into one another at the Kennedy
Space Center, Apollo 8 completed the first two orbits around the moon,
captured the breathtaking *Earthrise* photograph, and captivated the world
when astronauts Frank Borman, Jim Lovell, and William Anders took
turns reading excerpts from the Book of Genesis live from space on the
night before Christmas. One week later, after the astronauts' picture-perfect
splashdown in the Pacific Ocean south of Hawaii, Thomas Paine, NASA's
chief administrator, began echoing Buckley's praise for Apollo's technological achievements. The spacecraft worked exactly as planned, he explained to the media, by providing new navigation data necessary for the
upcoming moon landing. Paine then criticized young people such as Kunen
who dismissed the space race, and he praised those responsible for Apollo
8's success. The mission, argued NASA's chief, according to United Press
International, showed "restless students all over the world" the benefit of
those "who work with computers and slide rules, of engineers and scien-

tists, of men who read the Bible on Christmas Eve." Paine then concluded with a simple statement. Apollo 8, he declared, "was the triumph of the squares."[13]

Six months later, on June 7, 1970, Paine doubled down on such rhetoric in his commencement speech at the Worcester Polytechnic Institute (WPI). After thanking the university for inviting him to speak, Paine focused his entire address on what he believed to be "one of the most fascinating actions in the world today: the clash between two contemporary social worlds." One world, he explained to the graduating seniors, was the one they were raised in, the world in which their parents lived, a world led by "the pillars of society" housed in government offices, university administration buildings, corporate boardrooms, and church pulpits. The other was more difficult to describe because, according to Paine, it was "shadowy and shifting and partly underground." Its leaders included Harvard psychologist and LSD advocate Timothy Leary as well as musicians Arlo Guthrie, Jefferson Airplane, and John Lennon and Yoko Ono, among others, while its "home territories" comprised left-leaning locales such as Woodstock and the East Village in New York and San Francisco's Haight-Ashbury. The battle between these two social worlds, Paine concluded, was no less than a "war between 'Potland' and 'Squareland.'"[14]

Paine's portrayal of the counterculture squaring off against mainstream America became increasingly widespread among conservatives during the mid- to late 1960s. The controversial novelist and philosopher Ayn Rand, for instance, delivered a speech on November 9, 1969, at Boston's Suffolk University that similarly juxtaposed what she felt were the two most important events of the preceding summer: one involving a million space buffs traveling south to Cape Canaveral for Apollo 11's launch and the other comprising nearly a half million young people, whom she dismissed as "hippies," trekking north to Bethel, a small town in upstate New York, for the Woodstock music festival. Although Rand, like Paine, envisioned these worlds at odds with one another, for her the battle was philosophical rather than social. On one side, congregating at the Kennedy Space Center, was rationality and reason symbolized by Apollo, not the rocket but rather the Greek god of light. On the other side was Dionysus, the god of wine, standing in for the irrational and emotional throngs of youths "wallowing in the mud on an excrement-strewn hillside near Woodstock." "Apollo and Dionysus represent the fundamental conflict of our age," Rand argued in her talk. "Reality has offered two perfect, fiction-like dramatizations of these abstract symbols: at Cape Kennedy and at Woodstock."[15] For Rand the Greek's square god of light, symbolized by NASA's rocket, outshone the stoned hippies frolicking in Woodstock's mud.

As conservatives continued portraying Apollo 8 as a triumph of the squares, the mainstream media followed suit. Ernest Furgurson, the chief of the Baltimore Sun's Washington, DC, bureau, opened his column celebrating the July 1969 lunar landing by repeating Paine's "triumph of the squares" catchphrase from the preceding year. He then described Apollo 11's astronauts as the archetypal Squareland residents; all three came from small-town America, attended church regularly, wore their hair short, and refrained from smoking. "Squares, the whole corps of them," Furgurson boasted before contrasting them with the counterculture. "Nowhere among them was ever a mophead, a whiff of pot, a shout of doubt or whine of self-pity." Time magazine extended this portrait to Apollo 11's supporters, noting not only that they hailed from Middle America and supported the astronauts' return with patriotic flags and prideful bumper stickers, but also that there were "few love beads among them, fewer bell-bottom trousers and no disparaging words about the nation."[16] In other words, they included very few Kunens. Instead, concluded Furgurson, NASA's supporters "sprang from the classes Mr. Nixon catered to last fall when he praised the 'forgotten American'—the American who pays his taxes, does his job, sends his kids to school."[17] Thus while Potland's denizens criticized Apollo, the silent majority that Nixon would court as president remained adamant supporters of the space program.

During the early 1960s conservative support for NASA's rational, reasoned space triumph rested initially on military and economic grounds. Republican Arizona Senator Barry Goldwater, in a foreign policy paper distributed during his failed 1964 presidential bid, promoted America's space program as a means of countering Soviet "space power" as well as strengthening the security of the noncommunist free world.[18] Conservatives' economic arguments cohered around the aerospace industry's bottom line. The Wall Street Journal, Business Week, and Fortune repeatedly praised NASA for pumping $24 billion into the U.S. economy and creating hundreds of thousands of jobs during the mid-1960s.[19] Many commentators during this period merged these two attributes—one military, the other economic—by equating the growing aerospace business to the already well-established defense industry. "Even for someone accustomed to the so-called military-industrial complex, the space-industrial complex," explained one *Washington Star* reporter, "is something to behold." In many ways, he concluded, "the two are indistinguishable."[20]

As it became increasingly obvious during the late 1960s that the United States would beat the Soviet Union to the moon, Buckley, Rand, and others on the right also began praising the Apollo program as a means of supporting a new brand of conservative politics. According to such thinking, the Apollo rocket was not only scientifically "beautiful," as Buckley saw it, or philosophically rational, as Rand argued. Equally important was the manner in which this space technology was produced. "It is not coercion, not the physical force or threat of a gun, that created Apollo 11," Rand argued in the September 1969 issue of her periodical, *The Objectivist*. Rather, independent businesses and autonomous university faculty researched and developed these space technologies for the public good. "The various parts of the spacecraft were produced by private industrial concerns," she noted with obvious pride.[21] The business section of the *Washington Post*, NASA itself, and the guardian of traditional Middle American culture, *Life* magazine, all promoted similar thinking, explaining to the public that 93 percent of NASA's budget for research and development went to 20,000 private contractors and involved 4,000 researchers at more than 100 universities.[22] Unlike Soviet space technology, which, according to a July 1969 *Birmingham News* editorial, was developed mysteriously by a "closed, secrecy-shrouded society on the other side of the Iron Curtain," Apollo illustrated the fruits of a decentralized, free-market, "open American society."[23]

The relationship between NASA's space technology and the natural environment also informed this emerging conservative ideology. For Buckley, the beauty of the Saturn V rocket depended, to a large extent, on the nature of the Cape; with the arrival of dawn, he explained in his coverage of the early-morning Apollo 8 launch, "the rocket is seen in relief against a Homeric sky."[24] Rand embraced a similar aesthetic in her "Apollo 11" essay, which described the Kennedy Space Center as "an enormous place that looks like an untouched wilderness" with eagles' nests, alligators, and "wild subtropical growth." Yet while for Buckley wild nature enhanced the spectacle of NASA's technology, for Rand the relationship was more hierarchical. "Scattered at random, in the distance," she explained, are "a few vertical shafts rising from the jungle, slender structures of a shape peculiar to the technology of space, which do not belong to the age of the jungle."[25] Rand similarly placed space technology above nature in her "Apollo and Dionysus" essay when she quite consciously contrasted the Saturn V towering over the Cape with the lowly mud-strewn farm of the Woodstock music festival.[26] For these conservatives the space agency's technology was beautiful, rational, democratic, and sublime, in part, because it contrasted with, and rocketed high above, wild nature.[27]

Rand's hierarchy also appealed to fellow conservatives because it suggested that mankind, through modern technology, could tame wild, unruly, irrational nature. "Those dark red wings of fire," she wrote of Apollo 11's exhaust, represented "a cataclysm which, if unleashed by nature, would have wiped man out of existence." Fortunately, Rand explained, "this spectacle was not the product of inanimate nature, like some aurora borealis," but rather had been "planned, unleashed, and *controlled* by man" and "ruled by *his power*, and obediently serving his purpose." Rand concluded by arguing that at least during the seven-minute launch "the worst among those who saw it," by which she undoubtedly meant the hordes of hippies watching on television sets across the nation, were forced to feel "not 'How small is man by the side of the Grand Canyon!'—but 'How great is man and how safe is nature when he conquers it!'"[28] Paine extended such thinking about nature beyond Earth. The residents of Squareland, he told WPI's graduating seniors, probe the mysteries of nature "with microscopes, telescopes, computers, and spacecraft," the last of which would soon allow scientists "to intelligently monitor and manage the planet's biosphere."[29] Such "powerful forces," Paine argued in an earlier speech, are central to "our national effort to master the space environment, and use it for the benefit of mankind."[30]

The New Right enlisted this belief—that beautiful, rational, clean space technology could actually improve nature—in order to recycle an old idea regarding the American frontier. For Rand, Titusville, Florida, near Cape Canaveral, where NASA's engineers and scientists lived, was a "frontier settlement—the frontier of science."[31] President Nixon was more direct, noting in his 1971 annual foreign policy report to Congress that America's dominant role in space not only reflected its technological superiority over the Soviets, but also was "equally a measure of an older American tradition, the compulsion to cross the next mountain chain." The Apollo spacesuit, Nixon continued, was the technological equivalent of "the buckskin jacket," the spacecraft merely the latest "packhorse," and NASA's astronauts only "the most recent of a long line of American pioneers."[32] Whereas axes, ploughs, and railroads had helped these early settlers tame and utilize the old West's natural resources, so, too, would the Saturn V, the Apollo command module, and the lunar lander allow the United States to reap the benefits of the space environment. As the NASA rocketeer and spokesperson Wernher von Braun explained in a 1976 *Popular Science* editorial, "I see space as an endless frontier." Someday soon, he added, "minerals can be mined on the moon, in the asteroid belt, and on nearby planets."[33]

The association of outer space with America's western settlement had originated on the political left, on July 15, 1960, when John F. Kennedy gave his acceptance speech for the Democratic presidential nomination in Los Angeles. "We stand today on the edge of a New Frontier," he explained to the gathered delegates.[34] Many since have taken the young presidential hopeful to task for mythologizing America's frontier history, for not only erasing from it the social injustices inherent in conquering what was never a truly empty wilderness, but also ignoring the environmental destruction that occurred in western settlement's wake.[35] Yet Kennedy and his Democratic Party at least attempted to update this prized American myth, to modernize it for the late twentieth century, by replacing references to buckskin jackets, wagon trains, and transcontinental railroads with the future promise of space technology. "What was once the furthest outpost on the old frontier of the West," explained Kennedy a year and a half later while giving a speech at Houston's Rice University, "will be the furthest outpost on the new frontier of science and space."[36] Equally important, Kennedy also tried to secularize the myth. Although he referenced scripture once in his Los Angeles acceptance speech, he did so not to defend exploration on his New Frontier but rather to motivate a new generation of future-oriented "pioneers."[37]

As soon as he became president, Richard Nixon began consciously co-opting Kennedy's New Frontier rhetoric, with its references to both space technology and earthbound nature, in an effort to appeal to his silent majority. Well aware that conservatives remained supportive of NASA's space efforts while liberal interest waned, Nixon repeatedly linked his presidency with Apollo technology; he constantly phoned astronauts in space from his Oval Office, visited them in their quarantine facility soon after splashdown, sent Apollo command modules on national and global public viewing tours, and made sure his name, as president, appeared on the plaque left behind on the moon by Apollo 11. Yet Nixon's New Frontier rhetoric, unlike Kennedy's, faced backward, looking behind once again to the nineteenth century. Like his old frontier predecessors, the Republican president downplayed the environmental costs of exploration on the New Frontier by instead emphasizing that NASA's space technology, including its satellites, could help monitor pollution, assess hydrological and agricultural processes, and, most important, aid in what Nixon called "resource utilization planning," a quaint euphemism for natural-resource extraction.[38] Nixon likewise harked back to the former frontier by infusing space exploration, as well as "the technological expertise" that made it possible, with an old-fashioned religiosity that struck a deep chord with Squareland residents. "Can we look at the record of 24 men sent to circle the Moon or to stand upon it," he wondered in December of 1972, "and not see God's hand in it?"[39] As *Time* magazine reported soon after these remarks, President Nixon had recast NASA's Apollo program "as a kind of Manifest Destiny for the space age."[40]

Conservatives quickly embraced Nixon's older New Frontier, and the space agency's role in exploring and exploiting it, to shore up support for an increasingly nationalistic political outlook. Ayn Rand, for example, concluded her "Apollo 11" essay by claiming that the American flag, planted firmly on the lunar surface, was a "worthy monument" for a country that had developed highly sophisticated space technology both to rise above Cape Canaveral's jungle and to transport men beyond Earth through an extremely hazardous space environment.[41] Middle America seemed to agree. After the *Washington Post* ran an article that questioned NASA's right to plant the U.S. flag on the moon, several subsequent letters to the editor defended the action. "The U.S. taxpayers have a right to what they earn," argued E. A. Kendall in June of 1969. "The U.S. flag should be raised over that territory on the moon which U.S. nationals have seen or stepped upon or landed instruments upon."[42]

Such public support for planting the American flag on the moon arose early in 1969, when NASA officials began discussing the possibility of having the Apollo 11 astronauts plant a United Nations flag instead. The U.N. flag, some within NASA believed, with its symbolic olive branches encircling a map of the world, would better represent the peaceful goals of Apollo as well as the original intention of the 1958 Space Act, which created NASA and tasked it with promoting international cooperation while securing American leadership in space science and technology. To help settle the issue, on February 25 Thomas Paine, the space agency's acting administrator, appointed a Committee on Symbolic Activities for the First Lunar Landing, whose charge it was to select culturally meaningful actions undertaken by Apollo 11 astronauts that would neither interfere with the mission's objectives nor give the impression that the United States was taking possession of the moon, which was outlawed by the Outer Space Treaty of 1967. While the committee deliberated, conservative Americans like E. A. Kendall began writing letters not to the *Washington Post* but rather to Congress in support of Armstrong and Aldrin planting the American flag. Such grassroots activism bore political fruit just a few months later, when President Nixon signed legislation requiring that the U.S. flag, and no other flag, be planted on the lunar surface during all Apollo missions.[43]

Such thinking by conservatives also resulted in a host of editorial cartoons appearing soon after Apollo 11 landed on the moon. "Flag-Raising," which appeared in Alabama's *Birmingham News*, was a case in point (see Figure 5.1). Drawn by the conservative Southern cartoonist Charles Brooks, whom President Nixon hailed as the "Herblock of the right," the cartoon replicated the imagery of the famous World War II *Raising the Flag on Iwo Jima* photograph but set the scene on the lunar surface.[44] Brooks replaced soldiers with a chronological who's-who of pioneers from the old and new frontiers, including a pilgrim, a coonskin-capped settler, a California gold miner, and a short-haired NASA engineer or administrator sporting glasses and thin black tie, all helping two astronauts plant Old Glory firmly on a crater-pocked moon. The settler's powder horn, the miner's helmet, and the astronauts' extravehicular activity backpacks all suggested the central role played by technology in conquering extraterrestrial nature. Brooks even drew his American flag with the lunar environment in mind; while on Iwo Jima the flag hung downward in a mild breeze, in his cartoon it stood stiffly at attention, reminding readers that NASA engineers had to insert aluminum rods along the top of the flag to make it "fly" in the airless lunar atmosphere.[45]

Censuring Squareland

During his coverage of the Apollo 8 launch, James Kunen evidently agreed with conservatives such as Paine, Buckley, and Rand that NASA and the space race hailed from Squareland. Yet for Kunen this was not a good thing. Along with criticizing the breakfast attire of his fellow hotel guests at the Cape, who gobbled eggs, toast, and grits in chinos too short and dresses too long, Kunen teased space agency administrators for being prudish when it came to wearing no clothes at all. "The NASA people seem somewhat

Flag-Raising

5.1   Charles Brooks, "Flag-Raising," *Birmingham News*, July 22, 1969.

self-conscious," he explained while noting that a space agency press book, which included drawings of biomedical sensors on naked astronaut bodies, omitted the spaceman's phallus. "Perhaps it was deemed insignificant," Kunen joked to his readers, "beside the 363 foot rocket." He next set his sights on NASA's engineers, a small army of automaton-like men identified by an impersonal "eighteen digit identification tag." Kunen, however, saved his most straightforward critique for NASA's crewmen. "It's too bad the first men to the moon have to be astronauts," he lamented in his coverage of Apollo 8. "This country would never think of sending a poet up. It's intent on banging squares into round orbits."[46]

Six months later, at the Apollo 11 liftoff, other alternative writers repeated Kunen's mantra that the space agency was exceedingly unhip. Norman Mailer, for instance, whose "hipster" in his 1957 essay "The White Negro" was regarded by many in the counterculture as the precursor to the hippie, lambasted the identical dark pants, short-sleeved button-down shirts, somber narrow ties, crew-cut hair, and dark-rimmed glasses worn by nearly all of NASA's technicians.[47] He then censured space agency administrators for being overly courteous, expert at repeating the same information over and over and over again, and "subtly proud of their ability to serve interchangeably for one another." The entire Apollo mission, Mailer concluded in *Of a Fire on the Moon*, epitomized "the last chalice of Good Square Life," and he openly refused to drink.[48] While it had been *Women's Wear Daily* that called out Kunen as a "hippie" during Apollo 8, it was Mailer himself who voluntarily joined such company during Apollo 11. In an effort to distinguish himself from NASA's straight-laced engineers, administrators, astronauts, and fans flocking to the Cape from Middle America, the colorful Pulitzer Prize–winning author referred to himself throughout the book in the third person as "Aquarius."

A similar critique of the Apollo program as morbidly square quickly spread throughout the counterculture, which expressed such disdain through a wide variety of daily practices. Abbie Hoffman wrote about such lifestyle politics in his book *Woodstock Nation*, which detailed his own experiences at the three-day music festival. While Hoffman, who in 1967 cofounded the politically radical Youth International Party, often criticized the counterculture for its apparent lack of direct political engagement, at Woodstock this founder of what became known as the Yippies had an epiphany about the hippies. "I realized I had badly misjudged the event," he wrote. Although "during the past few years I have straddled the line between 'the movement' and 'the community,' between 'the left' and 'the hip,'" explained Hoffman, the Woodstock experience "made me have a clearer picture of myself as a cultural revolutionary—and not a political revolutionary." He then admitted that when he stood trial in Chicago later that year for his arrest during protests at the 1968 Democratic National Convention, he now wanted to be tried not because he supported South Vietnam's National Liberation Front, which he did, but rather because he had "long hair," liked to "smoke dope," and had "a good time."[49] Hoffman then went on to draw a stark distinction between those residing in Paine's Potland, which he called "Woodstock Nation," and the squares of NASA. Neil Armstrong was an "albino crewcut" with the personality of Arlington National Cemetery's unknown soldier, Hoffman argued before going after the astronaut's family. "I mean did you dig his parents," he asked, before referring to them, tongue in cheek, as "Mr. and Mrs. A-OK Armstrong—The Mom and Pop of Moonman."[50]

Just as conservatives supported Apollo on economic and military grounds, many in the counterculture refuted such beliefs. The cartoonist R. Crumb, who became famous among hippies for his underground comic strips such as "Keep on Truckin'," his characters including Mr. Natural, and his designs for dozens of album covers for bands from the Grateful Dead to Janis Joplin's Big Brother and the Holding Company, expressed such scorn for NASA while covering a mid-1970s "Space Day Symposium" in Los Angeles for Stewart Brand's *CoEvolution Quarterly*. "What was the purpose of all this talk about our 'destiny,' the 'thrust into space?'" Crumb wondered in his four-page strip about the symposium. His cynical answer: "To drum up business for the aerospace corporations, obviously!!" Several panels later, the cartoonist also critiqued the argument leveled by conservatives that a strong U.S. space program would ensure global security. "The military WILL use space technology," Crumb wrote under a tag line labeled "DANGEROUS!," "to create a whole new array of war toys in space." He then concluded the strip by dressing himself in space-travel attire and placing a ray gun in his left hand. "Don't be duped by foolish Buck Rogers dreams of glorious adventures among the planets!!" he warned readers. "Let's wait until we've learned to get along with each other on Earth before we go barging into the cosmos! Whataya say??"[51] (See Figure 5.2.) Unlike Goldwater and Rand, who praised the aerospace industry for fostering free-market democracy and making the planet safer, Crumb and other

5.2   R. Crumb dressed as a spaceman in R. Crumb, "R. Crumb on Assignment for
the *CoEvolution Quarterly* Goes to the Space Day Symposium (Or Whatever the Hell
It Was Called)," *CoEvolution Quarterly*, fall 1977.

countercultural critics lambasted NASA for wasting money and endangering the Earth.[52]

Hippies were also critical of the space agency's machinery, at least according to Paine, who argued in his WPI speech that all Potlanders viewed technology "as inhuman and responsible for the world's problems."[53] The statement was only partially correct; as Stewart Brand and his *Whole Earth Catalog* illustrated, many on the left opposed only certain types of technology. The hippie artist Chip Lord, for instance, who in 1968 founded the art collective known as Ant Farm on the outskirts of San Francisco, expressed this important distinction by likening NASA's space technology to that developed by the American auto industry. Each was large, impersonal, militaristic, and, according to Lord, nurtured an unhealthy habit "of machine idol." To draw attention to this addiction, in the mid-1970s Lord installed Cadillac Ranch, a public sculpture comprising ten of the General Motors cars partially buried in a field near Amarillo, Texas. Lord then publicly censured NASA for feeding this harmful habit. The space agency's promotion of space exploration, Lord wrote in *CoEvolution Quarterly*, "sounds like General Motors Futurama at the New York World's Fair of

1964." Not until the mid-1960s, concluded Lord, did we realize the "fallout from our frenzied technological growth."[54]

The counterculture was especially critical of the relationship between NASA's space technology and nature both on Earth and beyond. The rock band Jefferson Airplane, which performed at Woodstock, expressed this most directly in their song "Have You Seen the Saucers." Written by band member Paul Kantner around the Apollo 11 lunar landing before being released in 1970, the song opened with a lament regarding NASA's role in polluting the moon. "Tranquility Base, there goes the neighborhood," warned the Airplane's singers. "American garbage dumped in space and no room left for brotherhood." After a second chorus that accused the federal government of lying, Grace Slick crooned that Mother Earth "is getting old" before the entire band implored the "children of the forest, and child of the Woodstock nation," to "care for the needs of your planet."[55] Kantner and Slick returned to a similar space age critique the following year when their rechristened Jefferson Starship, with participation from musicians David Crosby and Graham Nash as well as members of the Grateful Dead and Santana, released *Blows against the Empire*, a concept album about hippies hijacking a spaceship in order to escape from a troubled world and find a more environmentally friendly planetary home.[56] Such artistic efforts highlighted themes expressed perhaps most popularly through David Bowie's "Space Oddity," which was released during the summer of Apollo 11 and featured the renegade astronaut Major Tom "sitting in a tin can / Far above the world."[57]

The counterculture's belief that space technology was harming the natural environment fostered an attitude toward Kennedy's New Frontier that was quite different from that espoused by conservatives during the late 1960s. Similar to their right-wing counterparts, hippies also viewed the New Frontier in space through the history of the old frontier out West. Yet while the likes of Rand and Nixon promoted NASA's hardware as evidence of national prestige and efficient "resource utilization," those in the counterculture saw such technology as the mechanism for colonial expansion into outer space. The author and activist Wendell Berry, whose poems and essays struck a loud chord with Hoffman's Woodstock Nation, was only one among many who voiced such concern. Space exploration, including the possibility of space colonies, Berry argued in the spring of 1976, represents "the renewal, in 'space-age' terms, of an old chauvinism: in order to make up for deficiencies of materials on earth we will 'exploit' (i.e., damage or destroy) the moon and the asteroids." Such thinking, he concluded, "is in absolute obedience to the moral law of the frontier."[58] His fellow countercultural poets Gary Snyder and Anne Waldman also questioned NASA's New Frontier ideology.[59]

This condemnation fed an alternative political philosophy that spread throughout the counterculture during the late 1960s. One expression of such thinking took center stage on the eve of the Apollo 11 launch in Fayetteville, Arkansas, during a guerrilla theater performance titled "A Loose Happening by Spaced-Out Free People." According to the local newspaper, this "space-in" involved neighborhood "hippies" dressed as an American astronaut purchasing the moon from several Indian women for $24 in trinkets and beads. After concluding the transaction, the astronaut explained that he would require more land by 1999 and that such efforts would most definitely lead to war. The hippie astronaut then "landed" on the moon and immediately planted a flag bearing red and white stripes and fifty skull-and-crossbone "stars."[60] Abbie Hoffman described similar political sentiment in *Woodstock Nation* when he faulted what he called "Amerika" for bringing its morality, symbolized by NASA's gravity-less flag, to the lunar surface. "What pig in the Pentagon ordered the project that would fix the flag so it would fly forever unfurled like some perpetual hard-on in space?" asked Hoffman. Unlike Rand, Nixon, and E. A. Kendall, all of whom wanted to see the Stars and Stripes "flying" next to Apollo 11, Hoffman was embarrassed and angry at such imperial pretensions. The residents of Woodstock Nation, he explained, want to "rip down that fuckin' flag on the Moon."[61]

Along with condemning NASA's space program and the conservative politics that supported it, hippies also promoted their own alternatives to both. "I would like to let you know that the young people here in WOOD-STOCK NATION are learning to fly in space," wrote Hoffman.[62] Yet for those gathered in upstate New York, like hippies living all over the country, "space" resided much closer to home. The *San Francisco Oracle*, for instance, an alternative newspaper published in Haight-Ashbury during the late 1960s, provided detailed instructions for what it called "inner space astronauts" interested in a different sort of trip. Obtain your "fuel supply" from a fellow inner astronaut, establish a "flight plan" to ensure a "successful orbit," and have an experienced "ground crew standing by" who can "relay valuable flight and weather information" so travelers can avoid "meteor showers or unexpected cosmic storms." The *Oracle* concluded by reminding readers that while "inner space has not been fully explored," and although the duration of a mission may vary from "a billion or more light years of inner space to 10 to 12 hours of external time," the counterculture's astronauts "will always land."[63] Hoffman concurred, adding, "If I had to sum up the totality of the Woodstock experience I would say it was the first attempt to land a man on the earth."[64]

Where they landed on Earth was also extremely important to the counterculture's inner space astronauts. As the 1960s wore on, and communities such as Haight-Ashbury became congested with youths seeking out the psychedelic scene, hippies began leaving cities such as San Francisco and New York for quieter, natural settings in rural regions. As early as October 1967, for instance, the Diggers of San Francisco warned their fellow hippies that the Haight was no longer "where it's at" and implored them to "gather into tribes: take it anywhere. Disperse."[65] Such sentiment was a large part of Woodstock's appeal. "Walk around for three days without seeing a skyscraper or a traffic light," explained a half-page advertisement for the music festival appearing in the *Village Voice* and other alternative newspapers in mid-summer 1969. "Hundreds of acres to roam on."[66] That same summer *Newsweek* magazine reported that the counterculture, inspired by LSD visions and repelled by "the stale 'straight' life," had "scattered across the country—to New Mexico's mesas and mountains, to lush valleys in Oregon, to Big Sur country in California, to remote corners of Arizona and Maine."[67] In other words, while conservatives such as Rand cheered NASA for building the Kennedy Space Center within the Cape's "untouched wilderness," hippies were embracing alternative living spaces by leapfrogging over their parents' suburbia to rural launchpads scattered across America.

Although Hoffman joked in *Woodstock Nation* that the counterculture would quite soon "fly off in some communal capsule," hippies by the tens of thousands during the late 1960s and early 1970s were serious about moving to rural communes where their lifestyle could include a heavy dose of inner space exploration.[68] Timothy Leary, for instance, always one step ahead of the countercultural curve, not to mention the law, fled in 1963 to a decrepit estate in upstate New York's Millbrook, where he created a commune dedicated to psychedelic drug research that quickly became a hippie mecca visited by the likes of Allen Ginsberg and Ken Kesey's Merry Pranksters. "On this space colony," explained Leary of his Millbrook commune, "we were attempting to create a new paganism and a new dedication to life as art."[69] Hippies established thousands of similar communes, often without the heavy hallucinogens, across rural America, including Drop City near Trinidad, Colorado; Morning Star Ranch north of San Francisco; and Packer Corner Farm in southern Vermont.[70] These countercultural communalists often consciously contrasted their own exploration of inner space in rural nature with NASA's technological adventures beyond Earth. "On the day two Americans harnessed technology to land on the moon," explained *Newsweek* one week after Apollo 11, "25 members of New Mexico's New Buffalo commune harvested wheat by hand—'the way Babylonians did 3,000 years ago.'"[71]

New Buffalo's hippie farmers were not the norm. While embracing the rural lifestyle, counterculture communalists had a much more ambivalent relationship to modern technology. As already noted, Stewart Brand promoted technology to both "citizens of planet Earth" and "hippie environmental spacemen" through his extremely popular *Whole Earth Catalog.*[72] Yet it was a specific type of technology—individualized, democratic, and, most important, small-scale—that appealed to his readers. Brand called them "tools," and NASA's enormous rockets and satellites simply did not fit this bill.[73] Smaller home technologies, however, did. *Whole Earth's* "Shelter" section, for example, which quickly became one of the most popular and enduring features of the publication, provided design plans and home-starter kits for Native American tipis, Buckminster Fuller geodesic domes, and what Brand called "Space Structures."[74] Additional home technologies hawked by *Whole Earth* included modest windmills to generate electricity, energy-efficient wood-burning stoves, and rudimentary solar-heat collectors. Such tools, Brand argued, represented a "movable education" for his counterculture friends "who were reconsidering the structure of modern life and building their own communes in the backwoods."[75]

One such group of friends was the New Alchemists, a band of hippie scientists who in the early 1970s moved with their families to a twelve-acre farm on Cape Cod where they constructed what they called "living machines." Inspired to a great extent by the bioregenerative life support experiments undertaken for NASA by the Odum brothers during the mid-1960s, the New Alchemists began, as their founder, John Todd, explained, by "simulating a variety of aquatic and terrestrial ecosystems in contained spaces or 'capsules.'"[76] These capsules, or "arks," included three greenhouse-covered ponds built on an incline that together produced energy from a windmill and food in the form of fish, while recycling waste and cleaning the system's water and air through ecological, rather than mechanical, processes. The cornerstone of these man-made ecosystems were so-called "solar ponds," fiberglass cylinders filled with water and plankton that absorbed the sun's energy and passed it on as heat to maintain the ark's living and growing climates year-round.[77] During the 1970s the New Alchemists built seven additional arks, often visited by the likes of Stewart Brand and astronaut Rusty Schweikart, from Costa Rica to Prince Edward Island, Canada.[78] Brand's nonprofit Portola Institute even helped fund these projects. "The Ark is a 'spaceship' concept, very far out," explained the *Atlantic Advocate* in a mid-1970s article titled "The Ark: Prince Edward Island's Spaceship to the Future." It represents a "simple return to where we all started, on the land with a few self-sustaining acres surrounding a humble abode."[79]

The chance encounter of James Kunen and William F. Buckley Jr. at the launch of Apollo 8 back in late December of 1968 thus symbolized more than a clash of cultures. In many respects it represented a political prize-fight. In one corner were Buckley and fellow conservatives, including Ayn Rand, cartoonist Charles Brooks, and even President Nixon himself, all of whom praised space technology for both taming wild nature on Earth and in outer space, and demonstrating at home and abroad the superiority of America's free-market political democracy. In the opposite corner were countercultural critics such as R. Crumb, Jefferson Airplane, and hippie actors playing astronauts and Indians, who warned instead that NASA's rockets not only endangered the planet and polluted the moon but also represented a more sinister political plot to extend the old western frontier into space. The counterculture promoted such politics not by marching with mules to the Cape, shutting down campus labs conducting research for Vietnam, or holding mock beauty pageants inside the Johnson Space Center, but rather by venturing into their own inner space with psychedelic drugs, loud music, radical art, guerrilla theater, and alternative living arrangements that incorporated small-scale technologies for both the home and the home planet. In other words, while the residents of Squareland supported NASA's New Frontier in space, which celebrated technology's control *over* nature, Potland's hippies instead installed solar-heat collectors on their roofs far out in the countryside as a means of living more closely *with* the natural environment.[80]

Siding with Squareland

In this culture war over space exploration, NASA sided squarely with Squareland. Such conservative favoritism was evident back in 1968 at the Apollo 8 launch, both when Paine's administrative office personally invited Buckley to attend, in the hopes that he would report favorably on the experience, and when the NASA official handing out press credentials reprimanded Kunen for his long hair.[81] Such animosity toward hippies only intensified in 1970 when Congress cut NASA's budget significantly and Paine responded by blaming the counterculture, among other left-wing groups, which he argued despised reason and science and had unduly influenced politicians in Washington, DC.[82] As a result, even though Paine admitted to WPI graduates in 1970 that it would be tough for his space agency to win what he called "the Squareland versus Potland war of the ages," NASA consciously began consolidating its alliance with conservatives like Buckley beyond Cape Canaveral's press room.[83]

The space agency nurtured its relationship with the New Right during the early 1970s by promoting its own back-to-the-land movement, one decidedly different from that embraced by rural, communal hippies. At the center of this movement was solar energy. Beginning in 1971, NASA teamed up with the National Science Foundation to create a Solar Energy Panel (SEP) to assess current solar technologies and to recommend possibilities for future development.[84] The space agency's involvement on the panel stemmed from its expertise developing solar technology to power Apollo spacecraft. Originally SEP supported such efforts by awarding approximately $200,000 to several NASA contractors, including the Grumman and Raytheon corporations, to explore the possibility of developing "satellite solar power stations" that could collect solar energy from Earth's orbit and beam it as microwave radiation to ground-based receiving stations that would convert it into electricity. Such satellites would, in effect, plug in the Earth to free energy from the sun. The following year NASA intensified such research by using funds from its own budget to commission additional studies on possible solar power satellite systems, which administrators believed could benefit not only the space agency but also the public at large.

President Nixon agreed, sort of. "The sun offers an almost unlimited supply of energy," argued Nixon in June 1971, for Americans who can "learn to use it economically."[85] This emphasis on the economic intensified two years later when the oil embargo by the Organization of Petroleum Exporting Countries (OPEC) ignited an international energy crisis. Nixon responded by immediately taking to the national airwaves, on both television and radio, to announce Project Independence, his federal initiative intended to help the United States achieve energy self-sufficiency by 1980. The president's project first and foremost asked ordinary Americans to make individual sacrifices by driving less and lowering their home thermostats, and then only vaguely referenced his administration's interest in developing alternative technologies for harnessing nuclear and solar power.[86] Perhaps most telling, Project Independence requested that Congress relax a slew of environmental regulations in order to boost domestic production of nonrenewable energy sources, including oil, gas, and coal. As Nixon explained, the United States must have "an appropriate balancing of our environmental interests" with "our energy requirements." To rally his silent majority, the president then compared Project Independence to the Apollo moon landing. "In the spirit of Apollo," he exhorted, "let us set as our national goal" that by the end of this decade "we will have developed the potential to meet our own energy needs without depending on any foreign energy sources."[87] To that end, Congress also requested that NASA prepare a report detailing how its space hardware might support Project Independence.[88]

Conservatives were quite prepared to back Nixon's halfhearted solar power pledge, and NASA's role in attaining it, as a means not only of achieving national energy sovereignty but also of exploring possible colonization along the New Frontier. Buckley's *National Review*, for instance, had already promoted solar settlements in space with language borrowed from the nineteenth-century eugenics movement. "The fate of man lies in man's ability to find new worlds to colonize," contended the conservative magazine in the spring of 1969. Unlike migrants from earlier eras, which the *Review* described as "malcontents or rejects of the old world," those selected for space settlement "will be chosen for both brain and brawn, and their pedigrees scrutinized," much like the process for selecting NASA's astronauts.[89] Although the *Wall Street Journal* and *Time* magazine refrained from such language, they, too, had celebrated the possibility of space colonies powered by solar satellites.[90] Such conservative backing was understandable, because NASA had been promoting the possibility of solar-powered space colonies since the dawn of the space age; during the mid-1950s

Wernher von Braun had widely publicized the idea of space settlements in three vividly illustrated special issues of *Colliers* magazine as well as through three television programs for Walt Disney.[91] As Disneyland's "Man in Space" told millions of television viewers in 1955, NASA's space-ships and space stations would contain "scientific apparatus" for "converting solar energy into electricity."[92]

During the early 1970s Walt Disney's dreams of solar settlements beyond Earth seemed on the verge of reality when Princeton physics professor Gerard O'Neill teamed up with NASA to undertake and publicize scientific research on the physical construction of space colonies. O'Neill, who had unsuccessfully applied to become an astronaut in 1966 when NASA opened up its corps to scientists, began undertaking mathematical calculations with his students concerning the construction and maintenance of self-supported space habitats floating at gravitationally stable points between Earth and the moon. In order to attract broader support for such work, O'Neill took what he called a more "conservative and pragmatic" approach based on the economics of NASA's proposed orbiting solar-power stations.[93] "The colonies," O'Neill wrote in a 1974 *Physics Today* article outlining his research, "can obtain all the energy they could ever need from clean solar power."[94] Due, in part, to his astute blending of hard science and economic optimism, NASA administrators, scientists, and even astronauts supported O'Neill's efforts. They funded and participated in two of his space-colonization conferences held at Princeton in 1974 and 1975, awarded him several grants for up to $500,000 per year for space-colony research, hired him to lead several studies from 1975 through 1977 at NASA's Ames Research Center that focused on developing solar energy to power lunar bases, and ultimately incorporated his concepts into the space agency's long-range planning.[95] During this period O'Neill's space-colony research also became front-page news in the *New York Times* and reached a national audience in 1977 with the publication of his extremely popular book, *The High Frontier: Human Colonies in Space*.[96]

The space colonies promoted by both O'Neill and NASA must have looked quite familiar to the residents of Squareland, at least from the inside. During one of his ten-week-long studies at Ames, which was also sponsored by Stanford University, O'Neill directed a research team of approximately thirty scientists to create designs for Stanford Torus, a cylindrical, donut-shaped space habitat that could accommodate 10,000 permanent residents. To promote the study's findings, NASA commissioned artists to depict in full color the space colony from the inside out, and it even produced a short film for the public that highlighted the colony's economically efficient use of solar power.[97] While the overall designs were radical, even within scientific circles, the aesthetics of the colonies were far from it. Detached single-family homes with green front lawns and neat backyard patios dominated the depictions. They "looked like New Jersey in orbit," argued science-fiction writer Allen Steel years later. Nothing less than "space as a giant suburb."[98] Social commenters back in the Apollo era agreed. "Will Tranquility Base become a Levittown," asked a July 1969 editorial in the Norfolk, Virginia, *Ledger-Star*, "and the Sea itself the midden for our packaged culture of pop-top cans, waxed milk cartons and broken bottles?"[99]

When in the mid-1970s NASA began abandoning its research on satellite solar-energy stations, along with the space colonies they would support, because developing such technology proved economically unfeasible, the space agency continued to court conservatives by bringing its solar energy research back down to Earth. Such ground-based efforts began in 1974 when NASA announced a solar energy heating-and-cooling initiative at three of its centers. While scientists at the Lewis Research Center in Cleveland began developing solar-heat collectors, those at NASA's Langley Research Center began testing these new technologies on the roof of a new engineering building being constructed at their Hampton, Virginia, campus. "This 53,000 square-foot building," explained NASA Administrator James Fletcher in a press release, "will be the first building of its size in the world for which solar energy will provide a significant part of the buildings' heating and cooling load."[100] Space agency scientists conducted similar solar tests at the Marshall Space Flight Center in Huntsville involving heating and air conditioning in a simulated private residence. "The Marshall demonstration unit," explained NASA in a 1974 press release titled "Solar Energy for Heating and Cooling of Buildings," "is representative of the size required for a single-family dwelling."[101]

The following year the space agency expanded such efforts by building Tech House, an acronym for "The Energy Conservation House," on the grounds of Langley. The 1,500-square-foot single-story home contained three bedrooms and, according to NASA officials, "a large living room with a fireplace, dining area, kitchen, two baths, laundry, and garage."[102] It also included an array of space age innovations, including eighteen solar collectors on the roof to supply the home with its heating and cooling requirements. The space agency publicized Tech House aggressively during the mid-1970s through press releases that garnered coverage in local newspapers and national magazines, by producing a short film on the experimental house, and by moving, with much public fanfare, a family of four into the home for one year.[103] According to such publicity, Tech House was an economic success, reducing hot water energy consumption by 70 percent and saving potential owners, over a twenty-year period, approximately $20,000 in utility costs. "Putting space technology to work in home construction," explained a June 1979 *Popular Mechanics* article on Tech House, "can save more than half the energy an average family uses in a conventional all-electric house."[104]

The space agency's Tech House could not have been more different from the New Alchemists' arks. Whereas the solar technology installed on Cape Cod and Prince Edward Island was small, relatively simple, and inexpensive, that at Langley included large solar panels as well as one hundred sensors placed strategically throughout the house, required two computers to run and monitor the system, and ultimately cost much more than NASA's stated price tag of $45,000. The computer equipment alone, lamented a *Popular Mechanics* article that questioned NASA's financial calculations, cost "as much as several Cadillacs."[105] More important, while the New Alchemists and the majority of hippies reading the *Whole Earth Catalog* used technologies such as solar-heat collectors to live more lightly on the land, NASA promoted Tech House instead as a means of using nature for financial gain. "As a home owner, how would you like to be relatively independent of rising utility costs," asked a NASA publicity report in the late 1970s, "and draw instead on nature's bounty for a significant part of your energy needs"?[106] Economic savings, not ecological salvation, was the foundation on which Tech House stood.

Perhaps most significant were the starkly different geographies of the solar technologies embraced by NASA and the counterculture. The space agency's Tech House was a grounded version of its space colonies; both reflected a suburban architectural ideal embraced most forcefully during the 1970s by the very conservatives who supported the space race. Not only could the Tech House's floor plan be easily mistaken for a Levittown ranch, but NASA administrators emphasized this aesthetic in promotional literature by consciously picturing the experimental solar home as a suburban haven, complete with grassy lawn, tree plantings and landscaped shrubbery, and a prominent garage.[107] To further emphasize this point the space agency moved the Swains, which NASA called a "typical" American family composed of a "husband, wife, and two children," along with their border collie, into Tech House for a year. It then photographed the wife, Elaine, performing most of the household chores, including cooking, dish-washing, and cleaning. "The point of the experiment," explained Elaine's husband, Charles, was "to see if an average middle-class family" could live in this house "and save energy."[108] The Swains as depicted by NASA, surrounded in their modern kitchen by suburban appliances, seemed a world apart from the New Alchemists, who often published in their journal photographs of their "small band" composed of multigenerational adults with no obvious partners, nearly a dozen children with no seemingly des-ignated parents, and visual references to farming and rural life (see Fig-ures 5.3 and 5.4).[109]

During the late 1970s NASA administrators tried to transplant the solar technologies it had tested in Tech House throughout America's suburban

5.3   NASA publicity photograph of the Swain family in the Tech House kitchen, 1978.

5.4   Self-portrait of the New Alchemists from their publication, *Journal of the New Alchemists*, 1974.

frontier. Such efforts entailed writing a "do-it-yourself" handbook for homeowners that explained how to build and install a solar heating system, publishing a "Guide to Space Heating through Solar Energy Technology" for contractors and engineers in the home-building industry, and by becoming a lead agency, along with the Department of Housing and Urban Development (HUD), in implementing the Solar Heating and Cooling Demonstration Act of 1974, which sought to illustrate to the American public the feasibility of solar heating and cooling technology developed by NASA.[110] One such demonstration took place twelve miles outside Washington, DC, in Greenbelt, Maryland, which since its creation during the Great Depression by New Deal planners had become one of the most famous, and studied, suburbs in America. For the Greenbelt demonstration NASA installed solar-heat collectors, similar to those at Tech House, onto four homes and monitored the energy use in each. The space agency then publicized the results, which indicated a 40–50 percent savings in oil needed for heating and hot water.[111] According to NASA, more than one hundred similar demonstrations involving space age solar technology were taking place by 1979.[112]

Although a short promotional film on the Tech House argued that "from outer space to inner space, NASA scientists have brought home some new ideas to solve the problems of diminishing resources here on Earth," the space agency's version of "inner space" was the antithesis of the counterculture's.[113] For hippies, inner space was a form of consciousness achieved by practicing alternative lifestyles, more often than not in rural regions. For NASA, Nixon, and their conservative supporters, inner space was instead located in the suburbs and required no lifestyle changes whatsoever. "Tech House represents an engineering test as to whether technology can achieve energy conservation without lifestyle adjustment," announced NASA administrators in 1978.[114] For hippies this was the ultimate cop-out. Such was the warning of "Beyond Solar Suburbia," a 1979 essay appearing in the alternative publication *RAIN*. Written by Peter Calthorpe, an architect and associate of the New Alchemists, the article criticized federal solar demonstration programs, such as those led by NASA and HUD, because they focused primarily on the suburban home. "Solar and other alternative energy sources may become a mechanism to perpetuate these inefficient patterns" sprawling across suburbia, Calthorpe warned, "rather than a means to a more environmentally sound culture."[115] In other words, while hippies embraced technologies such as solar collectors to forge a new, more ecological lifestyle, NASA, the Swains, and suburbanites in Greenbelt, Maryland, relied on them instead to maintain an older lifestyle predicated on using natural resources for economic gain.

NASA's New Research Parks

During the 1970s NASA succeeded in encouraging space age suburbs, just not the sort of suburbs promoted by its Tech House experiments. Similar to its energy-efficient heating and cooling systems, which engineers retooled from space capsules but installed in very few inner-city housing projects, NASA's solar suburban homes also failed to catch on with the American public due to financial constraints; as the OPEC embargo collapsed in the late 1970s public interest in, and congressional funding for, solar energy research dimmed.[116] Yet conservative support for NASA continued unabated, driven in part by long-standing economic optimism regarding the so-called space-industrial complex. Back in 1962, for instance, in a cover story titled "Space Crescent Transforms Gulf Area," *Business Week* described how NASA's five centers—arcing across the region from Cape Canaveral in the east through Alabama and Louisiana along the Gulf and on to Houston out west—had opened up a "new frontier" for prosperity.[117] The editors of *Fortune* extended such thinking across the entire nation, noting in a special-issue book titled *The Space Industry: America's Newest Giant* that NASA's centers also signed hundreds of thousands of contracts with more than 20,000 private companies from Northrop in Southern California to Grumman out on Long Island.[118] This burgeoning aerospace industry, concluded *Business Week*, was not only creating "greater expansion" and "more economic growth," but along with it "upheaval" and "a new kind of life" in America.[119]

A similar "upheaval" had created "a new kind of life" in the United States immediately after World War II. Prior to that conflict, American business owners had located the majority of their corporate offices in urban areas and had hired architects to build headquarters that symbolized corporate power. In most instances these designs had focused on height, with sky-scrapers such as New York's Woolworth and Chrysler buildings as prime examples.[120] After the war, however, changes in the American city radically transformed this corporate geography. Increasingly noxious pollution, densely packed business districts, and stifling traffic all weakened these businesses' bond with their prime downtown real estate. Rising anxiety regarding nuclear war, and federal directives to disperse for civil defense, also helped to push companies from the city in a corporate version of "duck and cover."[121] These businesses were similarly pulled beyond city limits by new federal highways, lower taxes, and most obviously more open space that would allow them to separate their factories, with their low-wage workers, from new corporate headquarters and research labs that would attract high-level managers and scientists. As a result, after World War II there was an exodus of corporate headquarters from the city center out into America's suburbs.[122]

Nature played a central role in this corporate migration. While an in-creasingly polluted urban environment helped to push these corporations out of the city, bucolic nature beyond city limits welcomed them to the sub-urbs. This suburban ideal had deep historical roots in the United States and blossomed in the mid-nineteenth century through the writings and landscapes of designers such as Andrew Jackson Downing, Calvert Vaux, and Frederick Law Olmsted, all of whom promoted pastoral suburbs as "middle grounds" between city and country that fostered social order, cultural uplift, and physical well-being. After World War II corporations extended this correlation between the pastoral and the positive to include increased productivity. The result was a new corporate aesthetic that included sprawling, modern campuses with sweeping lawns, decorative lakes, and winding roads all set within secluded woods.[123] Nature, not height, became the symbol of corporate power in the postwar era.[124] The rise of the military-industrial complex after World War II, in other words, was accompanied by, and helped foster, the birth of the "corporate park."

This new suburban ideal reached its apex in the mid-1950s with the construction of the General Motors Technical Center on the outskirts of Detroit. Although other businesses had paved the way into the suburbs, including AT&T, with its Bell Labs just west of New York City, and General Electric, which built its Electronics Park northwest of Syracuse, New York, it was GM's new headquarters and research laboratory north of the Motor City that set a new standard for corporate America.[125] Located on 900 acres that *Fortune* magazine described as only "thirty minutes by Cadillac or Chevrolet" from downtown Detroit, GM's Tech Center, as it came to be known, was designed with the company's automotive technology in mind. The Finnish architect Eero Saarinen used mass-produced steel panels for building exteriors and developed a neoprene adhesive process to hold plate-glass windows in place, much like that used for sealing car windshields.[126] He also limited the height of the Tech Center's buildings, which were separated into four research complexes and administrative offices, to three stories, giving the center a horizontal orientation quite different from its skyscraping, urban counterparts.

Yet while automotive technology informed the Tech Center's overall design, so, too, did nature. Saarinen designed the headquarters to resemble a university campus, connecting the Tech Center's buildings with 155 acres of lawn and arranging them around a central, rectangular, 22-acre lake, much like a campus green. The well-known landscape architect for the project, Thomas Church, not only incorporated a gardenlike outdoor "viewing terrace" for use by employees and visitors but also spent $1 million on 13,000 trees, 3,180 shrubs, and more than 50,000 ground-cover plants that were used in rows to direct views, as borders around buildings and lawns, to create four "weeping willow islands" in the lake, and to encircle the entire property with a forest of uniformly spaced trees.[127] The ultimate goal, according to GM publicity materials, was to integrate the corporation's technology-inspired architecture with the surrounding natural environment. As *Architectural Forum* explained, Saarinen designed the center's

5.5   General Motors postcard promoting its Technical Center, circa early 1960s.

large windows "not by a standard of 'foot candles at desk level' but by the desire that the innermost draftsman be aware of the leaves of the trees."[128] "The verdant landscaping," the magazine concluded, was aimed "inward, through glass, to the employees."[129]

Nature at the Tech Center was also aimed, quite consciously, outward toward the public. This is all too evident in the company's extensive promotional materials. *Where Today Meets Tomorrow*, GM's glossy twenty-page booklet introducing the new Tech Center, displayed a cover illustration that situated the center's central lake, shade trees, and expansive lawns within a larger pastoral landscape running far off to the horizon. The booklet's color photographs, as well as tourist postcards sold at the center, similarly framed the new headquarters' modern architecture with tall trees, shimmering water, and bright green turf (see Figure 5.5).[130] So bucolic was the Tech Center's setting that *Life* magazine, in a six-page photographic spread covering the facility's opening in May 1956, noted the "unusual beauty of this modern 'Industrial Versailles.'"[131] During the same opening ceremony President Eisenhower deployed a more American metaphor, comparing GM executives to western explorers and the new corporate headquarters, with its beautifully landscaped grounds, to a "technological frontier."[132]

While Ayn Rand used nearly identical language to praise the Kennedy Space Center during her visit to the Cape, it was the Manned Spacecraft Center, just twenty-two miles outside Houston, that was NASA's own version of GM's Tech Center. Opened in 1964, what was later renamed the Johnson Space Center sat on approximately 1,600 acres of former ranchland. The center's architect, Charles Luckman, was not only a contemporary of Saarinen but both men had designed different headquarters for CBS during the 1950s.[133] Similar to Saarinen, Luckman used metal, in this case glazed white, on building exteriors. He likewise physically separated each building according to different NASA departments, and kept them low and horizontal. Moreover, whereas Saarinen had designed GM's Tech Center with the corporation's automotive technology in mind, Luckman incorporated what has been called "aerospace modernism" into the Houston space center.[134] The "complex of white metal and marble chip concrete buildings," explained a *New York Times* reporter in 1969, resembled "a clean futuristic city."[135]

Also like Saarinen, Luckman relied on nature to unify NASA's newest headquarters. He arranged the center's buildings in a campus-like fashion, substituting three smaller geometric "lagoons" for Saarinen's central rectangular lake. He also connected the buildings with the same expansive green lawns, which in this case provided open views of the surrounding prairie, and he planted rows of trees, shrubs, and ground-cover plants, similar to those at GM's Tech Center, around parking lots and along entrance driveways.[136] "Meticulously cut evergreen shrubs and Rockford ferns," explained the *New York Times*, bordered the center's 1,620 acres, while "pines and oaks" scattered deliberately around the grounds "lie still in the languid heat."[137] Finally, NASA administrators, like their GM counterparts, also promoted their new campus extensively with glossy color postcards picturing the center's modern architecture framed by lawns, lagoons, and landscaped plantings (see Figure 5.6).[138]

During the early 1960s NASA located, designed, and promoted its other NASA centers similarly. Administrators first acquired land on the outskirts of cities—620 acres fifteen miles northeast of Washington, DC; nearly 2,000 acres ten miles southwest of Huntsville, Alabama; and nearly 14,000 acres an hour from New Orleans—as the sites for the Goddard Space Flight Center, Marshall Space Flight Center, and Mississippi Test Facility, respectively. At each of these suburban locations NASA then constructed pastoral corporate parks. The plan for Goddard, much like that in Houston and

5.6   NASA postcard of Building 2 at the Manned Spacecraft Center, circa late 1960s.

GM's Tech Center, included low-rise buildings that, as one historian explained, "blended inconspicuously into the landscape" and were physically separated by function to create "a campus-like atmosphere."[139] The architectural and landscape design plans for the Marshall Space Flight Center near Huntsville and the Mississippi Test Facility outside New Orleans also included horizontal buildings, extensive lawns, and tree plantings to screen parking lots and buildings and to direct views.[140] Administrators at NASA also promoted these pastoral campuses to the public, much as they did with the agency's Houston headquarters, with glossy postcards depicting both aerial photographs of the centers and hand-colored architectural and landscape drawings of the campuses' extensively landscaped grounds (see Figures 5.7 and 5.8).[141]

   Just as GM built its Tech Center, in part, to promote a positive corporate image, so, too, did NASA build and design its research centers as advertisements for space exploration. If, as the cultural geographer J. B. Jackson has argued, "greenery" is a "way of communicating with others," then the artificial lakes, sweeping lawns, and groves of planted trees covering the grounds of NASA's new research parks spoke eloquently to the public.[142] For instance, as mainstream reporters visited NASA's centers with increased frequency just prior to the Apollo 11 launch, they began to paint a

5.7   NASA postcard of an artist's rendering of the Goddard Space Flight Center.

5.8   NASA postcard of the Marshall Space Flight Center.

particularly rosy portrait of the aerospace business. "It is a quiet, clean-cut sort of industry, the kind that almost everyone likes to have next door," explained *Newsweek* in a special series titled "The Business of Space." For the *Newsweek* reporter, "clean-cut" cut two ways; NASA's centers were both square, like the managers and scientists who worked in them, as well as environmentally tidy. "There are hardly any belching smokestacks and no crashing assembly lines," *Newsweek* concluded.[143] Ayn Rand agreed, explaining during her own visit to the Cape that the Kennedy Space Center was crisscrossed by a network of "clean" roads and that NASA overall was the "cleanest" of all government programs.[144] It wasn't, obviously, but it appeared that way because administrators had physically separated the aerospace industry's messy production processes from NASA's clean, green centers. As one engineer who began working on Apollo during the mid-1960s said, in a moment of minor self-deception, "I'd hate to go back to industry."[145] Aerospace during the late 1960s was most definitely an industry, but one that NASA successfully masked with beautifully landscaped research parks.[146]

Spurring Aerospace Sprawl

The space agency's construction of pastoral centers on the outskirts of Washington, D.C., Houston, Huntsville, and New Orleans was not, initially, a migration onto the suburban frontier. Rather it was primarily a move into undeveloped territory. In 1960, one year before NASA announced its intention to locate a new space center in Texas, the area in the proposed vicinity was a sleepy community of just over 6,500 residents living in a half dozen unincorporated towns with just one bank, one savings and loan association, and no supermarket or dentist. The land in this region, most of which was devoted to grazing, sold for less than $750 an acre. "Five years ago, the 1,620-acre tract of land on which NASA's Manned Spacecraft Center now stands was a cow pasture" reached by a two-lane farm road, explained one reporter in December 1966. "It consisted of miles and miles of scrub grass, broken only by cattle and an occasional windmill."[147]

By 1966, just two years after the opening of the space center outside Houston, the cow pasture had been transformed into a space-age Levittown. Not only had the local population grown by almost 600 percent to more than 38,000, but these newcomers were now living in more than a dozen suburban developments.[148] The planned community of Clear Lake City, for instance, which broke ground less than a year after NASA announced its move to the area, sat on 15,000 acres of land adjacent to the center and entailed the construction of 40,000 new homes as well as shopping centers, office buildings, schools, churches, and even a country club with an eighteen-hole golf course.[149] Seven incorporated towns and eighteen suburban subdivisions within ten miles quickly followed. So did 125 aerospace firms, including North American Aviation, General Electric, Lockheed, IBM, and Sperry-Rand.[150] Just beyond the center's gates a "half dozen suburban satellites have sprung up with clusters of 'pure vanilla' ranch-style houses," reported the *Washington Post* in July 1969. "There is no single community or city-like center to give the area a focal point," but instead "a gaudy strip, motels, hamburger stands, restaurants and real estate offices have sprouted since the Center's site was announced."[151]

Additional suburban communities spread like weeds on the periphery of NASA's other research centers. The city of Huntsville, for instance, which had a population of 30,000 before the arrival of the Marshall Space Flight Center, boasted more than 160,000 on the eve of Apollo 11, an increase of more than 500 percent. In response to this population boom, politicians expanded the city's limits, from three to fifty-six square miles, to accommodate new suburban developments that required the construction of water, gas, and electrical systems; nineteen new shopping malls; and, during the mid-1960s, one school on average per week.[152] Picayune, a small town near NASA's Mississippi Test Facility, also experienced a housing boom.[153] Even the sleepy beach communities near Cape Canaveral became increasingly suburbanized with the arrival of the space agency in 1962. The General Development Corporation, for instance, which strategically acquired 43,000 acres near Cape Canaveral immediately after NASA announced the location of its new "Moonport," advertised the real estate in *Life* magazine with a full-page color illustration of a rocket launching above "Port Malabar," a brand-new suburb complete with "shopping facilities, schools, churches, movies, golf," and rows and rows of ranch-style homes with postage-stamp lawns.[154] Within four months the company, which also noted that the development was located in the "fastest growing county" in the nation, had sold 10,000 lots.[155]

Clear Lake City, Picayune, and Port Malabar were not unique; areas on the outskirts of cities across the nation were undergoing similar changes during the 1950s and 1960s. Yet as historian Greg Hise has shown in his study of Southern California, corporations did not always move to sub-urbia to appease their suburban workforce with shorter commutes to the office or a more bucolic work environment. Corporate flight, Hise persua-sively argues, did not always follow white flight from cities. Instead, many businesses were pioneers on the "crabgrass frontier," only to be followed later by their employees. This was especially true of defense industries such as airline manufacturing, which congregated near Los Angeles after World War II. Corporations relocating to the suburbs during the 1950s and 1960s, much like NASA's new research centers, were thus not only driving the military-industrial-space complex, but also causing suburban sprawl.[156]

This suburbanization intensified in the early 1960s as the aviation busi-ness gave way to the aerospace industry, and as NASA farmed out much of its production work to tens of thousands of subcontractors. In California, which garnered by far the most contract dollars of any state, nearly a dozen of the largest aerospace companies in the nation congregated between Los Angeles and San Diego where, much like GM and NASA, they built modern headquarters and research labs beyond city limits in part to promote their role in the space race.[157] These Southern California aerospace firms, in-cluding Northrop, Lockheed, Douglas, and North American, hired not only more than 250,000 workers but also none other than Charles Luckman and his partner, William Pereira, to design their facilities. In Southern Cal-ifornia the architects again integrated architectural references to these corporations' technological achievements with pastoral nature. For Gen-eral Dynamics' Convair Astronautics campus outside San Diego, Luckman and Pereira included an "Astro Café," a "Missile Park" complete with an au-thentic Atlas rocket, and a spiraling space age staircase ramp that dominated the facility's lobby, as well as an elegant oriental garden, reflecting pool, and strategically placed trees that provided, as one commentator noted, "the classic Southern California scene of buildings amid palm trees."[158] While the Convair campus, much like NASA's Houston headquarters, sat in the middle of undeveloped land when it opened in 1958, it, too, quickly became bounded by suburban subdivisions. Similar aerospace sprawl followed NASA contrac-tors such as Grumman on Long Island and Raytheon, RCA, and Avco along Route 128 outside Boston.[159]

Thus, while the sun was setting on NASA's suburban solar homes, aero-space suburbs began spreading across the space crescent and rising as well near NASA's contractors in Southern California and the Northeast. Square-land's "home territories," as Thomas Paine explained to WPI graduates, "the world you were born and raised in," were expanding, dramatically. Real estate salesmen immediately promoted this annexation. In the newly developed Clear Lake City on the outskirts of NASA's Houston center, for example, real estate signs began popping up on the newly seeded lawns of what were being advertised as "inner space homes."[160] Such residences, of course, would never interest Abbie Hoffman's "inner space astronauts," who had bypassed the growing suburbs of Houston, Huntsville, and Los Angeles to settle in more rural environs, including Big Sur and Woodstock. For such hippies, many of whom installed solar-heat collectors atop geodesic domes, the aerospace industry's "inner space homes" were just too square.

The Political Ecology of Space Age Suburbs

Such homes were also an environmental nightmare.[161] While many from the counterculture moved to the countryside to live more lightly on the land, NASA's suburbs devoured the land itself. What had been described in the early 1960s as "barren prairie" near Houston, "back country" orange groves at Cape Canaveral, and "sleepy hamlets" outside New Orleans became by the mid-1970s sprawling suburban subdivisions covering tens of thousands of acres.[162] In dry regions such as those encircling Houston's Manned Spacecraft Center, builders made way for tract housing by stripping the landscape of its natural flora, which increased erosion that damaged local soils and silted nearby streams, rivers, and lakes. In wetter areas, including those near Cape Canaveral's launch center, developers expanded their buildable acreage by filling in marshes, wetlands, and portions of coastal estuaries, a process that not only destroyed sensitive ecosystems but wildlife habitat as well.[163] To encourage even more building around the Houston center, explained *Business Week* in September 1965, "some woodland will soon be opened to development."[164]

As aerospace suburbs went up, so did the demand for water, which caused local water tables to drop precipitously. In Southern California this resulted in a permanent man-made drought, and even near Cape Canaveral such increased water use caused temporary shortages as local water systems failed to keep up with demand.[165] In the new suburbs of Houston this growing thirst also resulted in widespread subsidence, or the sinking of surface land, as aquifers underneath drained. Yet the worst impact these new suburbs had on local water supplies involved pollution. During the

1970s the mushrooming developments around NASA's center near Houston quickly overburdened local sewage-treatment plants, which were forced to disgorge hundreds of thousands of gallons of raw sewage into nearby water-courses, including the increasingly misnamed Clear Lake.[166] At the Cape, where most new suburbs relied instead on septic tanks located near private wells, water contamination was even more acute. "Soil conditions, high water tables plus an overabundance of septic tanks," warned a 1966 report by Florida State University's Institute for Social Research titled *NASA Impact on Brevard County*, "have created health hazards."[167]

The air high above, much like the land and water underneath, similarly suffered as NASA and its contractors spurred suburban sprawl. Unlike the space agency's energy-efficient Tech House or the New Alchemists' ecologically oriented arks, the tract homes that arose in the aerospace industry's shadow gobbled up fossil fuels, including coal, oil, and gas, that, when burned for heat, electricity, and lighting, fouled nearby air. Stripping the landscape of trees, and replanting it with lawns that needed constant cutting by woefully inefficient mowers, only added to these emissions while reducing the local environment's ability to sequester carbon from the atmosphere.[168] Yet it was the skyrocketing use of automobiles that made breathing most difficult for NASA's new neighbors. The influx of 250,000 aerospace workers into Southern California during the 1960s no doubt helped Los Angeles garner the ignominious achievement of having the worst air quality in the nation.[169] The following decade, as NASA's newly rechristened Johnson Space Center outside Houston pulled dozens of aerospace firms into its suburban orbit, the region averaged 2,000 additional cars per week, resulting in tens of thousands of additional workers driving to and from their homes each day. "Houston has a visible non-point air-pollution problem," concluded one report, "emanating from its over-crowded highways and streets."[170]

While hippies during the 1970s remained critical of what the folk singer Malvina Reynolds disparaged as "little boxes made of ticky tacky," especially because these new suburban subdivisions had an abysmal environmental record, conservatives moved into them in droves in part because they were, as Reynolds crooned, "little boxes all the same."[171] "If the area is physically fragmented," explained the *Washington Post* in a 1969 article describing daily life in Houston's sprawling new suburbs, "its citizens are rather homogeneous." On average they were thirty-five years old, married with at least two children, and owned two cars because, as the *Post* reminded its readers, "there are virtually no sidewalks."[172] They were also overwhelmingly middle class and laboring as engineers, scientists, and technicians for an industry that was openly hostile to working-class unions. Not only did aerospace businesses migrate to Southern California in part because it was an open-shop stronghold, but NASA also supported the federal government's efforts during the 1960s to break numerous strikes by machinists, ironworkers, and construction laborers building NASA facilities, including those at Cape Canaveral.[173] In the suburbs of sunny Southern California, as well as those outside Houston, Huntsville, and the Cape, there was thus no need to worry about an unruly working class; few from that economic stratum lived there.

Conservatives were also drawn to these communities' lily-white demographics and to the ability of their residents to maintain such racial uniformity. In Huntsville, for instance, whites moving into the city's peripheral suburbs were able to distance themselves quite successfully from racial minorities. "Expansion in Huntsville has done two things," explained Dr. John Cashin, an African American and local civil rights leader, on the eve of the Apollo 11 launch. It has "relocated blacks to the northwest side of town or moved black people out entirely." Cashin's wife, Joan, who had been arrested for trying to desegregate the local Walgreens lunch counter, had similar complaints, noting that "there is only one street for Negroes to live on that is not in a slum location."[174] The suburbs sprouting around Southern California's aerospace firms, as well as those at Cape Canaveral, were also strictly segregated.[175] Whereas advertisements for Port Malabar, with its high-end marina and eighteen-hole golf course, targeted the middle-class and predominantly white readers of *Life* magazine, others hawking "unimproved land" with "no roads or drainage" in what was being touted as Canaveral Grove Estates appeared in the pages of *Ebony*.[176] The nearby town of Cocoa Beach also built exclusionary public housing.[177]

Within these new suburbs, women were often sequestered within their "inner space homes," while men left each day for work. "I was amazed at the stubborn persistence of dated gender roles in the technical community," explained M. G. Lord in her memoir about growing up as the daughter of a rocket engineer at NASA's Jet Propulsion Laboratory in Southern California. While each morning her father "drove off to ply his brain," she and her mother stayed home. "If a wife had a brain to ply, she did so in the house," Lord added, "or volunteered with, say, the Girl Scouts, a paramilitary organization that promoted cleanliness, piety, and obedience to patriarchal authority." As a result, Lord witnessed her educated mother suffer from the very "malaise" identified in Betty Friedan's *The Feminist Mystique*. "It was 'the problem that has no name,'" Lord remembered, "chronic unhappiness and a lack of self-worth."[178] The wives of Houston's rocket engineers suffered similarly. "He is so pressurized and rushed for time," complained one wife from Clear Lake, "he expects me to have whatever he wants—food, clean clothes—ready in a minute."[179]

Class and racial homogeneity along with strict adherence to traditional gender roles, not to mention the practice of turning a blind eye to local environmental problems, resulted in communities that were ripe for right-wing politics. To a great extent, it was conservatives who were attracted to aerospace work in the first place; unlike many from the counterculture, the engineers, scientists, and technicians working for NASA and its contractors were quite comfortable with both the fuzzy line separating aerospace and military research as well as the security clearances necessary in order to undertake classified work.[180] The result was that conservatives quickly became the majority in these aerospace suburbs. "Sometimes I feel like I'm the only Democrat in town," explained the wife of a NASA engineer living in Houston during the summer of 1969.[181] Even the local clergy, who in most American suburbs stood guard over traditional mores, felt out of place on the outskirts of Houston. The Reverend John Elder, for instance, who served as pastor of the Clear Lake Presbyterian Church, acknowledged that many of the priests in town were "far to the left of their parishioners" and felt quite uncomfortable about it.[182]

Yet while the aerospace industry attracted conservatives into its suburban orbit, the residential developments encircling both NASA's centers and its contractors' office parks also tended to nudge newly arrived residents to the right. As historians Lisa McGirr and Matthew Lassiter have illustrated, the sprawling built environment of the postwar suburbs located in Southern California and across the Sunbelt South promoted privacy, individual property rights, homeownership, and isolation. These community characteristics, in turn, devalued public space and town centers that could have created a deeper sense of community responsibility, not to mention environmental sensitivity. "The physical landscape," McGirr argues persuasively in her history of Orange County, California, "contributed to creating a hospitable terrain for the Right."[183] For Lassiter, the rise of the New Right took root in local schools, where middle-class parents used spatial sprawl to escape classroom integration.[184] The result, for both students and their parents, was peer pressure to conform to a more conservative status quo.

This was especially true of NASA's aerospace suburbs. During the late 1960s, for instance, male students attending Clear Creek High School just beyond the gates of NASA's Houston space center had to wear collared shirts and were not allowed to sport long hair or sideburns. Three students who stenciled "Youth for Darwin" on their sweatshirts, because they felt outnumbered by fellow classmates in the Youth for Christ student group, were also expelled from the public school for being "improperly uniformed." Not surprisingly, female students could not wear skirts more than three inches above their knees.[185] As a result of such peer pressure, reported the *Washington Post* one week prior to the launch of Apollo 11, "politically the area tends to be conservative and Republican, and in religion, traditional."[186] Thus as NASA's aerospace suburbs slowly spread across the Sunbelt toward Southern California they not only caused a host of environmental problems, but also helped give rise to the New Right.

NASA's Southern Strategy

After bumping into one another in the viewing bleachers and then watching Apollo 8 roar into the early morning sky, both Kunen and Buckley boarded a bus and took a press tour of the Kennedy Space Center. Similar to the earlier Greyhound bus excursions that NASA initiated back in 1966, this one also made stops at Apollo's technological landmarks. The two reporters first visited the Apollo launchpad, charred black from the recent liftoff, and then moved on to the Vehicle Assembly Building (VAB) where the Apollo 8 spacecraft had been put together. It was while inside the VAB that a NASA representative educated the press about the space agency's technology, explaining that Apollo 8 contained 5.6 million individual parts that took more than 250,000 technicians to produce. Near the end of the tour Kunen and Buckley also walked through the Kennedy Space Center's launch control room, with its rows of now-dormant computers. Yet although they took the same bus tour, perhaps sitting only a few rows from one another in air-conditioned comfort, Kunen and Buckley came to dramatically different conclusions regarding its highlights, much as they produced quite contrasting coverage of the Apollo 8 launch.[187]

Buckley, not surprisingly, once again reveled in NASA's technological expertise. After complaining, only partially in jest, that Apollo 8's $361 million price tag did not include free coffee in the viewing area cafeteria for the press, Buckley marveled at the more than five and a half million components that went into NASA's spacecraft. "The figure is very nearly unfathomable," he wrote, "like the venture those parts combine to attempt." He similarly praised the success rate of those parts, noting that when Apollo 7 returned to Earth a little more than two months earlier, NASA engineers and scientists computed its success rate at 99.99 percent, or, as Buckley wrote, "as nearly perfect as our asymptotic universe permits." Technicians at the space agency expected no less for Apollo 8, he noted. "Science has done so much," Buckley concluded, "so then permit us O Lord our optimism."[188]

Kunen, of course, saw the sights rolling by his Greyhound bus window in a slightly different light. First, he once again paid homage to nature by noting that "a bird landed on the launch pad" where Apollo 8 had just taken flight. He then continued to criticize every stop along the tour. At the launchpad, Apollo 8's "mobile service structure," the scaffolding that had guided the Saturn rocket during liftoff and that was still visible, was "an erector set with thyroid trouble." Inside the VAB, the five and a half million parts constituting Apollo 8 suggested not technological achievement, but rather that the technicians responsible for those components knew very little regarding how the spacecraft functioned as a whole. "Of course the same is probably true of Wonder Bread, with a dozen Wonder Scientists each in charge of one 'way,'" Kunen wrote, tongue-in-cheek, of the brand's advertising slogan proclaiming that the bread's added nutrients "build strong bodies 12 ways." Finally, he compared NASA's launch control room, with its "endless line of gray computer consoles," to a graveyard.[189] He then concluded that "to see France's proudest works one goes to the Louvre. In the United States, one tours the Kennedy Space Center."[190] Buckley could not have agreed more, but he would have jettisoned the sarcasm.

Such contrasting coverage of Apollo 8 illustrates that the counterculture and the New Right not only had competing visions regarding space technology and the natural environment, but also a codependent, dysfunctional, love-hate relationship with the space race. For hippies, animosity toward Apollo began in the late 1960s when the counterculture criticized the American space program for being, as Thomas Paine explained, "too square, too disciplined, too rational." Too much a part of the Establishment, in other words. Yet rather than demonstrating in the streets, hippies instead drew cartoons that linked NASA to the military-industrial-complex, wrote poems that associated the space agency with extraterrestrial imperialism, listened to music that accused Apollo of degrading the moon and the Earth, and moved by the tens of thousands to rural communes where they relied on small-scale tools, rather than NASA's expensive and complicated spinoffs, to live more lightly on the land. The result was an alternative lifestyle politics that took aim at NASA and its conservative, flag-waving fans. In the end, though, just as hippies were unable to "rip down that fuckin' flag on the Moon," so, too, did NASA fail to substantially alter the counterculture. The space race did little more than provide creative fodder for the movement's "inner space astronauts."

The romance between the New Right and the space race was more mutual and impactful. Conservative support for Apollo by the likes of Buckley, Rand, and Nixon's silent majority encouraged NASA administrators to publicize their space technology as a "triumph of the squares," to transfer that hardware back down to Earth in Tech House, with its "typical American family," and to promote similar solar technology for suburban homes across the country. Yet in reorienting NASA earthward, conservative praise for the space race also altered the New Right, first and foremost by giving conservatives a tangible point of contrast with the counterculture. The New Frontier, not the New Alchemists, many on the right believed, was the key to America's future. Additionally, although NASA failed to convince suburbanites to install solar energy technology developed for spaceships on the roofs of their homes, it fostered suburban development around the space agency's new research centers and the headquarters of its contractors. While these elaborately landscaped corporate parks, much like GM's Tech Center, masked pollution by separating production from research, the suburban communities they spawned caused environmental problems of their own from water and air pollution to soil erosion and subsidence. Although alarming for hippies, for the New Right such complications were merely the unfortunate but necessary side effect of a surging, free-market aerospace economy.

Yet most important, the space race also dramatically altered the politics of the New Right. This was perhaps most evident in the so-called Southern strategy, a concerted effort by the Republican Party begun during the mid-1960s to capture the political support of the South, which prior to this time had historically voted Democratic. Nixon's 1968 "forgotten American" speech was part of this strategy, as was his appeal to the silent majority. While he and other politicians, including Barry Goldwater, orchestrated the strategy from above, it was enthusiastically joined from below by conservative middle-class whites, residing mostly in suburbs across the South, who opposed much of the recent progress made by those advocating for civil rights, peace in Vietnam, a clean environment, and women's liberation.[191] It was NASA and the aerospace industry that built many of these Southern suburbs, which in turn fueled a growing conservatism based on private property, a lack of public space, and security clearances on the one hand, and class homogeneity, racial exclusion, and strict gender roles on the other.[192] As a result, by the late 1970s the space crescent, which arced across the Sunbelt from Cape Canaveral in the east to Southern California out west, had also become a conservative crescent.

Conclusion

Grounding the Space Race

On June 5, 1970, the day after Apollo 13's James Lovell and John Swigert walked out of *Hair* in midperformance, a NASA spokesperson attempted to place the astronauts' abrupt departure in a broader context. They left Broadway's Biltmore Theater, explained Gene Marionetti to the *New York Times*, not because of the play's nudity or even due to a joke in the first act about a black astronaut. "That gag went by us so fast we missed it," admitted Marionetti, who had accompanied the astronauts to the show. Rather, they left because Lovell and Swigert were offended by the play's treatment of the American flag, which one of the characters used as a blanket before threatening to light it on fire at a be-in. "You've got to remember the plight these men were in a few weeks ago," Marionetti explained, by way of apology, referring to the astronauts' damaged Apollo spacecraft that had hobbled back to Earth in mid-April. "The flag was on the rocket when they left Cape Kennedy and on their shoulder patches during the flight." Those tribulations in space, Marionetti indicated, only amplified their patriotism. "Their country got them back when they were in trouble," he concluded, "and they're grateful."[1]

Those working on the Broadway production had a quite different take on the "*Hair* affair." The play's spokesperson, Mrs. Michael Gifford, noted that both the NASA public relations office and the Apollo 13 astronauts had been fully briefed on the content of the show several weeks before attending the June 4 performance. Gifford was thus understandably perplexed when Lovell and Swigert got up from their seats and bolted for the theater's lobby, where she asked them why they were leaving. When

Swigert answered by explaining his concern for Old Glory up on stage, Gifford countered at first by referencing the ongoing war in Vietnam. "If they can wrap that flag around dead men, I don't see why they can't wrap it around a living one," she stated. When such tactics fell flat, she tried a more traditional appeal. "It doesn't touch the ground," she explained of the flag in the play, "and it's not abused." As we already know, Lovell and Swigert would have none of it, and they quickly hailed a taxi uptown for a drink at the Sherry-Netherland Hotel.[2]

Spokespeople from NASA and for *Hair* were not the only ones expressing differing opinions regarding the Apollo 13 astronauts' midperformance departure. An editorial titled "Flag Respect," which appeared in the Spartanburg, South Carolina, *Herald-Journal* a week after Lovell and Swigert walked out of the Biltmore Theater, also used the incident to describe two contrasting etiquettes regarding the Stars and Stripes. On the one hand were those Americans who viewed the flag as symbolic of a more mainstream citizenship, which the editorial defined through practices such as "keeping informed on the affairs of government, voting and taking part in civic affairs, and giving of one's time and talents to make one's community and nation a better place in which to live." On the opposing side were others, including the creators of and performers in *Hair*, who used the flag instead to criticize what they felt was a corrupt political system. While the writers and cast members of *Hair* viewed their right to use the country's national symbol for such purposes as a centerpiece of American democracy, the *Herald-Journal*, much like Apollo 13's astronauts, found it "galling to see the flag misused and publicly desecrated in exercises of poor taste and bad manners." Not surprisingly, the editorial went on to praise the "quiet protest" registered by Lovell and Swigert, concluding, somewhat wishfully, that "if more people expressed their resentment in such quiet ways, productions like *Hair* would be in for hard times."[3]

But more people didn't, and the play wasn't. After it opened in April of 1968, *Hair* was nominated for two Tony awards and ran for nearly 2,000 performances into the summer of 1972. Even my parents snuck off to Broadway to see the play, and its nude scene, soon after it opened, most probably leaving my younger brother and me with my mother's sister, who was also raising two young sons in our small suburban town about forty-five minutes north of Manhattan. At the time, Martin Luther King Jr. and Robert Kennedy recently had been assassinated, the My Lai massacre and the Tet Offensive had just occurred, and there was buzz about a huge outdoor music festival planned in upstate New York about an hour and a half from our home. Three weeks before that concert, on July 24, 1969, my parents once again took me and my brother to my aunt's house, this time so we could all huddle together around the television set to watch Apollo 11 land on the moon at exactly 8:18 in the evening. Nearly seven hours later, at 2:56 in the morning, my parents woke us up to witness Neil Armstrong taking the first step on the lunar surface. As my mother remembers it, "While all those other things seemed to be happening very far away, the Apollo 11 landing was right in front of us on TV."[4]

My mother's two brothers did not join us that evening around our makeshift television hearth. They lived down in "the City," and although only about five years younger than their older sisters, they seemed to me, even at this early age, a generation apart. In the mid-1960s, for instance, one brother had decided to openly oppose the Vietnam War while studying at Le Moyne College in upstate New York; the anti-war priest Daniel Berrigan not only taught at Le Moyne and had several personal conversations with my uncle, but David Miller, the first American to protest the war by burning his draft card, was my uncle's classmate and friend. The other brother, the baby in their family, was between his sophomore and junior years at Holy Cross in Worcester, Massachusetts, during that summer of 1969. With several close friends in the mid-August heat, he piled into a station wagon and drove to a farm in upstate New York, where they got stoned for three days, sloshed around in the mud, and listened to Jefferson Airplane, John Sebastian, and The Who. "I wanted to go to Woodstock, not the moon," he explained. "Apollo pissed me off!" Perhaps because of such sentiment, he could not recall where he was at the exact moment Armstrong made history by taking "one giant step for mankind." He did remember, however, that he was "against *Hair*," which he refused to see because it was "trying to make a buck off the hippies."[5]

In the midst of this family dynamic, I often felt caught between the two worlds of my mother's generation. As a youngster born in 1964 and growing up in the suburbs, I remained woefully unaware of the enduring political fallout of the 1960s era; of the Vietnam War winding down, the women's movement gearing up, and the continuation of racial injustice in inner cities and pollution across the countryside. Yet such youthful ignorance subsided, just a bit, each time my two uncles visited for the holidays. Although at the time I was still too young to fully comprehend why, there was something about the way they dressed and the length of their hair, the friends and partners they brought along, and the heated discussions during dinner, especially after one too many cocktails, that brought the far-off issues of the 1960s into my suburban world of the 1970s. It was those dinner conversations that encouraged me to be a different sort of environmental historian, one who connects stories about nature to broader, more mainstream narratives in American history. Such table talk also shaped this book, and convinced me, on an almost instinctive level, that the race to the moon and the political problems back on Earth were interrelated, all part of the same family dialogue.

Americans during the 1960s era had similar conversations at dinner tables across the country, as well as in damp fields outside the gates of the Kennedy Space Center, on college campuses from MIT and Radcliffe to Stanford, and even in the mud at Woodstock. In the course of such discussions, and through the actions they sparked, the space race transformed these social and political movements. This impact was less pronounced during the early to mid-1960s, when Apollo's popularity allowed NASA administrators to turn a deaf ear to the criticism of these political activists. Yet such conscious disregard became increasingly difficult later in the decade, and nearly impossible during the early 1970s, as NASA fatigue set in and popular support for space exploration evaporated. It was then that administrators, engineers, scientists, and even some astronauts began listening to activists gathered at the national dinner table and reacting to them in order to salvage NASA's future.

In some instances, as with the civil rights movement and the counterculture, the space agency's engagement with grassroots politics was shallow, inadequately funded, highly publicized, and, in the end, ineffective. While the space agency ignored hippies, it did little more than provide civil rights activists with a global platform from which to voice their concerns. For other movements, however, the space race was more consequential. Administrators, scientists, and engineers at NASA produced cultural icons and scientific data that unified and strengthened environmentalism; provided second-wave feminism, albeit reluctantly, with new leaders, idols, and discourses that promoted gender equality through biological difference; and fostered the spread of the New Right's political ideology, quite unintentionally, across the Sunbelt South. The space agency's impact on the anti-war movement was different, yet again. Although NASA merely provided the New Left with yet another anti-war target, the space agency's deployment of Landsat technology across the developing world helped the U.S. government project its soft power abroad in more subtle, and ultimately more intrusive, ways. Thus while NASA was rarely, if ever, proactive when it came to these grassroots movements, its unintentional, belated, and often superficial efforts nevertheless became unexpected drivers of broader social change during the 1960s era.

By angering and then engaging these grassroots movements, the space race and NASA also inadvertently deepened social and political divisions within American society. There were many additional causes of this civic fragmentation, including the growing popularity during the 1970s of laissez-faire economic thinking, promoted by the likes of Buckley and Rand during their visits to Cape Canaveral, which endorsed individual liberty within the free market over social responsibility. For many Americans the civil rights and women's movements similarly weakened ties to a national culture by divvying up citizens by race and gender, as did the hippies, who rejected the mainstream "establishment" in favor of more personal lifestyle politics. "Imagined collectives shrank; notions of structure and power thinned out," argued one historian of the 1970s. The result was an "era of disaggregation, a great age of fracture."[6] Apollo during the Age of Aquarius encouraged such identity politics and thus aggravated this social splintering.

It was also no coincidence that the age of fracture overlapped with the era of the *Whole Earth* photograph. When in the late 1960s NASA began releasing to the public this and other images of the planet taken from space, they first became symbols of global unity. As already noted, Archibald MacLeish's short essay, written in December of 1968 after viewing a photograph taken by Apollo 8, spearheaded this notion of worldwide brotherhood.[7] Yet the astronaut's-eye view presented in these photographs also "flattened" Earth, obscuring important differences, discriminations, and disastrous problems affecting various groups of people living closer to the ground.[8] Grassroots activists in the 1960s era thus viewed *Whole Earth* and its photographic brethren with trepidation. The space agency's new global gaze, in other words, which often depicted white, wispy clouds encircling the Earth, threatened to obscure the racism, sexism, war, and pollution that divided communities across the planet.[9]

The acknowledgment of such divisions was part of a much broader conversation taking place at dinner tables across the country, including my own in the suburbs of New York, regarding the overall present and future course of the United States. For many hailing from Middle America,

Apollo illustrated the country's ability not only to agree on a specific national purpose—landing men on the moon by decade's end—but also to achieve it. Apollo 13's Lovell admitted as much. "It was a goal that you could go outside and see the moon and know that the United States had finally put a spacecraft around the moon," he reminisced years after walking out of the Biltmore Theater. It was "an achievement that everybody could look up to."[10] Well, perhaps not everybody. "I'm dissatisfied with the program," wrote J. N. Cooper in a letter to the editor of the *New York Times* in early January of 1969, one week after Lovell orbited the moon in Apollo 8. Such disappointment stemmed from the country's decision to spend ten long years on "this sort of glittering bauble of technological problem-solving" when there are so many other issues in need of resources. "For the really important national objectives," Cooper concluded, such as solving inner-city poverty, halting the Vietnam War, cleaning up our polluted environment, and providing equality for women, "we have hardly begun to define the questions."[11]

This national debate about America's national purpose, and the central role played by NASA and the space race within it, became widespread around the Apollo 11 lunar landing, and it took visual form in nearly one dozen editorial cartoons.[12] One of the most poignant examples was an illustration by Franklin Morse that appeared in the Los Angeles *Herald-Examiner* (see Figure C.1).[13] In it, Morse placed an obviously distraught Uncle Sam, dressed in what would no doubt be red and white striped pants if the cartoon were in color, directly between the two competing camps in this national dispute. Up above his head floated a crater-pocked moon with the words "U.S. Space Feats" written across its dark side, symbolizing what most Americans had wanted back in the early 1960s and what conservatives continued to celebrate during the summer of 1969. Below Uncle Sam's ladder was a quite different scene, represented by a sea of protest signs. The "URBAN CRISIS" of the civil rights movement, as well as other placards demanding that the federal government "END THE WAR" and halt "POLLUTION" together illustrated a quite different national priority. For Americans anxious about these issues, "HUMAN NEEDS" on Earth, also represented by a protest sign at the base of the stepladder, were much more important than the technological possibilities of outer space. Uncle Sam's frustrated query, "Didn't I Promise You the Moon?" neatly identifies the space race of the 1960s era as a catalyst for this disagreement over what kind of nation America was and wanted to be.

C.1 Franklin Morse, "Didn't I Promise You the Moon?," (Los Angeles) *Herald-Examiner*, May, 20, 1969.

As the space race transformed the grassroots political movements of the 1960s era, civil rights demonstrators, anti-war protestors, environmentalists, feminists, and hippies in turn altered the space race. This influence was possible because the leaders and activists of these disparate movements had a shared adversary in NASA; had successfully borrowed from one another similar strategies, including marches, sit-ins, boycotts, and letter-writing campaigns; and, perhaps most important, had embraced a common political language.[14] As already noted, for Ralph Abernathy this meant pressuring NASA technicians to "find ways to use their skills to tackle problems we face in society."[15] Students of the New Left, such as Paul Bunge, argued similarly, wondering why America was sending men to the moon when the Earth is "dying from hunger and war?"[16] Environmentalists also suggested that the United States turn its sophisticated space technology, as Stewart Brand explained, "180 degrees to look back," while Gloria Steinem noted that she and other feminists wished the nation would "conquer more inner space along with outer space."[17] Abbie Hoffman and the hippies, always wary of the established order, promoted a similar idea by taking matters into their own hands. Woodstock, he argued, "was the first attempt to land a man on the earth."[18] Thus while the civil rights, anti-war, environmental, feminist, and counterculture movements each had a distinct set of concerns, from racial and gender inequality to war and pollution, when it came to the space race the activists in these movements spoke with one voice: turn Apollo's space technology back around toward Earth!

In the late 1960s and early 1970s, as popular support for space exploration plummeted, NASA finally acquiesced, with mixed results. It began by retooling space technology, such as that used to heat and cool capsules in outer space, in an attempt to improve the living environment of African Americans in inner cities. It next redirected satellites to allow developing countries such as Vietnam to better manage their own natural resources, and it refocused space cameras and reformatted powerful computers to create both iconic images and environmental data that helped activists as well as scientists protect and assess the ecological health of the big blue marble. In the late 1970s the space agency even updated environmental simulators and spacesuits, originally designed to protect its all-male astronaut corps against the hazards of outer space, to accommodate women, such as Sally Ride, who had been grounded by sexism back on Earth. Finally, NASA tried, unsuccessfully, to reflect solar energy from its spaceships down onto homes along America's suburban frontier. In each of these efforts, space technology and earthbound nature together nurtured a new politics of space exploration.

The success of these grassroots movements in pressuring NASA to reorient much of its hardware toward Earth had a lasting, if fleeting, impact beyond the 1960s era. This became most evident in 1987, when Ride, who at the time was serving as a deputy administrator at NASA, presented to Congress the space agency's plans for Mission to Planet Earth. The proposed mission would include the development of an enormous orbiting observation system for measuring the environment of our home planet, as well as a sophisticated data and information system to predict future ecological changes on a global scale. "The scientific community has concluded that Earth is in the process of global change," explained Ride, "and scientists now believe that it is necessary to study the Earth as a synergistic system."[19] Two years later on the twentieth anniversary of the Apollo 11 lunar landing, when George H. W. Bush announced that his administration would undertake NASA's new mission, he sounded more like a 1960s activist than a Republican president. "Mission to Planet Earth," he explained, "reminds us of what the astronauts remember as the most stirring sight of all." Not the moon or the stars, Bush noted, but "the Earth—a tiny, fragile, precious, blue orb—rising above the arid desert of Tranquility Base."[20]

Unfortunately, much like NASA's efforts during the late 1960s and 1970s to address the concerns of the civil rights, anti-war, environmental, feminist, and counterculture movements, Mission to Planet Earth also fell short. Although the Bush administration announced in 1989 that the federal government would fund NASA's new mission with between $15 and $30 billion over the course of fifteen to twenty years, just four years later this budget had dwindled to a mere $8 billion over less than a decade.[21] Moreover, while NASA's efforts in the 1970s had at least attempted to train the agency's technology on alleviating the problems of racial discrimination, gender inequality, and war, Mission to Planet Earth was solely an environmental program. "This initiative," Bush explained, seeks "new solutions for ozone depletion and global warming and acid rain."[22]

More recently, the federal government and NASA have further relinquished responsibility for addressing these broader social problems by encouraging the privatization of space exploration. Although this shift began legislatively in 2004 with the Commercial Space Launch Amendments Act, which permitted private companies to fund and execute space exploration, its intellectual antecedents were evident as far back as the late 1960s in the free-market thinking of Buckley and Rand; both specifically praised NASA's nongovernmental contractors. Today it is the *Wall Street Journal*, the Cato Institute, and Buckley's *National Review* that lead this charge.[23]

"Private industry routinely takes technologies pioneered by the government—like air mail, computers and the Internet—and turns them into affordable, reliable and robust industries," argued the *Journal* in 2010. A similar privatization of space exploration would cultivate a "competitive market that will drive down the cost of getting you and me into orbit."[24] William Gerstenmaier, an associate NASA administrator, agreed, noting in 2015 that the space agency encourages "a private space station that is driven primarily by fundamental or basic research coming from the private sector." Administrators at NASA, he concluded, "can just buy these services."[25]

Perhaps, but such calls for privatizing the "space business" completely overlook the important history lessons of Apollo during the 1960s era. By outsourcing space exploration and removing it from the halls of government, we also distance it from the public sphere. During the 1960s this was definitely not the case. A damp field at Cape Canaveral, the hallways and elevator bays of the Johnson Space Center, the streets of Manhattan during an astronaut ticker-tape parade, and even a rock concert on a farm in upstate New York, all became town squares, of sorts, where grassroots activists rose up on their soapboxes and spoke out against the space race. Because of such public pronouncements in these public spaces, NASA, a civilian agency funded with taxpayer dollars, was forced to listen and pressured to respond. Private companies such as Space X and Virgin Galactic will be much more hard of hearing when it comes to a clamoring public. Ceding space exploration to such private interests thus jeopardizes our ability to influence federal policy and weakens the government's constitutional obligation to us, its citizens.[26] Doing so might also make it more difficult to look back and examine our own planet.

President Johnson was thus correct when he declared, less than a year after the *Hair* affair, that "space was the platform from which the social revolution of the 1960s was launched."[27] He failed to acknowledge, however, that this social revolution in turn brought the space race back down to Earth. While this technological turnaround left unresolved many of the problems associated with racial discrimination, gender inequality, unjust warfare, and an endangered environment, it both reflected the robust civic culture of the 1960s era and also strengthened it by sparking a national debate about national purpose. The space race, in other words, and grassroots opposition to it, invited more guests to the country's dinner table. The conversations that followed, like those taking place in my childhood home during the holidays, while often loud and contentious, ended with dessert and some semblance of deeper understanding. In our present age of bitter partisanship, government gridlock, and a host of new threats to both our planet and its diverse inhabitants, we would do well to remember Apollo in the Age of Aquarius.

NOTES

ACKNOWLEDGMENTS

ILLUSTRATION CREDITS

INDEX

Notes

Introduction

1. "Astronauts Find 'Hair' Offensive: Lovell and Swigert Walk Out after First Act of Musical," *New York Times*, 6 June 1970, 22.

2. This cultural divide was alluded to in "Astronauts Find 'Hair' Offensive." For additional news coverage that placed Apollo at the center of a gap between Middle America and 1960s liberals, see "The Moon and 'Middle America,'" *Time* 94, no. 5 (August 1969): 3; Ernest B. Furgurson, "Sons of the Forgotten," *Baltimore Sun*, 22 July 1969, 9; and Peter Collier, "Apollo 11: The Time Machine," *Ramparts*, October 1969, 56.

3. "Moon and 'Middle America.'"

4. There is an enormous literature linking the space race to the emergence of a global perspective. See especially Denis Cosgrove, "Contested Global Visions: One-World, Whole-Earth, and the Apollo Space Photographs," *Annals of the Association of American Geographers* 84, no. 2 (June 1994): 270–294; Sheila Jasanoff, "Image and Imagination: The Formation of Global Environmental Consciousness," in *Changing the Atmosphere: Expert Knowledge and Environmental Governance*, ed. Clark A. Miller and Paul N. Edwards (Cambridge, MA: MIT Press, 1996), 309–336; and Robin Kelsey, "Reverse Shot: Earthrise and Blue Marble in the American Imagination," in *New Geographies 4: Scales of the Earth* (Cambridge, MA: Harvard University Press, 2011), 10–16. Walter McDougall argues persuasively that this global perspective accompanied a new "technocracy" in the United States that augmented federal power at the expense of local freedoms. See McDougall, *The Heavens and the Earth: A Political History of the Space Age* (New York: Basic Books, 1985), 3–12. For a brief discussion of the role of the space race in this global turn, see James A. Vedda, "The Role of Space Development in

Globalization," in *Societal Impact of Spaceflight*, ed. Steven J. Dick and Roger D. Launius (Washington, DC: NASA History Division, 2007), 193–205; and Martin Collins, "The 1970s: Spaceflight and the Problem of Historically Interpreting the In-Between Decade," in *Limiting Outer Space: Astroculture After Apollo*, ed. Alexander Geppert (New York: Palgrave Macmillan, 2017): 35–55. On anxieties and concerns regarding globalization, see also Niall Ferguson, Charles Maier, Erez Manela, and Daniel Sargent, eds., *The Shock of the Global: The 1970s in Perspective* (Cambridge, MA: Belknap Press of Harvard University Press, 2010).

5. I coined this phrase and use it throughout the book to denote a drop in popular support for Apollo, which actually began in 1968 when Apollo 7's ten-day mission convinced Americans that the United States would beat the Russians to the lunar surface. See Herbert Krugman, "Public Attitudes toward the Apollo Space Program, 1965–1975," *Journal of Communication* (Autumn 1977): 87–93; and Stephanie A. Roy, Elaine C. Gresham, and Carissa Bryce Christensen, "The Complex Fabric of Public Opinion on Space," *Acta Astronautica* 47, issue 2–9 (July–November 2000): 665–675. On cuts to NASA's budget by Congress, see "The Plight of Apollo," *Newsweek*, 14 September 1970, 97.

6. The literature on the rise of identity politics during the 1960s era and the subsequent perception of increasing fragmentation within American society is vast. See especially Daniel T. Rodgers, *The Age of Fracture* (Cambridge, MA: Belknap Press of Harvard University Press, 2011); Mark Hamilton Lytle, *America's Uncivil Wars: The Sixties Era from Elvis to the Fall of Richard Nixon* (New York: Oxford University Press, 2006); and Bruce Schulman, *The Seventies: The Great Shift in American Culture, Society, and Politics* (New York: Free Press, 2001). For an overview of this history, see Cressida Heyes, "Identity Politics," in *The Stanford Encyclopedia of Philosophy* (Spring 2013 edition), ed. Edward N. Zalta, online at http://stanford.library.usyd.edu.au/archives/spr2013/entries /identity-politics/#Bib (accessed 5 May 2016).

7. Norman Mailer, *Of a Fire on the Moon* (London: Weidenfeld and Nicolson, 1970), 57.

8. E. F. Schumacher, *Small Is Beautiful: Economics as If People Mattered* (London: Blond and Briggs, 1973); and Lewis Herber (a.k.a. Murray Bookchin), *Our Synthetic Environment* (New York: Knopf, 1962). There is a rich literature on the history of the appropriate technology movement. See especially Fred Turner, *From Counterculture to Cyberculture: Stewart Brand, the Whole Earth Network, and the Rise of Digital Utopianism* (Chicago: University of Chicago Press, 2006); and Andy Kirk, *Counterculture Green: "The Whole Earth Catalogue" and American Environmentalism* (Lawrence: University Press of Kansas, 2007). See also Jordan Kleiman, "The Appropriate Technology Movement in American Political Culture" (Ph.D. dissertation, University of Rochester, 2000).

9. On this problem in women's history, see Joan Scott, "Gender: A Useful Category of Historical Analysis," *American Historical Review* 91, no. 5 (December 1986): 1053–1075.

10. The historian Mark Fiege has also championed an environmental history that analyzes broader, more mainstream historical moments and events. See his wonderful *Republic of Nature: An Environmental History of the United States* (Seattle: University of Washington Press, 2013). For less explicit calls for a more broad-based environmental history, see Theodore Steinberg, "Nature, Agency, and Power in History," *American Historical Review* 107, no. 3 (June 2002); and Adam Rome, "What Really Matters: Environmental Perspective on Modern America," *Environmental History* 7 (2002).

11. On this shift from technological determinism to the social construction of technology, see Wiebe E. Bijker, Thomas Park Hughes, and T. J. Pinch, *The Social Construction of Technological Systems: New Directions in the Sociology and History of Technology* (Cambridge, MA: MIT Press, 1987); Merritt Roe Smith and Leo Marx, *Does Technology Drive History? The Dilemma of Technological Determinism* (Cambridge, MA: MIT Press, 1994); and Langdon Winner, *The Whale and the Reactor: A Search for Limits in an Age of High Technology* (Chicago: University of Chicago Press, 1986).

12. Asif Siddiqi, "American Space History: Legacies, Questions, and Opportunities for Further Research," in *Critical Issues in the History of Spaceflight*, ed. Steven J. Dick and Roger D. Launius (Washington, DC: NASA History Division, 2006), 463–464, 480.

13. While the periodization of the 1960s era remains fluid among scholars, most agree that the era seeped beyond the decade proper. For expansive definitions of the period, see Lytle, *America's Uncivil Wars*, 1–12; Arthur Marwick, *The Sixties: Cultural Revolution in Britain, France, Italy, and the United States, c. 1958–c. 1974* (New York: Oxford University Press, 1998); and, to a lesser extent, Peter Braunstein and Michael William Doyle, "Introduction: Historicizing the American Counterculture of the 1960s and 1970s," in *Imagine Nation: The American Counterculture of the 1960s and 1970s*, ed. Peter Braunstein and Michael William Doyle (New York: Routledge, 2002), 5–14. For examples that trace the impact of the "long 1960s" into the 1970s, see Simon Hall, "Protest Movements in the 1970s: The Long 1960s," *Journal of Contemporary History* 43, no. 4 (October 2008): 655–672; and Dan Berber, ed., *The Hidden 1970s: Histories of Radicalism* (New Brunswick, NJ: Rutgers University Press, 2010).

14. Although the Apollo program officially ran from 1961 through 1972, and then continued into the mid-1970s with the Apollo-Soyuz Test Project, throughout this book I use the terms "Apollo" and "Apollo era" broadly to represent the period within NASA's history when the agency was focused on reaching the moon. I

thus include in my discussion of the Apollo era the missions of Project Mercury and Gemini. For an example of scholarship that similarly defines the Apollo era broadly, see Arnold S. Levine, *Managing NASA in the Apollo Era* (Washington, DC: NASA History Series, 1982).

15. This apology is mentioned in "Astronauts Find 'Hair' Offensive."

16. For an introduction to the concept of "envirotech," see the Envirotech Special Interest Group web page at http://www.envirotechweb.org/(accessed 27 April 2016); and Dolly Jorgensen, Finn Arne Jorgensen, and Sara Pritchard, eds., *New Natures: Joining Environmental History with Science and Technology Studies* (Pittsburgh: University of Pittsburgh Press, 2013), especially Sara B. Pritchard, "Joining Environmental History with Science and Technology Studies: Promises, Challenges, and Contributions," 1–20. On the idea of technopolitics, see Gabrielle Hecht and Paul N. Edwards, "The Technopolitics of Cold War: Toward a Transregional Perspective," in *Essays on Twentieth-Century History*, ed. Michael Adas (Philadelphia: Temple University Press, 2010), 271–314; and Gabrielle Hecht, *Radiance of France: Nuclear Power and National Identity after World War II* (Cambridge, MA: MIT Press, 2009), especially "Introduction," 1–20.

17. For a brief introduction to political ecology, see Paul Robbins, *Political Ecology: A Critical Introduction* (New York: Blackwell, 2004), especially "Part I: What Is Political Ecology"; and Paul Robbins, "Cultural Ecology," in *A Companion to Cultural Geography*, ed. James Duncan, Nuala C. Johnson, and Richard H. Schein (New York: Blackwell, 2004).

18. An important exception is Sara B. Pritchard, "An Envirotechnical Disaster: Nature, Technology, and Politics at Fukushima," *Environmental History* 17 (April 2012): 219–243.

19. This language was used by Langdon Winner to describe the audience of the *Whole Earth Catalog*. See Langdon Winner, "Building the Better Mousetrap: Appropriate Technology as a Social Movement," in *Appropriate Technology and Social Values—A Critical Appraisal*, ed. Franklin A. Long and Alexandra Oleson (Pensacola, FL: Ballinger, 1980), 31.

20. For these references, see "Walking in Space" and "Exanaplanetooch" in Gerome Ragni and James Rado, *Hair: The American Tribal Love-Rock Musical* (New York: Simon and Schuster, 1969), 143–155, 191–193.

21. Lyndon B. Johnson, *Vantage Point: Perspectives of the Presidency, 1963–1969* (New York: Holt, Rinehart and Winston, 1971), 285.

1. Spaceship Earth

1. Julian Scheer, "The 'Sunday of the Space Age,'" *Washington Post*, 8 December 1972, A26. For other descriptions of the gathering, see also Richard Lewis, "All Is Go for Trip to Moon," *Chicago Sun-Times*, 16 July 1969, 5; Bernard Weinraub,

"Hundreds of Thousands Flock to Be 'There,'" *New York Times*, 16 July 1969, 22; United Press International, "Space Head Tells Poor to Unite behind Apollo," *Boston Globe*, 16 July 1969, 12.

2. United Press International, "Space Head Tells Poor to Unite behind Apollo."

3. Scheer, "'Sunday of the Space Age.'"

4. Ibid.

5. William Greider, "Protesters, VIPs Flood Cape Area," *Washington Post*, 16 July 1969, A1.

6. Scheer, "'Sunday of the Space Age.'"

7. On Paine's impressive corporate career, see NASA, "Thomas O. Paine," online at http://history.nasa.gov/Biographies/paine.html (assessed 6 June 2016).

8. On Abernathy's early life and civil rights activism, see Ralph David Abernathy, *And the Walls Came Tumbling Down: An Autobiography* (New York: Harper & Row, 1989). On the Poor People's Campaign, see Gerald D. McKnight, *The Last Crusade: Martin Luther King, Jr., the FBI, and the Poor People's Campaign* (Boulder, CO: Westview, 1998).

9. For the history of NASA Headquarters in Washington, DC, which in 1968 was housed behind the current National Air and Space Museum, see Elizabeth Suckow (Archivist, NASA History Division) and Chris Jedrey (Executive Director, Office of Headquarters Operations), "Hidden Headquarters," 24 March 2009, online at http://hqoperations.hq.nasa.gov/docs/Hidden_Headquarters_March_24 _2009.pdf (accessed 6 June 2016).

10. For historiographical essays on this rich scholarship, see especially Jacquelyn Dowd Hall, "The Long Civil Rights Movement and the Political Uses of the Past," *Journal of American History* 91, no. 4 (2005): 1233–1263; Charles W. Eagles, "Toward New Histories of the Struggle for Black Equality since 1945," in *A Companion to Post-1945 America*, ed. Jean-Christophe Agnew and Roy Rosenzweig (Malden, MA: Wiley-Blackwell, 2002): 211–234; and Steven F. Lawson, "Freedom Then, Freedom Now: The Historiography of the Civil Rights Movement," *American Historical Review* 96, no. 2 (April 1991): 456–471. For a discussion of scholarship on the transnational history of the civil rights movement, see Mary L. Dudziak, *Cold War Civil Rights: Race and the Image of American Democracy* (Princeton, NJ: Princeton University Press, 2000).

11. For examples of this borrowing, see Todd Gitlin, *The Sixties: Years of Hope, Days of Rage* (New York: Bantam, 1987), especially chapter 6, 127–170; and David S. Meyer and Nancy Whittier, "Social Movement Spillover," *Social Problems* 41, no. 2 (May 1994): 277–298.

12. This paucity of scholarship works both ways; civil rights scholars have not analyzed the role of the space race in influencing the struggle for black freedom while space historians have similarly refrained from exploring the role of the civil rights movement in shaping the space race. The few exceptions include Kim

246 Notes to Pages 13–16

McQuaid, "'Racism, Sexism, and Space Ventures': Civil Rights at NASA in the Nixon Era and Beyond," in *Societal Impact of Spaceflight*, ed. Steven J. Dick and Roger D. Launius (Washington, DC: NASA Office of External Relations, 2007), 421–449; Steven L. Moss, "NASA and Racial Equality in The South, 1961–1968" (unpublished M.A. thesis, Texas Tech University, December 1997); and Lynn Spigel, "Outer Space and Inner Cities: African American Responses to NASA," in Lynn Spigel, *Welcome to the Dreamhouse: Popular Media and Postwar Suburbs* (Durham, NC: Duke University Press, 2001), 141–182.

13. There is a rich literature on the politics of technology. For an example of more recent work, see Gabrielle Hecht and Paul N. Edwards, "The Technopolitics of Cold War: Towards a Transregional Perspective," American Historical Association series in *Global and Comparative History* (2007). A classic along these lines remains Langdon Winner, "Do Artifacts Have Politics?," *Daedalus* 109, no. 1 (Winter 1980): 121–136, reprinted in *The Social Shaping of Technology*, ed. Donald MacKenzie and Judy Wajcman (London: Open University Press, 1985; 2nd ed., 1999).

14. For a brief discussion of these competing views, see David Nye, *American Technological Sublime* (Cambridge, MA: MIT Press, 1994), 225–256. See also Lewis Mumford, *The Pentagon of Power: The Myth of the Machine* (New York: Harcourt Brace Jovanovich, 1964), especially chapter 11.

15. Scheer, "'Sunday of the Space Age.'"

16. On NASA's Apollo rockets as symbols of power, see David Nye, *American Technological Sublime* (Cambridge, MA: MIT Press, 1996), 237–241; and Mumford, *Pentagon of Power*, 180. This distinction between Abernathy's mules and Paine's rocket was mentioned in "Lift for Poor Urged at Lift-Off," *Baltimore Sun*, 17 July 1969, A9.

17. The quote can be found in Gladwin Hill, "Apollo: Spacecraft Is Taking Shape," *New York Times*, 2 August 1962, 1.

18. "Project Apollo-1: The Saturn Launch Vehicle," *Evening Star* (Washington, DC), 12 May 1969, clipping from National Air and Space Museum (NASM) Archives Division, Technical Reference Files, Space History Series, Folder: 0A-250030-01 Project Apollo, Articles, 1969.

19. Cornelius Ryan, ed., "Man's Survival in Space: Picking the Men," *Colliers* (28 February 1953): 42.

20. Editorial, "History Is Watching," *Huntsville (AL) Times*, 21 December 1968.

21. Editorial, "We Can Do What We Set Our Minds to Do," *Miami News*, 22 July 1969, 14A.

22. For a full discussion of this symposium, see Frieda B. Taub, "Closed Ecological Systems," *Annual Review of Ecology and Systematics* 5 (1974): 151–152. For a transcript of Jack Myers's opening remarks at the symposium, see Jack Myers, "Introductory Remarks," *American Biology Teacher* 25, no. 6 (October 1963): 409–411.

23. In particular, Odum argued that no man-made mechanical system could be more efficient in converting photosynthetic power than natural ecosystems, which had evolved over millions of years. For a transcript of Odum's symposium paper, see Howard T. Odum, "Limits to Remote Ecosystems Containing Man," *American Biology Teacher* 25, no. 6 (October 1963): 429–430, especially 438.

24. All of the presentations made at the "Space Biology: Ecological Aspects" symposium were later published in two special issues of the *American Biology Teacher* 25, nos. 6 and 7 (October–November 1963).

25. Eugene Odum's discussion of Howard Odum's research on biologically regenerative systems at the 1963 Princeton conference is mentioned by Taub, "Closed Ecological Systems," 152.

26. On this statement at the 1963 Princeton conference by Eugene Odum, see the discussion titled "Regenerative Systems," in *Human Ecology in Space Flight: Proceedings of the First International Interdisciplinary Conference*, ed. Doris Howes Calloway (New York: New York Academy of Sciences Interdisciplinary Communications Program, 1966), 86, 88.

27. Of course not all scientists and engineers conducting bioregenerative life support research received NASA funding. On the space agency's funding for such research, see the NASA biologist Dale W. Jenkins, "Bioregenerative Life-Support Systems," in *Bioregenerative Systems: A Conference Held in Washington, D.C., November 15–16, 1966* (Washington, DC: NASA, 1968), 1. On the number of engineers and scientists in 1966 conducting similar research, see Jack Myers, "Space Biology," *American Biology Teacher* 25, nos. 6 and 7 (October–November 1963): 409. Such historical evidence, which clearly shows that ecologists were involved in professional discussions and scientific research sponsored by NASA as early as 1963, refutes Peder Anker's claim that ecologists did not participate in the engineering of the first space cabins. See Anker, "The Ecological Colonization of Space," *Environmental History* 10, no. 2 (April 2005): 243.

28. Very little has been written about Odum and Beyers's NASA grant. Frieda Taub mentions it only in passing in Taub, "Closed Ecological Systems," 153. Here I rely on E. P. Odum and R. J. Beyers, "Biodynamics of Microecosystems, First Semi-annual Report on NASA Grant NSG-706/11-003-001, Report to the National Aeronautics and Space Administration, March 1, 1965," Document ID 19650071169, Accession Number 65N83570, Report/Patent Number NASA-CR-62107, Contract/Grant/Task Number NSG-706; and Odum and Beyers, "Biodynamics of Microecosystems, Final Report on NASA Grant NSG-706, Report to the National Aeronautics and Space Administration, February 15, 1967," Document ID 19670016545, Accession Number 67N25874, Report/Patent Number NASA-CR-83884, Contract/Grant/Task Number NSG-706.

29. G. Denis Cooke, Robert J. Beyers, and Eugene P. Odum, "The Case for the Multispecies Ecological System, with Special Reference to Succession and Stability," in *Bioregenerative Systems*, 131–133.

30. Ibid., 136.

31. For examples of this language, see Howard T. Odum, "Limits to Remote Ecosystems Containing Man," 431, 439, 440, and 441.

32. For these comments in response to the paper by Odum, Beyers, and Cooke by Kraus, Repaske, and Kok, see "Comments," in *Bioregenerative Systems*, 138–139.

33. For an example of a similar conference promoting an engineering approach to space capsule life support systems, see W. B. Cassidy, ed., *Bioengineering and Cabin Ecology: Proceedings of a Symposium Sponsored by the American Astronautical Society and American Association for the Advancement of Science, Held December 30, 1968, Dallas, Texas* (Tarzana, CA: American Astronautical Society, 1969), volume 20 in the AAS Science and Technology Series.

34. For a detailed description of how the Environmental Control Subsystem regulated air, water, waste, and "climate" within the Apollo spacecraft, see "Apollo Spacecraft News Reference," National Air and Space Museum Archives Division, Technical Files, Space History Series, Folder: 0A-250500-01, Apollo Project, Plans and Training Manual, NASM, Washington, DC, 149–151.

35. After the tragic Apollo 1 launch-pad fire, NASA abandoned the practice of pressurizing the space capsule during launch with 100 percent pure oxygen, which created a highly flammable environment, and replaced it with a two-gas mixture of 60 percent oxygen and 40 percent nitrogen. For a discussion of the role of pure oxygen in the Apollo 1 fire, see *Aviation Week and Space Technology*, 6 February 1967, 29–36; and *Aviation Week and Space Technology*, 13 February 1967, 33–36.

36. For the shorter trip from the lunar orbit to the moon's surface, the lunar module transported water from Earth and stored it in tanks. For a detailed description of NASA's "Water Management Section," see "Apollo News Reference: Environmental Control, Quick Reference Data," Grumman Aircraft Engineering Corporation, EC-11–EC-13, located in the National Air and Space Museum Archives Division, Technical Reference Files, Space History Series, Folder: 0A-251220-01, Apollo Lunar Module (LEM, LM) Grumman Materials, NASM, Washington, DC, EC-4.

37. Sue Butler, "Of Men and Missiles: No Name for 'High C' Boys," *Daytona Beach (FL) Sunday News-Journal*, 15 August 1965, 7A.

38. On the "Atmosphere Revitalization Section" of NASA's Environmental Control Subsystem, see "Apollo News Reference: Environmental Control, Quick Reference Data," Grumman Aircraft Engineering Corporation, EC-11–EC-13, located in the National Air and Space Museum Archives Division, Technical Reference Files, Space History Series, Folder: 0A-251220-01, Apollo Lunar Module (LEM, LM) Grumman Materials, NASM, Washington, DC, EC-79. On the devel-

opment of these pneumatic toilets, see Amy Foster, "Sex in Space: The Politics and Logistics of Sexually Integrating NASA's Astronaut Corps" (unpublished Ph.D. dissertation, Auburn University, 2005), 176–178.

39. On the Heat Transport Section of NASA's Environmental Control Sub-system, see "Apollo News Reference: Environmental Control, Quick Reference Data," Grumman Aircraft Engineering Corporation, EC-1–EC25, located in the National Air and Space Museum Archives Division, Technical Reference Files, Space History Series, Folder: 0A-251220-01, Apollo Lunar Module (LEM, LM) Grumman Materials, NASM, Washington, DC, EC-13–EC-16.

40. Dr. Eugene B. Koneccl, "Human Factors Will Determine Space Flight Success," *New York Herald Tribune*, 28 June 1959, 116.

41. For examples of *American Biology Teacher* publishing articles by ecologists on space capsule life support systems, see "Space Biology: Part 1," special issue, *American Biology Teacher* 25, no. 6 (October 1963): 409–449; and "Space Biology: Part 2," special issue, *American Biology Teacher* 25, no. 7 (November 1963): 489–538. For examples of ecologists making this argument in books aimed at popular audiences, see Howard Odum, *Environment, Power, and Society* (Chapel Hill: University of North Carolina Press, 1971), 125; and Howard Odum and Elisabeth Odum, *Energy Basis for Man and Nature* (New York: McGraw-Hill, 1981), 116.

42. Dennis Cooke, "Ecology of Space Travel," in *Fundamentals of Ecology*, 3rd ed., ed. Eugene Odum (Philadelphia: Saunders, 1971), 498. Odum obviously approved of this language; not only was he the editor of the textbook, but Cooke was also one of his graduate students.

43. James Lovelock, *The Ages of Gaia* (New York: W. W. Norton, 1988), as quoted in Wolfgang Sachs, "The Blue Planet: An Ambiguous Modern Icon," *Ecologist* 24, no. 5 (September–October 1994): 173. For examples of Lovelock using the term "Spaceship Earth," see James Lovelock, *Gaia: A New Look at Life on Earth* (New York: Oxford University Press, 1979), 124.

44. John G. Mitchell, "The Good Earth?," *Newsweek*, 7 July 1969, 59.

45. For examples of reporters and editorialists using Apollo space capsules to reference the ecological health of "Spaceship Earth," see "Apollo's Return: Triumph over Failure," *Time*, 27 April 1970, 12–18; "Earth Day and Space Day," *New York Times*, 19 April 1970, 174; and "Our Troubled Spacecraft," *Sarasota (FL) Herald-Tribune*, 22 April 1970, 6A.

46. Marshall D. Ossey, "The Invalid Attacks on NASA," published in the "Speak Out!" column of the *Chicago Tribune*, 29 February 1972, 4.

47. Hill, "Apollo: Spacecraft Is Taking Shape."

48. On Abernathy's church service on the morning of the Apollo launch, see Weinraub, "Hundreds of Thousands," and Greider, "Protesters, VIPs Flood Cape Area."

49. Moss, "NASA and Racial Equality," 42.

50. On examples of hiring discrimination by NASA contractors, see James E. Clayton, "NASA Orbits into Rights Controversy," *Washington Post*, 31 October 1961, A2; and "NASA 'Didn't Know' about Bias," *Washington Post*, 2 November 1961, A3. The space agency did hire several prominent African American female mathematicians who were instrumental to the success of the Apollo program. See Alice A. Dunnigan, "Negro Women Technicians Help Chart Astronauts' Course," *Chicago Daily Defender*, 3 July 1963, 37. One of these women, Katherine Johnson, received the Medal of Freedom from President Obama in November 2015. She and several of her African American female colleagues are the subject of Margot Lee Shetterly, *Hidden Figures: The American Dream and the Untold Story of the Black Women Mathematicians Who Helped Win the Space Race* (New York: William Morrow, 2016). 20th Century Fox turned the book into a film titled *Hidden Figures*.

51. On NASA's claim regarding the dearth of technically trained African American scientists, engineers, and managers, see Moss, "NASA and Racial Equality," 42–43. For examples of the African American press rejecting such claims, and suggesting instead that racism restricted the hiring of blacks, see Steven Morris, "How Blacks View Mankind's 'Giant Step,'" *Ebony* 25 (September 1970): 33–42; and "In the Same Boat," *Ebony* 17 (October 1962): 72–73.

52. Dwight's story first broke in Charles L. Sanders, "The Troubles of 'Astronaut' Edward Dwight," *Ebony* 20 (June 1965): 29–36. For historical analysis of the discrimination faced by Dwight during his astronaut training, see Moss, "NASA and Racial Equality," 54–60; and Spigel, "Outer Space and Inner Cities," 155–156.

53. United Press International, "Seeks Investigation of Space Bias Charge," *Chicago Daily Defender*, 10 June 1965, 4. Later in 1974 NASA again faced widespread charges of racial discrimination after firing an African American, Ruth Bates Harris, the highest-ranking woman at NASA, after she filed an internal report critical of NASA's Equal Employment Opportunity Program. For a thorough analysis of the Ruth Bates Harris controversy, see McQuaid, "Racism, Sexism, and Space Ventures."

54. On Sputnik displacing Little Rock, see Moss, "NASA and Racial Equality," 31; and Taylor Branch, *Parting the Waters: American in the King Years, 1954–1963* (New York: Simon and Schuster, 1988), 224–225. On the Soviet Union using Sputnik to highlight racial unrest in Little Rock, Arkansas, see "Reds List Sputnik Time for Little Rock," *Washington Post*, 10 October 1957, A6.

55. Although Alan Shepard's historic Mercury flight took place several days before Lewis's Rock Hill beating, news of the mission dominated the news cycle the following week. On the Rock Hill Freedom Ride violence, see John Lewis and Michael D'Orso, *Walking with the Wind: A Memoir of the Movement* (New York: Simon and Schuster, 1998), 137–138. Supreme Court Justice Thurgood Marshall

reiterated such frustration in "Excerpts Remarks, Commencement Address Kalamazoo College," in *Unpublished Court Documents of Thurgood Marshall* (Alexandria, VA: Alexander Street Press, 2004), 1.

56. Jack Gould, "TV: Conveying the Depths of Feelings," *New York Times*, 26 March 1965, 70. For another example of the space race competing for media attention with the Selma-to-Montgomery march, see "Two Challenges: On Rights and Space," *New York Times*, 21 March 1965, E1.

57. Environmental historians have embraced the urban environment as a location of study. For historiographical essays on this subfield within environmental history, see Christine Meisner Rosen and Joel Tarr, "The Importance of an Urban Perspective in Environmental History," *Journal of Urban History* 20 (May 1994): 299–310; and Martin Melosi, "The Place of the City in Environmental History," *Environmental History Review* 17 (Spring 1993): 1–23.

58. Congress, Senate Committee on Government Operations, Subcommittee on the Executive Reorganization, *Federal Role in Urban Affairs*, "Testimony of the Reverend Martin Luther King, Jr.," 89th Cong., 2nd sess., 15 December 1966, 2970. For additional examples of civil rights leaders criticizing NASA for funneling funding away from urban problems, see also comments by NAACP executive director Roy Wilkins; Medgar Evers's brother, Charles; and the honorary "mayor" of Harlem, Dr. Benjamin W. Watkins, in Thomas A. Johnson, "Blacks and Apollo: Most Couldn't Have Cared Less," *New York Times*, 27 July 1969, E6; and also Donald Janson, "Webb Backs Cost of Space Program," *New York Times*, 6 December 1966, 34.

59. Alex Haley, *The Autobiography of Malcolm X* (New York: Ballantine, 1987), 273. See also Black Panther leader Eldridge Cleaver as quoted in Associated Press, "The Moon Landing as Eldridge Sees It," *Black Panther*, 2 August 1969, 17; and Eric Pace, "Cleaver Assails Apollo Program," *New York Times*, 21 July 1969, 40.

60. *Christian Science Monitor*, 5 August 1969, in *NASA Current News*, 5 August 1969, 27.

61. Whitney M. Young, "Men on the Moon," *Washington Daily News*, 28 July 1969.

62. For coverage of the 1966 congressional budget battle, see Marquis Childs, "How Far Can the U.S. Budget Be Stretched?," *St. Petersburg (FL) Times*, 2 February 1967, 12A.

63. Bernice Jones, "We Refuse to Allow Our Children to Live in Unsanitary and Hazardous Conditions," *Black Panther*, 1 August 1970, 9. Italics added; capitalization in original.

64. I am obviously oversimplifying this shift within the civil rights movement during the 1960s. For historiographical essays detailing this transformation, see Hall, "Long Civil Rights Movement"; Eagles, "Toward New Histories"; Lawson, "Freedom Then, Freedom Now"; Adam Fairclough, "Historians and the Civil

Rights Movement," *American Studies* 24, no. 3 (December 1990): 387–398; and Charles M. Payne, *I've Got the Light of Freedom* (Berkeley, CA: University of California Press, 1995), 413–441.

65. On these statistics regarding urban riots during this period, see Stephan Thernstrom and Abigail Thernstrom, *America in Black and White: One Nation, Indivisible* (New York: Simon and Schuster, 1999), 158–161; Hugh Davis Graham, "On Riots and Riot Commissions: Civil Disorders in the 1960s," *Public Historian* 2, no. 4 (Summer 1980): 12; and Jonathan J. Bean, "Burn, Baby, Burn: Small Business in the Urban Riots of the 1960s," *Independent Review* 5, no. 2 (Fall 2000): 165.

66. There is a rich literature on the role of poor housing conditions in sparking the urban riots of the 1960s era. For Detroit, see Thomas Sugrue, *The Origins of the Urban Crisis: Race and Inequality in Postwar Detroit* (Princeton, NJ: Princeton University Press, 1996). On housing and the Newark riots, see Max Herman, "Fighting in the Streets: Ethnic Succession and Urban Unrest in 20th Century America" (Ph.D. dissertation, University of Arizona, 1999).

67. "Statement by Dr. Martin Luther King, Jr. on Arrival in Los Angeles, August 17, 1965," Southern Christian Leadership Conference Records, Martin Luther King Jr. Center for Non-Violent Social Change, Atlanta, GA. Italics added. For additional examples of King citing poor living conditions as a primary cause for the Watts riots, see also Dr. Martin Luther King Jr., "Statement to the Press by Dr. Martin Luther King Jr., 20 August 1965, Los Angeles, California," Southern Christian Leadership Conference Records, Martin Luther King Jr. Center for Non-Violent Social Change, Atlanta, GA; and Associated Press, "King Denounces Riots, Hopes Won't Spread," *Park City Daily News* (Bowling Green, KY), 15 August 1965, 24, which quotes King as stating that "due to joblessness and housing conditions, every northern community is a potential powder keg."

68. "Return to 12th Street: A Follow-Up Survey of Attitudes of Detroit Negroes," *Detroit Free Press*, 7 October 1968; Governor's Select Commission on Civil Disorders, *Report for Action: An Investigation into the Causes and Events of the 1967 Newark Race Riots* (New York: Lemma, 1972), 55.

69. For descriptions of this sit-in, see "Hunger Protest Held at NASA: Welfare Group Sits by LM Mock-Up," *Toledo (OH) Blade*, 21 July 1969, 5; and Gertrude Wilson, "Home-Made Signs Impotent in the Roar of a Rocket," *New York Amsterdam News*, 2 August 1969, 1. On the history of the National Welfare Rights Organization and its involvement in the civil rights movement, see Premilla Nadasen, *Welfare Warriors: The Welfare Rights Movement in the United States* (New York: Routledge, 2004).

70. Associated Press, "Crew Rests: Apollo 13 Hailed," *Spokane (WA) Daily Chronicle*, 18 April 1970, Empire Edition, 1.

71. The quotes are from Hosea Williams and Joseph Hammonds, two SCLC leaders who helped to organize the march. For coverage of this protest, see Asso-

ciated Press, "Blacks Protest Apollo 14 Shot," *Rome (GA) News-Tribune*, 2 February 1971, 12; Associated Press, "Poor People's March Called on Launch," *Sarasota (FL) Journal*, 28 January 1971, 1B; and "SCLS [*sic*] Plans Cape March as Protest," *Daytona Beach (FL) Morning Journal*, 28 January 1971, 1. On the use of moon rocks as diplomatic gifts, see David Meerman Scott and Richard Jurek, *Marketing the Moon: The Selling of the Apollo* (Cambridge, MA: MIT Press, 2014), especially the chapter "The Apollo Roadshow: Moonwalkers and Moon Rocks," 100–110; and Teasel Muir-Harmony, "Project Apollo, Cold War Diplomacy and the American Framing of Global Interdependence" (Ph.D. dissertation, Massachusetts Institute of Technology, 2014), 210–240.

72. On Barry's call for this boycott, see Edward Ezell, "Apollo: So What? Earth Turmoil Dims Triumph," *Williamson (WV) Daily News*, 19 July 1979, 20; and Vincent Paka, "Barry Slams Apollo 11 Mission," *Washington Post*, 19 July 1969, A9.

73. Thomas Johnson, "Blacks and Apollo: Most Couldn't Have Cared Less," *New York Times*, 27 July 1969, E6.

74. "The Talk of the Town: The Moon Hours," *New Yorker*, 26 July 1969, 26.

75. Both the Mets fan's statement and the *New York Times* conclusion that African Americans ignored Apollo 11 can be found in Johnson, "Blacks and Apollo."

76. The Century Plaza protest was reported extensively in both local and national newspapers, and it included both civil rights and anti–Vietnam War activists. For examples, see James Wrightson, "1,600 Guests Will Cheer for Astronauts at Dinner Tonight," *Modesto (CA) Bee*, 13 August 1969, A2; Associated Press, "U.S. Leaders to Salute Astronauts at L.A. Fete," *Press-Courier* (Oxnard, CA), 13 August 1969, 1; and Don Oberdorfer, "Apollo Astronauts Hailed by Millions," *Washington Post*, 14 August 1969, A1.

77. Steven V. Roberts, "Astronauts Find Mixed Reactions: The Uninvited Hold Protest as Diners Hail Crew," *New York Times*, 15 August 1969, 14.

78. Paul Montgomery, "Protests Interrupt City Welcome for Astronauts," *New York Times*, 9 March 1971, 1. For an example of a smaller protest held in Union Square during New York City's ticker-tape parade for Apollo 11 astronauts, see Oberdorfer, "Apollo Astronauts Hailed." Civil rights activists planned, but canceled, a similar disruption in June 1965 during a parade for the Gemini astronauts James McDivitt and Edward White in Chicago to protest that city's segregated schools. On this Chicago protest, see Associated Press, "Astronauts Parade May Be Picketed," *Saskatoon* (Canada) *Star-Phoenix*, 12 June 1965, 7.

79. *Jet* editorial comment, "Moon Probe Laudable—But Blacks Need Help," *Jet*, 31 July 1969, 10; Diane Perry, "Moonshot Missed Most Blacks," *New Pittsburgh (PA) Courier*, 2 August 1969, 13; and Ronald Taylor, "'Only Way I Could Get on the Moon Is to Be the First Janitor There,'" *Washington Daily News*, 21 July 1969, 5. For other examples of editorials from African American newspapers

criticizing the space race for siphoning resources from urban housing, see also the *Ebony* photo editorial, "Giant Leap for Mankind?," *Ebony*, September 1969, 58; and "Moon Conquest: The Progress of Man," *Los Angeles Sentinel*, 24 July 1969, B6.

80. Morris, "How Blacks View Mankind's 'Giant Step,'" 33. *Jet* ran a similar editorial arguing that "landing an astronaut on the moon has more priority in America than putting a black man on his feet." See *Jet*, "Moon Probe Laudable"; and also Frank L. Stanley, "People, Places, Problems," *Chicago Defender*, 9 August 1969, 18.

81. *Ebony* photo editorial, "A Plan for Pioneer Eleven," *Ebony*, February 1974, 98.

82. Chester Commodore, "What about the Space between Races of Man," *Chicago Defender*, 12 July 1969, 8.

83. Chester Commodore, "An Expensive Rerun," *Chicago Defender*, 19 November 1969, 17.

84. For additional examples of these cartoons, see Chester Commodore, "No Fairy Tale," *Chicago Defender*, 19 July 1969, 8; Chester Commodore, "Another Space Problem," *Chicago Defender*, 20 April 1970, 13; and Chester Commodore, "A Couple of No No's," *Chicago Defender*, 20 November 1969, 21.

85. Gil Scott-Heron, "Whitey on the Moon," on *Small Talk at 125th and Lenox* (Ace Records, 1970).

86. Ibid. For an interesting analysis of the role of "Whitey on the Moon" in mobility studies, see Jenna M. Loyd, "'Whitey on the Moon': Space, Race, and the Crisis of Black Mobility," in *Mobile Desires: The Politics and Erotics of Mobility Justice*, ed. Liz Montegary and Melissa Autumn White (New York: Palgrave Macmillan, 2015), 41–52.

87. Herb Block, "Transported," *Washington Post*, 18 July 1969, A22.

88. John Hamilton, "Meanwhile, Back on Earth . . . ," *New York Times*, 28 July 1969, 30. For additional *New York Times* editorials comparing the well-funded space race to the meager amounts allocated to urban housing, see also "NASA's Reducing Pains," *New York Times*, 22 April 1968, 46; and Tom Wicker, "In the Nation: Reflections on Apollo," *New York Times*, 18 March 1969, 44.

89. The phrase "project low-cost housing" was used in Roscoe and Geoffrey Drummond, "Matching of NASA Success on Earth Requires Clear Goal, No Obstruction," *Washington Post*, 28 July 1969, A21. Similar arguments also appeared in Joseph Morgenstern, "What's It to Us?," *Newsweek* (special issue, *The Moon Age*), 7 July 1969, 67; and "Beyond Moon and Earth," *Christian Science Monitor*, 5 August 1969, 16.

90. Richard Wilson, "Will It Be Mars—or Housing?," *Minneapolis Tribune*, 25 July 1969.

91. On Vice President Agnew's proposal to explore Mars, see "Agnew Urges Flight to Mars," (Hopkinsville) *Kentucky New Era*, 16 July 1969, 9.

92. Bill Sanders, "On to Mars!," *Milwaukee Journal*, 20 July 1969, 1.

93. Hugh Haynie, "American Know-How," *Courier-Journal* (Louisville, KY), 17 July 1969, 10.

94. George Gallup, "Public Cool to Manned Mars Landing," *Washington Post*, 7 August 1969, F4. According to David Nye, less than one in four African Americans supported spending $4 billion a year on the Apollo program. See David Nye, *Narrative Spaces: Technology and the Construction of American Culture* (New York: Columbia University Press, 1997), 151.

95. For discussions of this drop in popular support for Apollo, see Herbert Krugman, "Public Attitudes toward the Apollo Space Program, 1965–1975," *Journal of Communication* (Autumn 1977): 87–93; and Stephanie A. Roy, Elaine C. Gresham, and Carissa Bryce Christensen, "The Complex Fabric of Public Opinion on Space," *Acta Astronautica* 47, issue 2-9 (July–November 2000): 665–675. On cuts to NASA's budget by Congress, see "The Plight of Apollo," *Newsweek*, 14 September 1970, 97.

96. Gallup, "Public Cool to Manned Mars Landing."

97. Paul Bunge, "Voice of Youth: Do We Really Need to Land on the Moon?," *Denver Post*, 1 February 1969, 12. For similar comments, see Joan Hanauer, "People Everywhere Eye America's Moon Effort," United Press International, 19 July 1969, article from Apollo 11 clipping file, National Air and Space Museum Library, National Air and Space Museum, Washington, DC, 40.

98. On Charles Grigsby's concerns about the space race's impact on urban housing, see Victor Chen, "Some Attack Cost of Trip," *Boston Sunday Globe*, 13 July 1969, 54. On Congressman Koch's concerns, see "Science: Post-Mortem on Apollo 13," *Time*, 4 May 1970, 82.

99. Senator Edward Kennedy, "Goddard Library Dedication, March 19, 1969," Box D5-16, Records of the Goddard Library Project, Clark University Archives, Clark University, Worcester, MA. Kennedy's Clark speech made news in dozens of newspapers and magazines nationwide, each of which listed urban housing as one of his primary concerns. For examples, see Robert Reinhold, "Kennedy Puts Earth Needs Ahead of Space Program," *New York Times*, 20 May 1969, 1; "Cutbacks in Space Race Appears [*sic*] in Order," *Miami News*, 21 May 1969; and "Space and Earth: They're Different," *New Haven (CT) Register*, 21 May 1969.

100. On this liberal attempt in Congress, see Warren Weaver Jr., "Senate Rejects Space Cuts: Adds to Funds for Renewal: Liberals Fail to Reduce the $3.3-Billion NASA Budget on Two Close Votes but Win Increase for Urban Aid," *New York Times*, 8 July 1970, 1.

101. For both Senator Proxmire's quote and a description of the seven-hour debate over "national priorities," see Philip D. Carter, "NASA Cuts Rejected by Senate," *Washington Post*, 8 July 1970, A1.

102. For historical data on NASA's total budget for these years in both real and in 2008 inflation-adjusted dollars, see National Aeronautics and Space Council,

Aeronautics and Space Report of the President: Fiscal Year 2008 Activities (Washington, DC: U.S. Government Printing Office, 2008), "Appendix D-1A: Space Activities of the U.S. Government, Historical Table of Budget Authority (in Millions of Real-Year Dollars)," 146, and "Appendix D-1B: Space Activities of the U.S. Government, Historical Table of Budget Authority (in Millions of Inflation-Adjusted FY 2008 Dollars)," 147.

103. "Plight of Apollo."

104. On Johnson's Cape Canaveral luncheon, see Associated Press, "Hails NASA: LBJ Arrives at Cape," *Chicago Sun-Times*, 16 July 1969, 5; Greider, "Protesters, VIPs Flood Cape Area"; and Weinraub, "Hundreds of Thousands." On Johnson sitting in a separate VIP section from where Abernathy was located, see "Lift for Poor Urged at Lift-Off," A9. Johnson's former vice president, Hubert Humphrey, sat just three seats away from Ralph Abernathy at the Apollo 11 launch.

105. Johnson made these remarks at a White House ceremony in honor of the Apollo 8 astronauts in early January of 1969. See the text of President Johnson's remarks, 9 January 1969, President's Appointment file, box 120, as quoted in Robert A. Divine, "Lyndon B. Johnson and the Politics of Space," in *The Johnson Years: Vietnam, The Environment and Science*, ed. Robert A. Divine (Lawrence: University of Kansas Press, 1987), 247. Divine's chapter wonderfully examines Johnson's extensive involvement in promoting the space race during the 1960s.

106. *Public Papers of the Presidents of the United States: Lyndon B. Johnson, 1963–1969*, 11 vols. (Washington, DC: Government Printing Office, 1964–1969), 1: 114.

107. For a general overview of the War on Poverty and Johnson's Great Society programs affecting urban neighborhoods, see Bruce Schulman, *Lyndon B. Johnson and American Liberalism: A Brief Biography with Documents* (Boston: Bedford / St. Martins, 1995).

108. Lyndon B. Johnson, "The New World of Space" (speech), in *Proceedings of the Second National Conference on the Peaceful Uses of Space: Seattle, Washington, May 8–10, 1962* (Washington, DC: U.S. Government Printing Office, 1962), 30.

109. For an analysis of these efforts by Johnson, see Foster, "Sex in Space," 107–109; and Moss, "NASA and Racial Equality," 3–4.

110. On Johnson's concern that the space race was taking financial resources away from his Great Society programs, see Divine, "Lyndon B. Johnson and the Politics of Space," 236.

111. On Johnson's acceptance of cuts to NASA's budget in a conscious effort to limit reductions in his Great Society programs, see Divine, "Lyndon B. Johnson and the Politics of Space," 242–245. Johnson did similarly in negotiations during the following two years, paving the way in 1970 for liberals, including Koch, Kennedy, and Proxmire, to fight publicly for the reallocation of NASA funding toward urban renewal programs.

112. NASA Administrator James Webb to President Lyndon B. Johnson, 30 November 1964, LBJ files, NASA History Office, as quoted in W. Henry Lambright, *Powering Apollo: James E. Webb of NASA* (Baltimore, MD: Johns Hopkins University Press, 1998), 136.

113. On Johnson promoting space technology as potentially helpful for his Great Society programs, see Kim McQuaid, "Selling the Space Age: NASA and Earth's Environment, 1958–1990," *Environment and History* 12, no. 2 (May 2006): 135; and W. Henry Lambright, *Presidential Management of Science and Technology: The Johnson Presidency* (Austin: University of Texas Press, 1985), 81–84.

114. For the most thorough assessment of NASA's Technology Utilization Program, see Samuel I. Doctors, *The Role of Federal Agencies in Technology Transfer* (Cambridge, MA: MIT Press, 1969), 61–155; and Samuel I. Doctors, *The NASA Technology Transfer Program: An Evaluation of the Dissemination System* (New York: Praeger, 1971). On the number of companies that were using the services of NASA's Technology Utilization Program by 1974, see Todd Anuskiewicz, William Thompson, and Sandra O'Hara, *Technology Utilization Program Report 1974*, NASA SP-5120 (Washington, DC: Technology Utilization Office, NASA, 1975), 4.

115. Hubert Humphrey, "HHH: On the Space Program," *Aerospace Technology*, 7 May 1967, 19.

116. For the General Electric study, see M. L. Feldman, L. A. Gonzalez, and A. B. Nadel, *Application of Aerospace Technologies to Urban Community Problems*, 23 September 1965, Document ID 19660022604, Accession ID 66N31894, Report Number NASA-CR-76524, RM-65TMP-53, Contract-Grant-Task Number NASA Order R-5177, NASA Technical Report Server, 2. For similar NASA-sponsored studies by Drexel University, see *Management Technology Applied to Urban Systems: Proceedings of a Symposium on the Application of NASA Management Technology to the Management of Urban Systems* (Philadelphia: Drexel Institute, Center for Urban Research and Environmental Studies, 1972). On NASA's partnerships with the International City Management Association and the American Institute of Aeronautics and Astronautics and its conferences, see Jennifer Light, *From Warfare to Welfare: Defense Intellectuals and Urban Problems in Cold War America* (Baltimore, MD: Johns Hopkins University Press, 2003), 122.

117. Thompson's comment was recorded in *Space, Science, and Urban Life: Proceedings of a Conference Held in Oakland, California, March 28–30, 1963* (Washington, DC: National Aeronautics and Space Administration, 1963), 1. For Webb's statement, see "Address by James E. Webb, Administrator National Aeronautics and Space Administration, Space, *Science and Urban Life Conference*, Oakland, California, March 30, 1963," NASA news release, 30 March 1963, Folder: 3755: Webb-Space, Science and Urban Life Conference, Oakland, California, March 30, 1963, Washington, D.C., NASA History Collection, NASA Headquarters Archive, Washington, DC, pp. 14–15.

118. Feldman, Gonzalez, and Nadel, *Application of Aerospace Technologies*, 17.

119. Anuskiewicz, Thompson, and O'Hara, *Technology Utilization Program Report 1974*, 40. On this water study by NASA, see also Technology Application Group, George Washington University Medical Center, Biological Communications Project, *Applications of Aerospace Technology in the Public Sector* (Washington, DC: Technological Utilization Office, NASA, 1972), 35.

120. Feldman, Gonzalez, and Nadel, *Application of Aerospace Technologies*, 21.

121. "Satellite Photos to Aid in Predicting Snow-Melt Runoff Volume in West," *Roundup* (newspaper of the Johnson Space Center), 19 December 1975, 2.

122. Feldman, Gonzalez, and Nadel, *Application of Aerospace Technologies*, 24–27.

123. Virginia P. Dawson, *Engines and Innovation: Lewis Laboratory and American Propulsion Technology*, NASA History Series (Washington, DC: NASA Scientific and Technical Information Division, 1991), 205. Such efforts at Lewis began when researchers at the NASA center formed the NASA Volunteer Air Conservation Committee in order to respond to air-quality problems in Cleveland.

124. NASA's "air pollution detection program" is described in detail in Anuskiewicz, Thompson, and O'Hara, *Technology Utilization Program Report 1974*, 37–39.

125. "Space-Derived Sewer Monitor," in *Spinoff: An Annual Report* (Washington, DC: NASA Technology Utilization Office, 1982), 90.

126. On NASA's ACTS technology, see "Sewage Treatment," in *Spinoff: An Annual Report* (Washington, DC: NASA Technology Utilization Office, 1976), 77; and "Technology for Societal Needs," in *Spinoff: An Annual Report* (Washington, DC: NASA Technology Utilization Office, 1977), 29. On the waste-treatment system installed in a Jacksonville, Florida, apartment complex, see *Spinoff: An Annual Report* (1977), 71.

127. "An Environmental Innovation: The Sewer Mouse," in *Spinoff: An Annual Report* (Washington, DC: NASA Technology Utilization Office, 1979), 37; and "Space-Derived Sewer Monitor," 91.

128. Workshop participants recommended that NASA and HUD work together on the urban housing crisis in a final report. See Volta Torrey, *Science and the City* (Washington, DC: U.S. Government Printing Office, 1967), 5.

129. "H. B. Finger Nominated to HUD Post," *Washington Post*, 29 March 1969, D19. For a discussion of how Finger's prior work at NASA would help him in his new position at HUD, see also United Press International, "NASA Aide Named to Housing Post: Scientist Will Direct Urban Research and Technology," *New York Times*, 16 March 1969, 39. For additional background on Finger's role at HUD, see Light, *From Warfare to Welfare*, 12, 120.

130. Paine discusses his consulting work with HUD on what he calls "slum housing" in Oral Interview of Dr. Thomas O. Paine by T. Harri Baker, 10 April 1969, Folder 4185: Paine Interviews Conducted by Baker, Logsdon, and Burke, NASA

Historical Materials, NASA Headquarters Archives, Washington, DC, pp. 12–13. On this meeting between Paine and Romney, see William K. Stevens, "Space Official Talks about Problems on Earth," *New York Times*, 19 May 1969, 31.

131. NASA's Ted Hays is quoted extensively in "USPO Conducts Technical Studies for MIUS Project," *Roundup*, 20 December 1974, 4. On the application of the USPO program for urban apartment complexes and urban development areas, see also "To Use Space Technology on Earth: Urban Systems Project Office Set Up Here: Hays Heads It," *Roundup*, 14 April 1972, 4.

132. HUD Assistant Secretary Harold Finger references the housing crisis as a cause for the creation of Operation Breakthrough in his 6 August 1969 speech before the Portland Cement Association in Chicago, Illinois. See Harold Finger, "The Housing Shortage: Operation Breakthrough," *Vital Speeches of the Day* 35, no. 23 (15 September 1969): 709–712. On the prefabricated components of Operation Breakthrough construction, see U.S. Department of Housing and Urban Development, "Operation Breakthrough," circa 1972, clippings file, Jersey City Public Library, Jersey City, NJ.

133. United Press International, "Romney Bids U.S. Put Housing First," *New York Times*, 23 July 1969, 23.

134. On the involvement of aerospace companies in Operation Breakthrough, see R. H., "No Easy Money in Housing," *Astronautics and Aeronautics*, April 1970, 10; and *Newsweek*, 21 July 1969, 78.

135. On the production goals of HUD's Operation Breakthrough, see R. McCutcheon, "The Role of Industrialised Building in Low-Income Housing Policy in the USA," *Habitat International* 14, no. 1 (1990): 168–170. For an analysis of Operation Breakthrough in general, see also National Research Council, *Rebuilding the Research Capacity at HUD* (Washington, DC: National Academies Press, 2008), 12, 65–67. For a bibliography of reports and articles on MIUS, see J. D. Ryan and B. Reznek, eds., *Abstracted Reports and Articles of the HUD Modular Integrated Utility Systems (MIUS) Program* (Washington, DC: U.S. Government Printing Office, 1977).

136. "USPO Conducts Technical Studies for MIUS Project," 4.

137. On the overall goals of the MIUS program, see "Technology Evaluation of Control / Monitoring Systems for MIUS Application," NASA Technical Memorandum, JSC-08973, NASA TM X-58135 (Houston, TX: National Aeronautics and Space Administration, 1974), v; Naomi Lede, Hortense Dixon, and Donald Hill, "Social Costs Considerations and Legal Constraints in Implementing Modular Integrated Utility Systems: Final Report, Covering the Period December 1973–December 1974 (Washington, DC: National Aeronautic and Space Administration, 1974), 28; and M. R. Mixon, "The Modular Integrated Utility System (MIUS) as a Potential Influence on Community Development," *Water, Air, and Soil Pollution* 7 (1977): 262.

138. Robert V. Gordon, "Urban Systems Project Office," NASA news press release no. 72-75, 6 April 1972, 2.

139. Private industries, including gas utilities, had experimented with "total energy" systems since 1962, but it was not until HUD and NASA's Urban Systems Project Office began working together on MIUS that the federal housing department began testing total energy systems based on MIUS technology in its Operation Breakthrough sites. On this history, see C. W. Hurley, J. D. Ryan, and C. W. Phillips, "Performance Analysis of the Jersey City Total Energy Site: Final Report," HUD Utilities Demonstration Series, Department of Housing and Urban Development, vol. 13 (NBSIR 82-2474), (Washington, DC: U.S. Government Printing Office, 1982), 1–3.

140. On United Aircraft Corporation's MIST lab, see "USPO Conducts Technical Studies for MIUS Project," 4. On the electrical and thermal subsystems of the MIUS system, see also Mixon, "Modular Integrated Utility System," 263.

141. For a detailed summary of the MIST lab results, see Tony E. Redding, "Application of the Integrated Utilities Concept to Community-Size Developments," in *9th Intersociety Energy Conversion Engineering Conference Proceedings Held in San Francisco, California, August 26–30, 1974* (New York: American Society of Mechanical Engineers, 1974), 493–498. See also "USPO Conducts Technical Studies for MIUS Project," 4.

142. "WATER News and Views," *Journal of the American Water Resources Association* 11 (1975): 401–402.

143. This photograph is reprinted, and NASA's total energy system discussed, in Richard Wright, *Building and Fire Research at NBS/NIST, 1975–2000* (Washington, DC: U.S. Department of Commerce, 2003), 137, online at http://fire.nist .gov/bfrlpubs/fire04/PDF/f04019.pdf (accessed 24 August 2016). On federal rent subsidies for the Jersey City Operation Breakthrough site, see Rudy Johnson, "Power Plant Using Waste Heat Is Inspected by Federal Officials," *New York Times*, 25 March 1974, 67.

144. Walter H. Waggoner, "Housing Plan Set for Jersey City," *New York Times*, 27 July 1970, 27. While the U.S. Census indicates that minorities accounted for 22 percent of the Jersey City population in 1970, the percentage of minorities was undoubtedly higher in and around the St. John's Urban Renewal Area, which was one of the poorer sections of the city. See *1970 Census of Population*, vol. 1: *Characteristics of the Population, Part 32: New Jersey, Section 1, Chapter B: General Population Characteristics, New Jersey* (Washington, DC: U.S. Government Printing Office, 1970), 32–52.

145. On HUD working with the local Jersey City redevelopment agency, see "Operation Breakthrough: New Housing Approaches," press release by Jersey City Mayor Thomas J. Whelan, 9 April 1970, Jersey City Public Library, Jersey City, NJ. For an example of a local housing contractor, the Townland Marketing and

Development Corporation of Cherry Hill, New Jersey, working on the Jersey City Operation Breakthrough site, see "Breakthrough Project Stalled by Addition of Fourth Firm," *Jersey Journal*, 1 April 1971, 9s; and Waggoner, "Housing Plan Set for New Jersey," 27.

146. The Summit Plaza heating and cooling system is described in detail in Hurley, Ryan, and Phillips, "Performance Analysis of the Jersey City Total Energy Site," 13: 15. It is also still functioning and was toured by the author during the summer of 2010.

147. Rudy Johnson, "New Furnace Recaptures 'Waste' Energy," *New York Times*, 1 January 1978, NJ10.

148. W. T. Davis and J. O. Kolb, *Environmental Assessment of Air Quality, Noise and Cooling Tower Drift from Jersey City Total Energy Demonstration*, HUD Utilities Demonstration Series, Department of Housing and Urban Development, vol. 17 (NTIS Issue Number 198313) (Washington, DC: Department of Housing and Urban Development, 1980), section 1.2.3, "Combustion Emission Discharge Conditions," and section 2.6.2, "Evaluation of the Effect of the TE Plant on Local Air Quality."

149. On the Summit Plaza being the site of the nation's first residential pneumatic trash collection system, see Jack Preston Overman and Terry G. Statt, *Evaluation of the Refuse Management System at the Jersey City Operation Breakthrough Site*, HUD Utilities Demonstration Series, Department of Housing and Urban Development, vol. 3 (Record ID: 30878) (Washington, DC: Department of Housing and Urban Development, 1978), iv.

150. Terrence G. Reese and Richard C. Wadle, "General Survey of Solid-Waste Management," JSC-06696, NASA TM X-58133, Johnson Space Center, NASA, 1974, pp. 2, 8–11.

151. For a detailed description of this data acquisition system, see C. W. Hurley and J. D. Ryan, "Performance Analysis of the Jersey City Total Energy Site: Executive Summary," HUD Utilities Demonstration Series, Department of Housing and Urban Development, vol. 18 (HUD IAA-H-37-72) (Washington, DC: U.S. Government Printing Office, 1982), 7–33.

152. On these savings, see Gamze-Korobkin-Caloger, Inc., "Final Report: Design and Installation, Total Energy Plant—Central Equipment Building, Summit Plaza Apartments, 'Operation Breakthrough Site,' Jersey City, New Jersey," HUD Utilities Demonstration Series, Department of Housing and Urban Development, vol. 12 (Washington, DC: U.S. Government Printing Office, 1982), 1.

153. For an example of published conference proceedings, see *Space, Science, and Urban Life: Proceedings of a Conference Held in Oakland, California, March 28–30, 1963*. For an example of such news releases, see "Address by James E. Webb, Administrator National Aeronautics and Space Administration, Space, *Science and Urban Life Conference*, Oakland, California, March 30, 1963." For examples of such coverage in the popular media, see David Bird, "Pollution Fight May Get NASA Aid,"

New York Times, 8 October 1967, 49; and Arthur Fisher, "Science Newsfront: Last-Minute News and Notes to Keep You Up-to-Date," *Popular Science* (November 1970): 12.

154. "Housing: Operation Breakthrough," *Newsweek*, 21 July 1969, 78. For additional coverage of HUD's Operation Breakthrough national tour, kicked off to coincide with the Apollo 11 launch, see Richard L. Strout, "Mars or Housing?," *Christian Science Monitor*, 24 July 1969, 3; and "Mortgage Bankers to Hear Romney on 'Breakthrough,'" *Chicago Defender*, 22 August 1970, 31. For examples of Finger giving speeches on Operation Breakthrough, see also Assistant Secretary for Research and Technology Harold Finger, "Research and Development: The Housing Industry, delivered before the Forest Products Research Society, San Francisco, California, 7 July 1969," published in *Vital Speeches of the Day* 35, no. 21 (8 August 1969): 652–654; and Finger, "Housing Shortage."

155. On Romney's speech before the NAACP, see "Romney Hails Blacks' Efforts in Solving Housing Problem," *Chicago Defender*, 17 May 1969, 14. For his speech at the NAACP's annual meeting announcing HUD's grant to the civil rights organization later that summer, see Thomas A. Johnson, "N.A.A.C.P. Gets Grant to Study Minority Builders," *New York Times*, 2 July 1969, 21.

156. Johnson, "Power Plant Using Waste Heat," 67.

157. Bruce Hicks, "New Communities Designed to Use 38% Less Energy," *Atlanta Daily World*, 4 January 1973, 3.

158. The *Chicago Defender* describes the Summit Plaza Operation Breakthrough housing project in "Unique Waste," *Chicago Defender*, 27 February 1971, 11. For *Jet* magazine's coverage of Operation Breakthrough, see "Black Builders to Be in U.S. 'Operation,'" *Jet* 37, no. 13 (1 January 1970): 3. For additional coverage of Operation Breakthrough by the *Chicago Defender*, see also "HUD Task Force to Take Part in Black Housing Producers Meet," *Chicago Defender*, 10 May 1969, 12; "Mortgage Bankers to Hear Romney on 'Breakthrough'"; "U.S. to Help Reduce Inequities," *Chicago Defender*, 6 November 1971, 18; "Inland Home Units Called 'Breakthrough,'" *Chicago Defender*, 27 November 1971, 6; and "Slate New Housing," *Chicago Defender*, 24 November 1973, 1.

159. Rudy Johnson, "Romney's Housing Plan Praised at Conference of Urban League," *New York Times*, 30 July 1969, 24.

160. Abernathy's quote and photos of Abernathy hanging a noose around Paine's neck, with newspaper photographers in the background, appear in Simeon Booker, "Blacks Scarce as Men on Moon at Launch," *Jet*, 31 July 1969, 6, 9.

161. On NASA's intention to apply MIUS technology to a 110,000-person community, see Tony Redding (of NASA's Urban Systems Project Office), "Application of the Integrated Utilities Concept to Community-Size Developments," presented at the 9th Intersociety Energy Conversion Engineering Conference, held in San Francisco, California, August 26–30, 1974, and published in *9th Intersociety*

Energy Conversion Engineering Conference Proceedings (New York: American Society of Mechanical Engineers, 1974), 493–498. For the *New York Times* quote, see "Decay in Jersey City Yielding to 'Springtime' of Renewal," *New York Times*, 12 November 1972, 98. On Operation Breakthrough's role in helping to revitalize Jersey City, see also Paul Goldberger, "Jersey City Plans Broad Renewal," *New York Times*, 7 January 1974, 33.

162. Operation Breakthrough also failed to alter significantly the U.S. housing industry. On this history, see National Research Council, *Rebuilding the Research Capacity at HUD* (Washington, DC: National Academies Press, 2008), 67. The Operation Breakthrough program ended in the late 1970s.

163. This statement by Abernathy was reported repeatedly in coverage of the Apollo 11 launch. See "The Bright Side of the Moon Mission . . . ," *Washington Post*, 20 July 1969, 40; and "Stop the World . . . ," *Washington Daily News*, 17 July 1969. For examples of other VIPs being similarly distracted from civil rights issues by the Apollo 11 launch, see Gertrude Wilson, "Mule Train Is in the Past," *New York Amsterdam News*, 26 July 1969, 17.

164. The term "whitewashing" has several meanings historically. Here I use the term in its political and racial contexts to suggest a glossing over of racial inequalities. The term "scrub" has an equally complex history, meaning both to clean as well as to cancel, call off, dismiss, or abort. I use the term "scrub" in both contexts. On the origin and use of both these terms, see "whitewash, v.," Oxford English Dictionary Online, http://www.oed.com.ezp-prod1.hul.harvard.edu/view/Entry /228644 (accessed 27 June 2016); and "scrub, v.1," Oxford English Dictionary Online, http://www.oed.com.ezp-prod1.hul.harvard.edu/view/Entry/173696? (accessed 27 June 2016).

165. Peter Flanigan (a Nixon administration presidential assistant) to NASA Administrator Thomas Paine, 22 July 1969, Box 5, Nixon Presidential Materials Project, White House Central Files, Subject Files, Outer Space, EX OS 3 7/22 / 69 through EX OS 3 8/4/69–8/6/69, Box 5, Richard Nixon Presidential Library and Museum, Yorba Linda, CA.

166. On Abernathy's experiences in the Kennedy Space Center VIP viewing section, see "Lift for Poor Urged at Lift-Off," A9; "Bright Side of the Moon Mission . . . ," 40; and David Streitfeld, "Footprints in the Cosmic Dust," Washington Post, 20 July 1989, available online at: https://www.washingtonpost.com /archive/lifestyle/1989/07/20/footprints-in-the-cosmic-dust/b23f9258-8f8f-4b89 -b2b0-5b49bae6e156/ (accessed 25 October 2016).

167. The photograph of Mrs. Mattie Gray and her daughter appeared in Booker, "Blacks Scarce as Men on Moon at Launch," 7. For the photograph of Hosea Williams giving a black power salute, see United Press International, "Space Head Tells Poor to Unite behind Apollo," *Boston Globe*, 16 July 1969, 12.

168. Booker, "Blacks Scarce as Men on Moon at Launch," 6.

169. For international coverage of Abernathy's Poor People's Campaign dem-onstration at the Apollo 11 launch, see "Astronauts," *Daily Gleaner* (Kingston, Ja-maica), 17 July 1969, 9; United Press International, "Irony: 'Rockets, Rickets,'" *Pacific Stars and Stripes* (Tokyo, Japan), 17 July 1969, 3; Associated Press, "Moon Voyage Begins," *Brandon Sun* (Canada), 16 July 1969, 1; "A Noose for NASA," *Winnipeg Free Press* (Canada), 25 July 1969, 40; George W. Eberl, "Moon Mission Puts Onlookers into Orbit: Laughter, Concern Mingle at Cape Kennedy," *Stars and Stripes* (Darmstadt, Germany), 22 July 1969, 8; and "Countdown Ahead of Schedule for Apollo Astronauts Today," *Times* (London), 16 July 1969, 1.

170. "Meanwhile, Back Home . . . ," United Press International (BWP 071512—Premium Service, 16 July 1969. For examples of this photograph also appearing domestically, see *Boston Globe*, 16 July 1969, 12.

171. Melor Sturua, "Soviet Media, Scientists Hail Apollo Success," *Izvestiya*, 18 July 1969, A6 (translated into English by the Foreign Broadcast Information Ser-vice on 22 July 1969). The Moscow Domestic New Service also reported on the Abernathy Apollo 11 protest, and the coverage was picked up by both *Izvestiya* and *Pravda*. For the text of this report, see Moscow Domestic Service, "*Pravda, Iz-vestiya* Tribute," 18 July 1969 (translated into English by the Foreign Broadcast Information Service on 18 July 1969).

2. Shooting (from) the Moon

1. Or at least for all Vietnamese who had access to television sets. Viet-namese News Agency, "Cosmonaut Addresses Vietnamese People from Space" (translated into English by the Foreign Broadcast Information Service [FBIS]), 31 July 1980, Folder #002345, Title: "Tuan, Pham (Lt. Col.) Soyuz 37 North Viet-namese," NASA History Archives, NASA History Division, NASA Headquar-ters, Washington, DC. On Tuan's broadcast being translated and relayed to other Asian countries, see James P. Sterba, "Hanoi Touts Its Astronaut as a Benefit of Moscow Ties," *New York Times*, 2 August 1980, 3.

2. For a discussion of space science and technology being used by the two superpowers to promote national identity, see Asif Siddiqi, "Spaceflight in the Na-tional Imagination," in *Remembering the Space Age*, ed. Steven Dick (Washington, DC: NASA Office of External Relations, History Division, 2008), 17–35; and Asif Siddiqi, "Competing Technologies, National(ist) Narratives, and Universal Claims: Toward a Global History of Space Exploration," *Technology and Culture* 51, no. 2 (April 2010): 427. An insightful example of this sort of space history is Walter Mc-Dougall, *The Heavens and the Earth: A Political History of the Space Age* (Balti-more, MD: Johns Hopkins University Press, 1986). For more general discussions of this relationship between a nation's technoscientific capabilities and its political

ideology during the Cold War era, see Carol E. Harrison and Ann Johnson, "Introduction: Science and National Identity," in "National Identity: The Role of Science and Technology," ed. Carol E. Harrison and Ann Johnson, special issue, *Osiris*, 2nd ser., 24, no. 1 (2009): 1–14. Also see the thirteen other essays on this theme in this special issue. This relationship is also discussed in Siddiqi, "Competing Technologies."

3. For an introduction to the historical literature on the links between science and technology and the Cold War, see Mark Solovey, ed., "Science in the Cold War," special issue, *Social Studies of Science* 31, no. 2 (April 2001), especially David Hounshell, "Epilogue: Rethinking the Cold War; Rethinking Science and Technology in the Cold War: Rethinking the Social Study of Science and Technology" (289–297); and John Cloud and Judith Reppy, eds., "Earth Sciences in the Cold War," special issue, *Social Studies of Science* 33, no. 5 (October 2003).

4. Vietnamese News Agency, "Cosmonaut Addresses Vietnamese People from Space."

5. Nhan Dan, "We Fly into Space," Hanoi Domestic Service (translated into English by FBIS), 23 July 1980, Folder #002345, Title: "Tuan, Pham (Lt. Col.) Soyuz 37 North Vietnamese," NASA History Archives, NASA History Division, NASA Headquarters, Washington, DC.

6. The literature on the role played by the natural environment in forging national identity is vast. For a classic introduction to this literature regarding the United States, see Barbara Novak, *Nature and Culture: American Landscape and Painting, 1825–1865* (New York: Oxford University Press, 1980). Other more recent examples include Franz-Josef Bruggemeier, Mark Cioc, and Thomas Zeller, eds., *How Green Were the Nazis? Nature, Environment, and Nation in the Third Reich* (Athens: Ohio University Press, 2005); Arja Rosenholm and Sari Aultio-Sarasmo, eds., *Understanding Russian Nature: Representations, Values, and Concepts* (Helsinki: University of Helsinki Press, 2005); and Gunnel Cederlof and K. Sivaramakrishnan, eds., *Ecological Nationalisms: Nature, Livelihoods, and Identities in South Asia* (Seattle: University of Washington Press, 2006).

7. On this statement by Nguyen Van Hieu, Vietnam's National Scientific Research Center deputy chairman, see "Further on Joint Soviet-Vietnamese Spaceflight, First Experiment," Moscow Domestic Service in Russian, 26 July 1980, U3, as transcribed by FBIS, Folder #014776, Title: "Vietnam," NASA History Reference Collection, NASA History Division, NASA Headquarters, Washington, DC.

8. There is a dearth of literature on the environmental history of the Cold War, and even less that examines this relationship between the natural environment and the Cold War across the developing world. For discussions of this historiographical gap, see John R. McNeill and Corinna R. Unger, "Introduction: The Big Picture," in *Environmental Histories of the Cold War*, ed. John R.

McNeill and Corinna R. Unger (New York: Cambridge University Press, 2010), 3–4, 15.

9. Vietnamese News Agency, "Cosmonaut Addresses Vietnamese People."

10. Nhan Dan, "We Fly into Space." There is a rich literature on the role of science and technology in helping to forge empire. See Daniel Headrick, *The Tools of Empire: Technology and European Imperialism in the Nineteenth Century* (New York: Oxford University Press, 1981); Michael Adas, *Machines as the Measure of Men: Science, Technology, and Ideologies of Western Dominance* (Ithaca, NY: Cornell University Press, 1989); and Richard Grove, *Green Imperialism: Colonial Expansion, Tropical Island Edens and the Origins of Environmentalism, 1600–1860* (New York: Cambridge University Press, 1995). For a review of literature on the topic, see John Krige and Kai-Henrik Barth, "Introduction: Science, Technology and International Affairs," in John Krige and Kai-Henrik Barth, "Global Power Knowledge: Science and Technology in International Affairs," special issue, *Osiris*, 21, no. 21 (2006); and Roy MacLeod, "Introduction," in Roy MacLeod, ed., "Nature and Empire: Science and the Colonial Enterprise," special issue, *Osiris*, 2nd ser., 15 (2000): 1–13.

11. The few exceptions include institutional histories, undertaken often by NASA's own history division, that describe the use of various technologies by the U.S. military during the Vietnam War. For examples, see Andrew J. Butrica, *Beyond the Ionosphere: Fifty Years of Satellite Communication* (Washington, DC: NASA History Series, 1997), 65–69.

12. While the concerns and activism of the New Left overlapped with other social and political movements of the period, especially regarding civil rights, it was most vocal and effective in its efforts to halt the Vietnam War and ultimately became the central player in the anti-war movement. For examples of early scholarship on the New Left, often written by participants, see Kirkpatrick Sale, *SDS: The Rise and Development of the Students for a Democratic Society* (New York: Random House, 1973); Todd Gitlin, *The Sixties: Years of Hope, Days of Rage* (New York: Bantam, 1987); James Miller, *"Democracy Is in the Streets": From Port Huron to the Siege of Chicago* (Cambridge, MA: Harvard University Press, 1987); and Maurice Isserman, *If I Had a Hammer . . . : The Death of the Old Left and the Birth of the New Left* (New York: Basic Books, 1987). For a more contemporary account of the movement, see John McMillian, " 'You Didn't Have to Be There': Revisiting the New Left Consensus," in John Campbell McMillian and Paul Buhle, *New Left Revisited* (Philadelphia, PA: Temple University Press, 2003), 1–8; and Rick Perlstein, "Who Owns the Sixties? The Opening of a Scholarly Generation Gap," *Lingua Franca* (May–June 1996): 30–37.

13. As quoted in Sterba, "Hanoi Touts Its Astronaut."

14. For further details on this incident, see Steven Hurst, "Vietnamese-Soviet Team Links Up with Space Lab," Associated Press, 24 July 1980; and As-

sociated Press, "Asian and Russian Board Salyut 6," *New York Times*, 25 July 1980, A13.

15. Gordon R. Hooper, "The Cosmonauts—Part 20," clipping from Folder #002345, Title: "Tuan, Pham (Lt. Col.) Soyuz 37 North Vietnamese," NASA History Archives, NASA History Division, NASA Headquarters, Washington, DC.

16. For a discussion of Vietnamese communists' use of nature during the First Indochina War, see David Biggs, *Quagmire: Nation-Building and Nature in the Mekong Delta* (Seattle: University of Washington Press, 2010), 133, 144–150, 204.

17. Curtis Jordan, "Uncovering Charlie: Spray Destroys Hiding Places of Viet Cong," *Air Force Times*, 11 May 1966, 54. For a similar conclusion by a scientist visiting Vietnam, see also Arthur Westing, "II. Leveling the Jungle," *Environment* 13, no. 9 (November 1971): 9.

18. Charles V. Collins, "Herbicide Operations in Southeast Asia, July 1961–June 1967," Department of the Air Force, Headquarters Pacific Air Forces, Directorate, Tactical Evaluation, CHECO Division, 11 October 1967, 16, distributed by National Technical Information Service, U.S. Department of Commerce.

19. On the history of the tunnels of Cu Chi, see Tom Mangold and John Penycate, *The Tunnels of Cu Chi: A Harrowing Account of America's "Tunnel Rats" in the Underground Battlefields of Vietnam* (New York: Random House, 1985).

20. On nighttime fighting along the Ho Chi Minh Trail, see James Gibson, *The Perfect War: Technowar in Vietnam* (Boston, MA: Atlantic Monthly Press, 1986), 396; and Terrence Smith, "American Squadron Bombs the Ho Chi Minh Trail Only after Dark," *New York Times*, 16 December 1968, 3. For other newspaper accounts of the widespread use of nighttime darkness by the communist guerrillas during the Vietnam War, see also Charles Mohrs, "G.I.'s Contesting Vietcong at Night: Camp in Areas Previously Held by Reds after Dark," *New York Times*, 2 August 1965, 1; and Bob Considine, "Reds Own the Country at Night in S. Viet," *Evening News* (Newburgh, NY), 11 August 1967, 3A.

21. Horst Faas for the Associated Press, "For Both Americans and Viet Cong Night Alters Face of War," *Eugene (OR) Register-Guard*, 15 September 1966, 8A.

22. Here I am engaging the concept of "legibility" as discussed by James Scott. While Scott does not focus his analysis specifically on natural-resource use, local knowledge of nature is central to his concept of "metis," meaning practical knowledge, which he juxtaposes with the "high modernist" practices of more industrialized cultures. On the role of nature in "metis," see James Scott, *Seeing Like a State: How Certain Schemes to Improve the Human Condition Have Failed* (New Haven, CT: Yale University Press, 1998), especially chapter 9.

23. Jordan, "Uncovering Charlie"; Peter R. Kann, "The Invisible Foe: New Intelligence Push Attempts to Wipe Out Vietcong Underground," *Wall Street Journal*, 5 September 1968, 1; Associated Press, "U.S. Airborne Device Sniffs for Foe under Jungle Canopy," *New York Times*, 28 May 1967, 7.

24. Lt. Col. Stanley D. Fair, "No Place to Hide: How Defoliants Expose the Viet Cong," *Army*, September 1963, 54.

25. On French colonial authorities using dredges to expand colonial control over the Mekong Delta, and Vietnamese locals reacting by attacking this technology, see Biggs, *Quagmire*, 41–48, 202.

26. Westing, "II. Leveling the Jungle."

27. Edward Miguel and Gerard Roland, "The Long Run Impact of Bombing Vietnam," *National Bureau of Economic Research Working Paper No. 11954* (January 2006): 2. For a careful review of the ecological and health impact of bomb craters on Vietnam, see E. W. Pfeiffer, "I. Craters," *Environment* 13, no. 9 (November 1971): 4.

28. In 1969 and 1970 ad hoc groups of international scientists, led by zoologists E. W. Pfeiffer and G. H. Orians, pressured the American Association for the Advancement of Science and the U.S. National Academy of Sciences to undertake field studies to measure chemical defoliant usage by the U.S. military in Vietnam as well as its impact on human health and the natural environment. On these field studies, see G. H. Orians, E. W. Pfeiffer, and Clarence Leuba, "Defoliants: Orange, White, and Blue," *Science, New Series* 165, no. 3892 (1 August 1969): 442–443; E. W. Pfeiffer, "Final Word on Defoliation Effects," *Science, New Series* 171, no. 3972 (19 February 1971): 625–626; and E. W. Pfeiffer, "Operation Ranch Hand: The U.S. Herbicide Program," *Bulletin of the Atomic Scientists* 38, no. 5 (May 1982): 20–24.

29. As quoted in R. W. Apple Jr., "U.S. Planes in Action," *New York Times*, 16 February 1966, 4.

30. Jack Foisie, "Defoliation Pressed along Supply Trails," *Washington Post*, 16 February 1966, A1. On Agent Orange being used in Vietnam to make the jungle more legible, see also Scott, *Seeing Like a State*, 189; and Jacob Darwin Hamblin, *Arming Mother Nature: The Birth of Catastrophic Environmentalism* (New York: Oxford University Press, 2013), 179–196.

31. Fair, "No Place to Hide," 54.

32. Westing, "II. Leveling the Jungle."

33. For details of this meeting, see Admiral W. Fred Boone, "NASA Office of Defense Affairs: The First Five Years, December 1, 1962, to January 1, 1968," Historical Division, Office of Policy, National Aeronautics and Space Administration, December 1970, 249, NASA Headquarters Historical Reference Collection, Washington, DC; and NASA Deputy Administrator Robert C. Seamans Jr. to Distribution List, memorandum, 31 March 1965, Box: Office of Defense Affairs, Office of D.D. and Interagency Affairs, 1958–1990, #18518, Box 1, Folder: "Chron File 1965/1966 Coulter," NASA History Division, Historical Reference Collection, NASA Headquarters Archives, Washington, DC.

34. On the creation of "NASA's Limited War Committee" and its work, see Boone, "NASA Office of Defense Affairs," 249–251; and Thomas O'Toole, "NASA's

Role in War Grows," *Washington Post*, 4 December 1967, A1. On the request to establish the Limited Warfare Committee's "Special Support Projects" account, see D. D. Wyatt to NASA Assistant Administrator of Programs William Fleming, memorandum, 10 February 1966, Record Number 701, Series: Biographies-NASA Employees, Subject: Fleming, William, A., Folder Title: Fleming-Chron File: 1/66–6/67 (7), Period Covered: January, 1966–June, 1967, Location: LEK1/6/5, National Aeronautics and Space Administration, Headquarters, Washington, DC, 1.

35. Seamans to Distribution List, memorandum, 2.

36. On the long relationship between NASA and the military, see Peter Hays, *The U.S. Military and Outer Space: Perspectives, Plans and Programs* (New York: Routledge, 2011).

37. On NASA's involvement with the CORONA program, see John Cloud, "Imaging the World in a Barrel: CORONA and the Clandestine Convergence of the Earth Sciences," *Social Studies of Science* 31, no. 2, special issue on "Science in the Cold War" (April 2011): 231–251; Dwayne A. Day, "CORONA: America's First Spy Satellite Program," *Quest: History of Spaceflight Quarterly* 4, no. 2 (Summer 1995): 4–21; and Dwayne A. Day, "CORONA: America's First Spy Satellite Program, Part II," *Quest: History of Spaceflight Quarterly* 4, no. 2 (Fall 1995): 28–36.

38. On the use of NASA's TIROS for military maneuvers, see Henry Hertzfeld and Ray Williamson, "The Social and Economic Impact of Earth Observing Satellites," in *Societal Impact of Spaceflight*, ed. Steven Dick and Roger Launius (Washington, DC: NASA Office of External Relations, History Division, 2007), 238.

39. *National Aeronautics and Space Act of 1958*, Public Law #85-568, 85th Cong., 2nd sess. (7 January 1958), 1.

40. O'Toole, "NASA's Role in War Grows."

41. Between 1967 and 1968 Congress cut NASA's budget from $4.966 billion to $4.578 billion, its lowest amount since 1964. When adjusted for inflation according to 2008 dollars, this represents a decrease of 10.5 percent. On this, see National Aeronautics and Space Council, *Aeronautics and Space Report of the President: Fiscal Year 2008 Activities* (Washington, DC: U.S. Government Printing Office, 2008), "Appendix D-1A: Space Activities of the U.S. Government, Historical Table of Budget Authority (in Millions of Real-Year Dollars)," 146, and "Appendix D-1B: Space Activities of the U.S. Government, Historical Table of Budget Authority (in Millions of Inflation-Adjusted FY 2008 Dollars)," 147.

42. On the inverse federal funding relationship between NASA and the Vietnam War, see J. V. Reistrup, "War Could Cut Space Funds," *Washington Post*, 4 December 1967, A3; and Murray L. Weidenbaum, "Aerospace Technology and the Federal Budget," NASA Technical Report #AIAA Paper 68-915, 28–30 August 1968, 1, 6.

43. On NASA's concern that its military research for the Vietnam War might damage its relations with foreign countries, see O'Toole, "NASA Cut $282 Million More," *Washington Post*, August 19, 1967, A4.

44. This internal memo is discussed in Boone, "NASA Office of Defense Affairs," 251.

45. The quote comes from Seamans to Distribution List, memorandum, 1. On the expanded role of NASA's Limited Warfare Committee, see Boone, "NASA Office of Defense Affairs," 251; and O'Toole, "NASA's Role in War Grows."

46. On the various research proposals undertaken by NASA's Limited Warfare Committee, see "NASA Limited Warfare Projects Supporting DOD," undated, Box 1: Office of Defense Affairs, Office of D.D. and Interagency Affairs, 1958–1990, #18518, Folder: "Chron File, 1965/1966 Coulter," NASA History Division, Historical Reference Collection, NASA Headquarters Archive, Washington, DC; Freitag to Mueller, memorandum, 20 January 1966, Folder #002276, Title: Teague (1963–67-NASA Oversight Comm.), NASA History Archives, NASA History Division, NASA Headquarters, Washington, DC; and William Schimandle (JPL) to James O. Spriggs, 19 June 1967, Folder: #002175, Title: "SA Chron 1963–70 Spriggs," NASA History Archives, NASA History Division, NASA Headquarters, Washington, DC.

47. "Summary of Suggestions by NASA Headquarters Personnel as to Ideas That May Have Application to the War in Southeast Asia," 10 December 1965, Folder: #014776: "Vietnam," NASA History Reference Collection, NASA History Division, NASA Headquarters, Washington, DC, 1.

48. On the development by NASA's Limited Warfare Committee of an ATS satellite, "Cloud Camera," "to provide meteorological reports of Vietnam to the U.S. Air Force," see James Spriggs, "Log for Week of 28 Nov," Folder #002175, Title: "SA Chron-1963–70 Spriggs," NASA History Archives, NASA History Division, NASA Headquarters, Washington, DC; James Spriggs to William Schimandle, 13 April 1966, Folder #002175, Title: "SA Chron-1963–70 Spriggs," NASA History Archives, NASA History Division, NASA Headquarters, Washington, DC; and especially, Leonard Jaffe (NASA Space Applications Programs Director) to Edgar Cortright (Director of NASA's Langley Research Center), memorandum, 16 May 1966, Folder #005628, Title: "ATS Documentation," NASA Archives, NASA History Division, NASA Headquarters, Washington, DC.

49. For a brief overview of "Project Able," see "Administrator's Back-Up Book, New Item—PROJECT ABLE," Folder #011535, "Project Able / Mirror Illumination," NASA Headquarters, NASA Archives, Washington, DC.

50. Mueller was quoted in Norman Carlisle, "We Can Turn Night into Day," *Sun* (Baltimore, MD), 8 December 1968, 381. On Mueller's congressional testimony regarding Project Able, see also Associated Press, "Illuminating Satellite Underway," *Victoria (TX) Advocate*, 11 August 1966, 6A. There is a rich literature on the history of illumination and civic control. See especially Chris Otter, *The Victorian Eye: A Political History of Light and Vision in Britain, 1800–1910* (Chicago:

University of Chicago Press, 2008); and the forum on Otter's book in *History and Technology* 26, no. 2 (June 2010): 147–185.

51. For a brief chronology of NASA's involvement with Project Able, see "Administrator's Back-Up Book, New Item—Project Able," 24 July 1967, Folder #011535: "Project Able / Mirror Illuminator," NASA Headquarters, NASA Archives, Washington, DC, 2; and "Project Moonlight, Edward Z. Gray to Associate Administrator for Manned Space Flight George Mueller," memorandum, 18 March 1966, Folder #011538: "Project Brilliant / Moonlight / Reflector," NASA History Reference Collection, NASA History Divisions, NASA Headquarters, Washington, DC.

52. Seismometers were part of the Apollo 11 Early Apollo Surface Experiments Package (EASEP), and they were on the five subsequent Apollo missions as part of the Apollo Lunar Surface Experiments Package (ALSEP). On the history of EASEP and ALSEP, see Donald A. Beattie, *Taking Science to the Moon: Lunar Experiments and the Apollo Program* (Baltimore, MD: Johns Hopkins University Press, 2001), 132–133, 144–145.

53. On NASA's work with the Air Force on "seismic detectors," see "Summary of Suggestions by NASA Headquarters Personnel," 2.

54. On Project Igloo White and the electronic battlefield in Vietnam, see George Weiss, "Battle for Control of Ho Chi Minh Trail," *Armed Forces Journal,* 15 January 1971, 19–22; and Paul N. Edwards, *The Closed World: Computers and the Politics of Discourse in Cold War America* (Cambridge, MA: MIT Press, 1996), 3–8, 142–143.

55. Weiss, "Battle for Control of Ho Chi Minh Trail," 19. In a coordinated effort, NASA also worked with the Air Force to develop for dispersal along the Ho Chi Minh Trail piezoelectric crystals, which, when crushed, emitted an electric charge that could be sensed by nearby radio receiving equipment. On these crystals, see "Summary of Suggestions by NASA Headquarters Personnel," 3.

56. Kenneth W. Gatland, "Soviet Satellites Eye Viet Military Scene," *Christian Science Monitor,* 23 April 1975, 4.

57. William Broad, "Soviet Studies Satellite to Convert Solar Energy for Relay to Earth," *New York Times,* 14 June 1987, 1. While "Star Electricity" was never deployed over Vietnam, when it was unfurled in space in 1993 it cast a 2.5-mile-wide beam of light across Western Europe.

58. On opposition to Project Able from astronomers, see "Scientists Oppose Orbiting Mirrors," *New York Times,* 26 May 1967, 4. On the American Astronomical Society's letter-writing campaign opposing Project Able, see American Astronomical Society Secretary G. C. McVittie to NASA Administrator James Webb, 2 September 1966, Folder #011535, Folder Title: "Project Able / Mirror Illumination," NASA Headquarters, NASA Archives, Washington, DC; Arizona Senator

Carl Hayden to NASA Administrator James Webb, 20 October 1966, Folder #011535, Folder Title: "Project Able / Mirror Illumination," NASA Headquarters, NASA Archives, Washington, DC; and Richard L. Callaghan to Kentucky Senator John S. Cooper, 4 April 1967, Folder #011535, Folder Title: "Project Able / Mirror Illumination," NASA Headquarters, NASA Archives, Washington, DC.

59. On the New Left's critique of the American university as supporting the rise of technocracy during the 1960s, see Leo Marx, "Reflections on the Neo-Romantic Critique of Science," *Daedalus* 107, no. 2 (Spring, 1978), 72. The argument that an aversion to technology unified the 1960s counterculture, of which the New Left was a part, originated with Theodore Roszak, *The Making of a Counter Culture: Reflections on the Technocratic Society* (New York: Doubleday, 1969), 8.

60. For the transcription of this speech by Mario Savio on the Sproul Hall steps in Berkeley, California, 2 December 1964, see http://www.americanrhetoric .com/speeches/mariosaviosproulhallsitin.htm.

61. There is a rich literature on academia's role in the military-industrial complex. See especially Stuart Leslie, *The Cold War and American Science: The Military-Industrial-Academic Complex at MIT and Stanford* (New York: Columbia University Press, 1993); Margaret Pugh O'Mara, *Cities of Knowledge: Cold War Science and the Search for the Next Silicon Valley* (Princeton, NJ: Princeton University Press, 1005); and Matt Wisnioski, "Inside 'the System': Engineers, Scientists, and the Boundaries of Social Protest in the Long 1960s," *History and Technology* 19, no. 4 (2003).

62. On the naval recruitment protest at Berkeley, see Harold V. Streeter, "Even Peace Can Lead to Controversy in Berkeley," *Gazette* (Emporia, KS), 20 July 1967, 14; and Roszak, *Making of a Counter Culture*, 30. On the Free Speech Movement at Berkeley transforming into the anti-war movement, see Jeremi Suri, *Power and Protest: Global Revolution and the Rise of Détente* (Cambridge, MA: Harvard University Press, 2003), 166–172.

63. On this outpouring of student activism in opposition to university research for the Vietnam War, see Paul Goodman, "Can Technology Be Humane?," *New York Review of Books* 13, no. 9 (20 November 1969): 1. On student sit-ins at Princeton, see Wisnioski, "Inside 'the System,'" 319–320.

64. "The College Poll: Students Would Cut New Space Program," *San Francisco Examiner*, 29 July 1969, 2.

65. Paul Bunge, "Voice of Youth: Do We Really Need to Land on the Moon?," *Denver Post*, 1 February 1969, 12.

66. For reference to the militarization of space, see the "Deterrence Policy" section of *The Port Huron Statement of the Students for a Democratic Society* (1962) (no page numbers given). For SDS attitudes toward Vietnam, see the "Colonial Revolution" section of *Port Huron Statement*. Years later Tom Hayden, one of the *Statement* authors, stated that "those who were really engaged with the

struggle against . . . war, probably had a jaundiced view of [the space race] as a kind of escapism." For Hayden's comment, see James Sterngold, "Bound Once More for the Moon, without the Stressful 60's Cargo: Mini-Series Casts the Mission as the Defining Spirit of a Decade," *New York Times*, 5 March 1998, E1.

67. "College Poll."

68. O'Toole, "NASA's Role in War Grows"; "NASA to Study Military Satellites," *Technology Week*, 25 July 1966, 16.

69. "Table B. Federal Obligations for Research and Development, by Major Agency and Performer: Fiscal Years 1956–1994," in National Science Foundation, *Federal Funds for Research and Development Detailed Historical Tables: Fiscal Years 1956–1994*, NSF 94-331 (Bethesda, MD: Quantum Research, 1994).

70. Kenneth J. Heineman, *Campus Wars: The Peace Movement at American State Universities in the Vietnam Era* (New York: New York University Press, 1994), 14.

71. "The Moon Age, a Special Section: How We Got There, Where We're Going," *Newsweek*, 7 July 1969, 64.

72. Victor Cohn, "MIT Study Due," *Washington Post*, 31 May 1969, A1.

73. On student demonstrations at MIT against the university's Instrumentation Laboratory, see Cohn, "MIT Study Due"; Associated Press, "Scientist at MIT Says He Was Fired," *St. Joseph (MO) News-Press*, 17 October 1969, 4B; and Crocker Snow Jr., "800 in Antiwar March at MIT," *Boston Globe*, 5 November 1969, 1. On MIT's decision to "divest" itself of the Instrumentation Laboratory, see Robert Reinhold, "Panel Urges Shift in MIT Research," *New York Times*, 3 June 1969, 1; and John Noble Wilford, "MIT Weapons Lab Shifts to Urban Problems," *New York Times*, 17 October 1969, 49.

74. Joseph Alsop, "New Left's Assault on I-Lab Is Uglier Than Yet Revealed," *Washington Post*, 21 November 1969, A27. On student activism at MIT against the Draper lab, see also Cohn, "MIT Study Due"; Reinhold, "Panel Urges Shift in M.I.T. Research"; and Wilford, "MIT Weapons Lab."

75. NASA's Cloud Camera was also canceled because its implementation would interfere with other experiments already scheduled for the space agency's Advanced Technology Satellite. On this scheduling conflict, see Leonard Jaffe (NASA Space Applications Programs Director) to Edgar Cortright (Director of NASA's Langley Research Center), memorandum, 16 May 1966, Folder #005628, Title: "ATS Documentation," NASA Archives, NASA History Division, NASA Headquarters, Washington, DC.

76. NASA Deputy Administrator Robert C. Seamans Jr. to Heads, Program and Staff Offices and Center Directors, memorandum, 10 November 1969, Folder #014777: "Vietnam, South," NASA History Reference Collection, NASA History Division, NASA Headquarters, Washington, DC. Italics in the original.

77. On Senator Proxmire's leak, see United Press International, "'Automated Battlefield' Rapped," *News and Courier* (Charleston, SC), 6 July 1970, 10B; and Associated Press, "Goldwater Raps Proxmire Security Leak," *Pittsburgh Press*, 8 July 1970, 11.

78. The Pentagon director of defense, John Foster, as quoted in Larry Lempert, "On Classified Research," *Michigan Daily*, 16 March 1971, 4. On student activism at the University of Michigan opposing faculty research on the electronic battlefield, see Sara Fitzgerald, "Faculty to Vote on Classified Research Tomorrow," *Michigan Daily*, 21 March 1971, 1; and Alan Lenhoff, "Research Panel Report," *Michigan Daily*, 19 May 1971, 4. On NASA contracts at the University of Michigan during the 1960s, see Heineman, *Campus Wars*, 13.

79. United Press International, "Pennsylvania Students Hit Nixon Talk," *Beaver County (PA) Times*, 26 April 1972, B8.

80. For details of the Columbia building take-over, see John Darton, "Lab Occupation Ends," *New York Times*, 29 April 1972, 1. On the JASON involvement in the electronic battlefield, see Edwards, *Closed World*, 142. On the history of the JASON, see Ann Finkbeiner, *The JASONS: The Secret History of Science's Postwar Elite* (New York: Penguin, 2006).

81. Martin Arnold, "Columbia Classes Are Held under Makeshift Conditions," *New York Times*, 27 April 1972, 45.

82. On students finding materials linking Columbia to the "war machine" in Vietnam, see Martin Arnold, "Lawsuit Is Threatened," *New York Times*, 29 April 1972, 1; and Darton, "Lab Occupation Ends."

83. Don Hesse, "Could I Interest You in Some Earthly Problems?," *St. Louis (MO) Globe*, 28 December 1968.

84. Gene Bassett, "What Have They Been Feeding You?," *Washington Daily News*, 28 July 1969.

85. John Fischetti, "Horizons," *Chicago Daily News*, 26 December 1968.

86. A breakthrough in this impasse occurred on 8 May 1972, one month after this cartoon appeared, when President Nixon made a major concession to the North Vietnamese by announcing that the United States would accept a cease-fire as a precondition for its military withdrawal from South Vietnam.

87. Bruce Shanks, "Space Progress, Peace Progress," as reprinted in the *Los Angeles Times*, 9 April 1972.

88. Associated Press, "Automated Battlefield Object of Anti-War Group Anger," *Morning Record* (Meriden, CT), 20 December 1971, 13.

89. For photographs of this anti-NASA protest, see National Archives and Records Administration, NAIL NWDNS-428-KN-193335, as reprinted in Mitchell K. Hall, "The Vietnam Era Antiwar Movement," *OAH Magazine of History* 18, no. 5 (2004): 13.

90. On the American Friends Service Committee's slide show on the electronic battlefield, see "Citizens for Peace to Hear Retired Brigadier General," *Me-*

riden (CT) Journal, 7 July 1972, 17; and Associated Press, "Automated Battlefield Object," 13. On the proposed amendment by members of the American Physical Society, see Associated Press, "Amendment Urged," *New York Times*, 26 April 1972, 11. For a description of the fast by Berrigan and other inmates at the Federal Correctional Institution at Danbury, Connecticut, see Raymond Pontier, "A Fast Continues," *New York Times*, 30 August 1972, 37.

91. Constance Holden, "Vietnam Land Devastation Detailed," *Science, News Series* 175, no. 4023 (18 February 1972): 737.

92. For statements by both Indira Gandhi and Olof Palme, see Jon Tinker, "Indochina: Ecology Which Stockholm Forgot," *New Scientist* (22 June 1972): 694.

93. On the termination of the electronic battlefield, see Edwards, *Closed World*, 3–8, 142–143.

94. My thinking here regarding the causes of détente has been influenced both by Jeremi Suri's work and by its critics. While Suri argues that détente around 1968 both deflected domestic political pressures by movements such as the New Left and also reasserted federal control through stronger international ties, critics fault him for overemphasizing politicians' reactions to these grassroots movements and for failing to adequately address the economic causes of détente as well as the impact of the nuclear arms race. For a discussion of Suri's work, see Carolyn Eisenberg, Greg Grandin, T. Christopher Jespersen, and David Painter, H-Diplo Roundtable on Jeremi Suri, *Power and Protest: Global Revolution and the Rise of Détente*, online at https://networks.h-net.org/pdf-h-diplo-roundtable-eisenberg-suri (accessed 11 July 2016). See also Jeremi Suri, *Power and Protest: Global Revolution and the Rise of Détente* (Cambridge, MA: Harvard University Press, 2005).

95. The quote comes from "Cambodia Protests 'Consideration' of Project Able," Folder #011535: "Project Able / Mirror Illuminator," NASA Headquarters, NASA Archive, Washington, DC. On the opposition of the Cambodian government, see G. Ostruoumov, "U.S. Mirror Satellite Research Denounced," *Izvestiya* (Moscow), 11 May 1968, Folder #011535: "Project Able / Mirror Illuminator," NASA Headquarters, NASA Archive, Washington, DC.

96. Jonathan Spivak, "The First Space Handshake," *Wall Street Journal*, 22 July 1975, 16. Even *Aviation Week and Space Technology*, historically a staunch supporter of space exploration, called such experiments "pseudo-scientific." See William H. Gregory, "Special Report: ASTP Mission—Airmanship, Science Highlight Project," *Aviation Week and Space Technology*, 28 July 1975, 16.

97. On the ASTP mission's $4 million television camera, see Spivak, "First Space Handshake."

98. Thomas P. Stafford with Michael Cassutt, *We Have Capture: Tom Stafford and the Space Race* (Washington, DC: Smithsonian Institution Press, 2002), 190, as quoted in Jennifer Ross-Nazzal, "Détente on Earth and in Space: The Apollo-Soyuz Test Project," *OAH Magazine of History*, July 2010, 29.

99. James Haggerty, "U.S.-Soviet Détente in Space," *Saturday Evening Post* 247 (July / August 1975): 8.

100. United Press International, "Space Flight Hailed as Effort to Peace," *Rome (GA) News-Tribune*, 20 July 1975, 10A.

101. Joseph Kraft, "Space Shots Are Boosters for Economic System," *Sunday Bulletin* (Philadelphia), 25 May 1969.

102. David Streitfeld, quoting the San Francisco writer Barry Malzberg, in "Footprints in the Cosmic Dust: Twenty Years Later, Six Voices on the Lost Promise of the Apollo Mission," *Washington Post*, 20 July 1989, D1. For similar comments see also Frank Joyce, "Moonism," *Fifth Estate Newspaper* (Detroit, MI) 4, no. 6 (July–August 1969), 5; and Mary Clarke and Varda Ullman, "Escalation," *L.A. Wisp* (Women Strike for Peace) (February 1971), 1.

103. As the *Wall Street Journal* reported, "The earth resources program owes most of its technology to the highly classified military programs." See William Burrows, "Sizing Up the Planet: Satellites Will Seek to Inventory Resources of Earth from Orbit," *Wall Street Journal*, 8 June 1970, 1. On the evolution of NASA technology from TIROS to ATS to Landsat, see Hertzfeld and Williamson, "Social and Economic Impact of Earth Observing Satellites," 240–238. Landsat was originally called the Earth Resources Technology Satellite (ERTS).

104. United Press International, "NASA Satellite to Be Launched: 2d Earth Resources Craft to Relay Environmental Data," *New York Times*, 15 January 1975, 53. See also "Photography from Space to Help Solve Problems on Earth: NASA Earth Resources Technology Satellite," pamphlet published by NASA's Goddard Space Flight Center, circa 1972, NASA Headquarters Archives, NASA Historical Materials, Folder 5745: ERTS Photos and Booklets, Washington, DC, 2.

105. M. Mitchel Waldrop, "Imaging the Earth (I): The Troubled First Decade of Landsat," *Science* 215 (26 March 1982): 1601.

106. Ibid.

107. For an overview of the history of Landsat, see Pamela Mack, *Viewing the Earth: The Social Construction of the Landsat Satellite System* (Cambridge, MA: MIT Press, 1990). See also "The Landsat Program," http://www.earth.nasa.gov /history/landsat/landsat.html; and Waldrop, "Imaging the Earth (I)."

108. "Ecological Surveys from Space," Office of Technological Utilization, NASA SP-230, 1970, NASA Headquarters Archives, NASA Historical Materials, Folder 5754: Earth Resources Satellite, 1970, Washington, DC; NASA, *Improving Our Environment* (Washington, DC: U.S. Government Printing Office, 1973); "Photography From Space to Help Solve Problems on Earth."

109. Jeffrey St. John, "Space Effort: No Apologies Necessary," *Los Angeles Times*, 9 April 1972, C3. See also "Space: Can NASA Keep Its Programs Aloft?," *Business-Week*, 13 February 1971, 23.

110. NASA's budget during these years decreased from $3.9 billion to $3.2 billion annually. For historical data on NASA's total budget for these years in both real and in 2008 inflation-adjusted dollars, see National Aeronautics and Space Council, *Aeronautics and Space Report of the President: Fiscal Year 2008 Activities*, "Appendix D-1A," 146, and "Appendix D-1B," 147.

111. George Low Papers #27, 18 July 1970, "Fiscal Year 1972 Budget and Programmatic Discussions," 12–14, Box 70, Low, Rensselaer Polytech Institute.

112. On the economic benefits to American farmers of NASA's LACIE program, see Hertzfeld and Williamson, "Social and Economic Impact of Earth Observing Satellites," 240–241, 262. On the economic benefits for the United States of other satellite earth observations, see "Aerospace Research Profits Earth," *Roundup* (newspaper of the Johnson Space Center), 18 February 1972, 2.

113. Between 1975 and 1980 Congress increased NASA's total budget from $3.22 billion to $5.24 billion. When adjusted for inflation according to 2008 dollars, this represents an increase of 10 percent. For historical data on NASA's total budget in both real and inflation-adjusted dollars, see National Aeronautics and Space Council, *Aeronautics and Space Report of the President: Fiscal Year 2008 Activities*, "Appendix D-1A," 146, and "Appendix D-1B," 147.

114. "Text of Address by President Nixon to General Assembly of the United Nations," *New York Times*, 19 September 1969, 16.

115. "NASA Position Paper on the Remote Sensing of Planetary Surfaces (Earth, Moon, Mars, Venus, Etc) from Orbital and Fly-By Spacecraft," paper attached to memorandum by NASA Advanced Missions Program Chief Peter Badgley, 8 October 1965, Box#075-14, Series: Apollo, Johnson Space Center History Collection, University of Houston at Clear Lake, Houston, TX.

116. For an example of this concern, see Edward Keating, "Hard Times: World Spy," *Ramparts* 9, no. 8 (March 1971): n.p.

117. The proposal, which ultimately stalled in the UN's Scientific and Technical Sub-Committee, was actually stricter than a similar United Nations proposal put forth by France and the Soviet Union the year before. On both these proposals to the United Nations, see Hamilton DeSassure, "Remote Sensing by Satellite: What Future for an International Regime?," *American Journal of International Law* 71, no. 4 (October 1977): 714, 720. On Landsat possibly being used by developed countries to exploit natural resources in developing countries, see also John Hanessian, "International Aspects of Earth Resources Survey Satellite Programs," *Journal of the British Interplanetary Society* 23 (Spring 1970): 548.

118. "ERTS-EREP Proposal Review," memorandum by Verl Wilmarth to TA / Director of Science and Applications, 7 July 1971, Record Number 14994, Box 529, Johnson Space Center Archives, University of Houston at Clear Lake, Houston, TX.

119. On the lack of trained photo interpreters in the developing world, see Hanessian, "International Aspects," 545. For additional concerns within NASA regarding a lack of skilled scientists in developing countries to take advantage of space technology, see "Practical Applications of Space Systems" (NASA-CR-145434), National Academy of Sciences, Washington, DC, 1975.

120. For examples of these press releases by NASA, see "Earth Resources Experiments RFP," Press Release No 70-117, 14 July 1970, Folder 5754: Earth Resources Satellite, 1970, NASA Headquarters Archive, NASA Historical Materials, Washington, DC; and "Skylab Experimenters Sought," Press Release No 71-5, 19 January 1971, Record Number 142778, Box 502, Johnson Space Center Archives, University of Houston at Clear Lake, Houston, TX.

121. In 1971 the United Nations created a Space Applications Program specifically to promote the use of earth-observing remote-sensing data throughout the developing world. On such efforts, see V. Klemas and D. J. Leu, "Applicability of Spacecraft Remote Sensing to the Management of Food Resources in Developing Countries," Center for Remote Sensing, University of Delaware, report prepared for the School of Engineering and Applied Science, George Washington University, Washington, DC, and the Division of International Programs, National Science Foundation, Washington, DC, 31 March 1977, 31–48.

122. On NASA's Landsat conference at the University of Michigan, see "Earth Resources Survey Workshop," NASA Press Release No. 70-215, 23 December 1970, NASA Headquarters Archive, NASA Historical Materials, Folder 5754: Earth Resources Satellite, 1970, Washington, DC. On the Johnson Space Center conference, see "All You Wanted to Know about Earth Resources," *Roundup* 14, no. 12 (6 June 1975): 1. Lady Bird Johnson had NASA administrator Thomas Paine give a similar lecture on remote sensing of earth resources at a cocktail party she threw for a group of foreign correspondents covering one of the early Apollo launches. On this cocktail party lecture, see Oral Interview of Dr. Thomas O. Paine by T. Harri Baker (Tape 2), Folder 4185: Paine Interviews Conducted by Baker, Lodsdon, Cohen, and Burke, NASA Historical Materials, NASA Headquarters Archives, Washington, DC.

123. For a description of Landsat conferences in Africa in Ghana and Mali, see "Landsat May Help Bridge Technological Gap," *Roundup*, 23 May 1975, 2; and "W. Africans Confer on Uses of Remote Sensing Data," *Roundup*, 6 June 1975, 2. On Landsat data being used in response to the Sahel drought, see "Aid to W. Africa Aim of US Profs.," *Chicago Defender*, 8 September 1973, 25.

124. On similar Landsat conferences in the Philippines, see "Earth Resources Team Visits to the Philippines," Record Number 210333, Report Number SRE, 21 September 1971, Box 546, Johnson Space Center History Collection, University of Houston at Clear Lake, Houston, TX. On efforts to promote Landsat across Latin America, see "Inter American Geodetic Survey Proposal for Multi-National

ERTS (Earth Resources Technology Satellite) Cartographic Experiments," Record Number 213145, Report Number IAGS-EROS, 7 April 1972, Box 563, Johnson Space Center History Collection, University of Houston at Clear Lake, Houston, TX.

125. On the expansion of NASA's international fellowship program to include study at U.S. universities on Landsat, see John Hanessian Jr. and John Logsdon, "Earth Resources Technology Satellite: Securing International Participation," *Astronautics and Aeronautics* (August 1970): 60.

126. On NASA's training of Brazilian and Mexican scientists and engineers at the Johnson Space Center, see Hanessian and Logsdon, "Earth Resources Technology Satellite," 59; and Hanessian, "International Aspects," 546. In this particular case the remote-sensing data were obtained from aircraft circling above those countries rather than from satellites.

127. On NASA encouraging developing countries to build their own receiving stations, see Hertzfeld and Williamson, "Social and Economic Impact of Earth Observing Satellites," 239. On developing nations building Landsat receiving stations, see Klemas and Leu, "Applicability of Spacecraft Remote Sensing," 42. On Zaire's Landsat receiving station in particular, see "Landsat May Help Bridge Technological Gap," *Roundup*, 23 May 1975, 2.

128. On this extensive use of Landsat data by fifty countries worldwide, see Klemas and Leu, "Applicability of Spacecraft Remote Sensing," 41.

129. For a detailed description of Landsat data use in South America, especially in Brazil, see Allen Hammond, "Remote Sensing (I): Landsat Takes Hold in South America," *Science* 196, no. 4289 (29 April 1977): 511–512.

130. For descriptions of Landsat data being used in Central and South America, see Klemas and Leu, "Applicability of Spacecraft Remote Sensing," 31–48.

131. On Landsat data being applied to Sudan's Kordofan Province and Botswana's Okavango Delta region, see Klemas and Leu, "Applicability of Spacecraft Remote Sensing," 7, 33. On Landsat data being used for range management in Kenya, see "Landsat May Help Bridge Technological Gap," 2.

132. On NASA's Landsat program across the Sahel region of Africa, see Klemas and Leu, "Applicability of Spacecraft Remote Sensing," 41.

133. Klemas and Leu, "Applicability of Spacecraft Remote Sensing," 7. For a list of African countries undertaking experiments with Landsat data, see Klemas and Leu, "Applicability of Spacecraft Remote Sensing," 47.

134. On NASA's use of Landsat in cooperation with the World Bank in Asia, see Klemas and Leu, "Applicability of Spacecraft Remote Sensing," 35–36.

135. NASA's Goddard Space Flight Center in Greenbelt, Maryland, oversaw this Landsat project, which was a cooperative effort with the Mekong Committee, an international planning organization sponsored by the United Nations' Economic and Social Commission for Asia and the Pacific. On this effort, see

Jeffrey W. Jacobs, "Mekong Committee History and Lessons for River Basin Development," *Geographical Journal* 161, no. 2 (July 1995): 142–143. For the early history of the committee, see also W. R. Derrick Sewell, "The Mekong Scheme: Guideline for a Solution to Strife in Southeast Asia," *Asian Survey* 8, no. 6 (June 1968): 448–455.

136. These three maps are described in detail in Mekong Committee Secretariat Mr. W. J. van der Oord and Mr. Frederick Gordon, Technical Monitor, Goddard Space Flight Center, "Agriculture / Forestry Hydrology," Quarterly Report, Dec. 1975–Feb. 1976, Mekong Committee Secretariat c / o ESCAP Sala Santitham Bangkok, Thailand, 1 April 1976, NASA Technical Reports Server, Document ID 19760016569, Accession ID 76N23657, Report Number E76-10330, NASA-CR-147211, REPT-2, 1–6.

137. Mekong Committee Secretariat Willem J. van Liere, "Applications of Multispectral Photography to Water Resources Development Planning in the Lower Mekong Basin (Khmer Republic, Laos, Thailand and Viet-Nam," 9 March 1973, NASA Technical Report Document ID 19730008739, Accession ID 73N17466, Report Number E73-10257, PAPER-W3, NASA Headquarters, Washington, DC, 76.

138. On diplomatic difficulties between China and the Soviet Union during this period, see Lorenz Luthi, *The Sino-Soviet Split: Cold War in the Communist World* (Princeton, NJ: Princeton University Press, 2008); and Chen Jian, *Mao's China and the Cold War* (Chapel Hill: University of North Carolina Press, 2000).

139. On the Intercosmos Council space program, see "Socialist Countries Cooperation in Space," *TASS* (Moscow), 16 September 1977, Folder 001693, Title: "Petrov, Boris N. (USSR Academician—Bio)," NASA History Archives, NASA History Division, NASA Headquarters, Washington, DC.

140. B. Petrov, "New Stage of Cooperation," *Pravda* (Moscow), 5 March 1978, translated from Russian by FBIS, Folder 001693, Title: "Petrov, Boris, N (USSR Academician—Bio)," NASA Archives, NASA Headquarters, Washington, DC, 1. For additional material on the central role played by the Salyut space station in the efforts of the Intercosmos Council, see Boris Petrov, "Orbital Stations: A Soviet Scientist Reviews the Scientific and Economic Benefits," *Spaceflight* (August 1973): 288–290; and Boris Petrov, "The USSR and International Cooperation in Space Research," *Moscow News*, no. 15 (1162) (21–28 April 1973): 3, 14.

141. As quoted in Marc Rosenwasser, "Soviet Space Program for Foreigners Reaps Public Relations Benefit," *Association Press*, 22 September 1980.

142. On the Soyuz 37 mission being used to promote the benefits of Soviet-style communism, see "Vietnamese Cable to Soviet Leaders," Moscow Domestic Service in Russian, 1 August 1980, translated from Russian by FBIS, Folder #002345, Title: "Tuan, Pham (Lt. Col.) Soyuz 37 North Vietnamese,"

NASA History Archives, NASA History Division, NASA Headquarters, Washington, DC.

143. On the Laotian government's praise for the Soyuz 37 mission, see "Leaders Congratulate SRV Counterparts on Space Shot," Vientiane KPL in English, 26 July 1980, as transcribed by FBIS, Folder #002345, Title: "Tuan, Pham (Lt. Col.) Soyuz 37 North Vietnamese," NASA History Archives, NASA History Division, NASA Headquarters, Washington, DC. On similar praise from Sri Lanka, see "Sri Lanka CP Leader Congratulates Le Duan on Space Flight," Hanoi VNA (Vietnamese News Agency) in English, 31 July 1980, as transcribed by FBIS, Folder #002345, Title: "Tuan, Pham (Lt. Col.) Soyuz 37 North Vietnamese," NASA History Archives, NASA History Division, NASA Headquarters, Washington, DC. For information on the Cuban reception, see "USSR-SRV Space Flight, Cooperation Lauded, Havana, Phnom Penh Meetings," Hanoi VNA (Vietnam News Agency) in English, 9 August 1980, as transcribed by FBIS, Folder #014776, Title: "Vietnam," NASA History Reference Collection, NASA History Division, NASA Headquarters, Washington, DC.

144. Sterba, "Hanoi Touts Its Astronaut." For additional examples of western concern regarding the impact of the Soyuz 37 across the developing world, see Anthony Austin, "Vietnam-Soviet Space Team Down; Political Impact Is Called Extensive," *New York Times*, 1 August 1980, B4, who reported that the mission was "bound to have an impact on people's minds throughout Asia."

145. On the Intercosmos Council chairman Boris Petrov's discussion of the MKF-6 multizone camera, see Petrov, "New Stage of Cooperation."

146. On Soyuz 37–Salyut 6 experiments regarding Vietnamese natural resources, see "Further on Joint Soviet-Vietnamese Spaceflight"; and Neville Kidger, "Salyut 6 Mission Report—Part 6," *Spaceflight* 23, no. 3 (March 1981): 75. On the mission's experiments forecasting typhoons and hurricanes, see "Academician on Soyuz-37 Program," Moscow Domestic Service in Russian, 25 July 1980, as transcribed by FBIS, Folder #001693, Title: "Petrov, Boris N (USSR Academician—Bio)," NASA History Archives, NASA History Division, NASA Headquarters, Washington, DC. Regarding the mission's hydrological studies of the Mekong and Red River basins, see "Further Details of Soyuz-37 Mission," Telegraph Agency of the Soviet Union in English and Vietnamese News Agency in English, as reported by the BBC Summary of World Broadcasts, 6 August 1980, Section: Part 1—The USSR; D. Space Research; SU/6490/D/1.

147. On this statement by Vietnam's National Scientific Research Center deputy chairman, Nguyen Van Hieu, see "Co-operation in Cosmic Studies with USSR and GDR," Hanoi in English, as reported by the BBC Summary of World Broadcasts, 6 August 1980, Section: Part 3—The Far East, Weekly Economic Report; B, Broadcasting, Telecommunications, Space, Vietnam, FE/W1096/B/2.

148. On the mapping of natural resources in Cuba during the Intercosmos Council's Soyuz-38 mission, see United Press International, "Cuban Cosmonaut and His Soviet Comrade Return to Earth," 26 September 1980. For a similar mapping of Mongolia during Soyuz-39, see Associated Press, "Soviet Spacemen Return to Earth," *Tri City Herald* (Pasco, Richland, and Kennewick, WA), 27 March 1981, 4. For the mapping of Afghanistan's natural resources during the Soyuz TM-6 mission, see Esther B. Fein, "Soviets Send Doctor to Space Station," *New York Times*, 30 August 1988, C10.

149. "Further on Joint Soviet-Vietnamese Spaceflight."

150. Kidger, "Salyut 6 Mission Report—Part 6."

151. The Soviets made this explicit in a joint public message from Soviet leader Leonard Brezhnev and his Vietnamese counterpart, Le Duan. On this statement see "Brezhnev–Le Duan Congratulations," *Pravda* (Moscow), 26 July 1980, as translated by FBIS, Folder #002345, Title: "Tuan, Pham (Lt. Col.) Soyuz 37 North Vietnamese," NASA History Archives, NASA History Division, NASA Headquarters, Washington, DC.

152. David K. Willis, "Soviets Bid for Third-World Prestige with Viet Cosmonaut," *Christian Science Monitor*, 15 July 1980, 3.

153. Sterba, "Hanoi Touts Its Astronaut."

154. On Pham Tuan returning home with photographs of Vietnam from space, undertaking a national tour of the country, and planting trees as "keepsakes," see "Hero Pham Tuan Talks about Historic Flight—Part IV: Seven Days in Space and a Lifelong Brotherhood," *People's Army Newspaper Online*, 8 May 2008, online at http://www.qdnd.vn/qdndsite/en-us/75/72/184/164/207/61322/Default.aspx (accessed 3 January 2013).

155. While it has been somewhat difficult to draw clear distinctions between the New Left's anti-NASA activism versus its anti-war activism, the combination of the two in the public arena during the late 1960s and early 1970s undoubtedly convinced NASA administrators to redirect their space technology back down toward Earth.

156. On the porous boundaries between science and technology on the one hand and politics on the other, especially during the Cold War, see Solovey, "Science in the Cold War," especially the introduction by Mark Sloveny, "Science and the State during the Cold War: Blurred Boundaries and a Contested Legacy." For other examples see Leslie, *Cold War and American Science*; and Paul Forman, "Behind Quantum Electronics: National Security as Basis for Physical Research in the United States, 1940–1960," *Historical Studies in the Physical Sciences* 18, pt. 1 (1987): 149–229.

157. For a similar discussion of local indigenous people using Landsat data for their own purposes, see Karen Litfin, "The Gendered Eye in the Sky: A Feminist

Perspective on Earth Observation Satellites," *Frontiers: A Journal of Women Studies* (Fall 1997): 41.

158. On NASA and the U.S. government retaining ultimate control over Landsat experiments, see Hanessian, "International Aspects," 552. In reviewing Landsat proposals from developing countries, NASA administrators were able to list an experiment as "N," meaning "Negotiation Required." Doing so gave NASA more control over the experiment being proposed. For an example of this process, see "Additional EREP Investigations," memorandum by NASA Associate Administrator for Applications Charles Mathews to Manned Spacecraft Center (Johnson Space Flight Center) Director Chris Kraft, 21 April 1972, Record Number 146924, Box 535, Johnson Space Center Archives, University of Houston at Clear Lake, Houston, TX.

159. My thinking here has been influenced by John Krige, *American Hegemony and the Postwar Reconstruction of Science in Europe* (Cambridge, MA: MIT Press, 2006), especially its introduction. For examples of the rich literature on coproduction within science studies, see Sheila Jasanoff, *States of Knowledge: The Co-Production of Science and the Social Order* (London: Routledge, 2004).

160. There is a rich literature on the concept of "soft power." See especially Joseph Nye, *Bound to Lead: The Changing Nature of American Power* (New York: Basic Books, 1990), and *Soft Power: The Means to Success in World Politics* (New York: Foreign Affairs, 2004).

161. These satellite data were compiled by both Landsat 1 and Landsat 2. For a description of this involvement by local scientists and government officials in corroborating Landsat data, see Mekong Committee Secretariat Mr. W. J. van der Oord and Mr. Frederick Gordon, Technical Monitor, Goddard Space Flight Center, "Agriculture/Forestry Hydrology," Quarterly Report, Dec. 1975–Feb. 1976, Mekong Committee Secretariat c/o ESCAP Sala Santitham Bangkok, Thailand, 1 April 1976, NASA Technical Reports Server, Document ID 19760016569, Accession ID 76N23657, Report Number E76-10330, NASA-CR-147211, REPT-2, 3 and page 2 of the report's "Annex II: Note on the Land Use Map of the Lower Mekong Basin."

162. For references to "South Viet-Nam," see ibid., 2. The Landsat map, which is titled "Landsat-2 Ground Track Coverage of Lower Mekong Basin," appears in the same publication on what is labeled by hand "5a," which is actually the eighth page in the report.

163. On the devastating impact of this ban on the Mekong Committee's Landsat work in Vietnam, see Jeffrey W. Jacobs, "The United States and the Mekong Project," *Water Policy* 1 (1998): 592.

164. Vietnamese News Agency, "Cosmonaut Addresses Vietnamese People from Space."

165. Hanessian, "International Aspects," 555.

3. Thinking Globally, Acting Locally

1. "Stewart Brand: Why Haven't We Seen the Whole Earth?," in *The Sixties: The Decade Remembered Now, by the People Who Lived It Then* (New York: Random House / Rolling Stone, 1977), 168–169; Stewart Brand, "The First Whole Earth Photograph," in *Earth's Answer: Explorations of Planetary Culture at the Lindisfarne Conferences*, ed. Michael Katz, William P. Marsh, and Gail Gordon Thompson (New York: Harper and Row / Lindisfarne, 1977), 185–189. NASA distributed the *Whole Earth* photograph to the national and international press just twelve hours after Apollo 17's splashdown on 19 December 1972.

2. An astronaut supposedly told Brand that the buttons were being worn at NASA Headquarters. For a reference to this, see Vicki Goldberg, *Power of Photography: How Photographs Changed Our Lives* (New York: Abbeville, 1993), 54.

3. On the importance of *Whole Earth* and other Earth photographs on environmental thought, see Denis Cosgrove, "Contested Global Visions: One-World, Whole-Earth, and the Apollo Space Photographs," *Annals of the Association of American Geographers* 84, no. 2 (June 1994): 270–294; Goldberg, *Power of Photography*, 52–57; Wolfgang Sachs, "An Ambiguous Modern Icon," *Ecologist* 24, no. 5 (September / October 1994): 170–175; Yaakov Jerome Garb, "Perspective or Escape? Ecofeminist Musings on Contemporary Earth Imagery," in *Reweaving the World: The Emergence of Ecofeminism*, ed. Irene Diamond and Gloria Feman Orenstein (San Francisco: Sierra Club, 1990), 264–278; Holly Henry and Amanda Taylor, "Re-Thinking Apollo: Envisioning Environmentalism in Space," *Sociological Review* 57, supp. S1 (May 2009): 190–203; Sheila Jasanoff, "Image and Imagination: The Formation of Global Environmental Consciousness," in *Changing the Atmosphere: Expert Knowledge and Environmental Governance*, ed. Clark A. Miller and Paul N. Edwards (Cambridge, MA: MIT Press, 1996), 309–336; Nina Edwards Anker and Peder Anker, "Viewing the Earth from Without or from Within," in *New Geographies 4: Scales of the Earth* (Cambridge, MA: Harvard University Press, 2011), 89–94; Robin Kelsey, "Reverse Shot: Earthrise and Blue Marble in the American Imagination," in *New Geographies 4: Scales of the Earth* (Cambridge, MA: Harvard University Press, 2011), 10–16; and Robert Poole, *Earthrise: How Man First Saw the Earth* (New Haven, CT: Yale University Press, 2008).

4. Although the St. Louis *Post-Dispatch* first used the phrase "bright marble" to describe the view of Earth from space in general, the name "blue marble" for 22727 arose in popular culture around 1974, when the Public Broadcasting System aired a syndicated half-hour children's television series titled *Big Blue Marble*. See the editorial "As the Apollo Missions End," *Post-Dispatch* (St. Louis, MO), 5 December 1972.

5. Stewart Brand, "Photography Changes Our Relationship to Our Planet," Smithsonian Photography Initiative, online at http://click.si.edu/Story.aspx?story =31#top (accessed 14 April 2014). *Click! Photography Changes Everything* is an online initiative by the Smithsonian Institution Archives.

6. NASA, *Apollo 17: Preliminary Science Report* (NASA SP-330) (Washington, DC: National Aeronautics and Space Administration, 1973), 4–9. This library of photographs of Earth from space is mentioned in Harold M. Schmeck Jr., "Apollo to Study Theory of Drift," *New York Times*, 12 October 1968, 20.

7. The quote comes from Schmeck Jr., "Apollo to Study Theory of Drift," 20. According to this article, scientists at the Environmental Science Services Administration prepared at least sixty to eighty weather targets for photography for the Apollo 7 astronauts. For other reporting on scientists using early NASA photographs of Earth from space, see "Space Shots Yield Knowledge of Seas," *New York Times*, 1 June 1969, 82.

8. There is a rich literature on the fragmentation of the early environmental movement, as well as the role of scientists within that movement. My thinking here is most influenced by Adam Rome, *The Genius of Earth Day: How a 1970s Teach-In Unexpectedly Made the First Green Generation* (New York: Hill and Wang, 2013), 20–29; Robert Gottlieb, *Forcing the Spring: The Transformation of the American Environmental Movement* (Washington, DC: Island Press, 1993), 81–88; and Michael Egan, *Barry Commoner and the Science of Survival: The Remaking of American Environmentalism* (Cambridge, MA: MIT Press, 2007), especially chapters 1 and 2.

9. On this decision by NASA, see Nicholas de Monchaux, *Spacesuit: Fashioning Apollo* (Cambridge, MA: MIT Press, 2011), 337.

10. There is very little scholarship by environmental historians about outer space. For examples, see Neil M. Maher, "Gallery: Neil Maher on Shooting the Moon," *Environmental History* 9 (July 2004): 526–531; and Michael Rawson, "Discovering the Final Frontier: The Seventeenth-Century Encounter with the Lunar Environment," *Environmental History* 20, no. 2 (2015): 194–216. Peder Anker has also published on the connections between space exploration and architecture: see "The Ecological Colonization of Space," *Environmental History* 10, no. 2 (2005): 239–268; "The Closed World of Ecological Architecture," *Journal of Architecture* 10, no. 5 (2005): 527–552; and *From Bauhouse to Ecohouse: A History of Ecological Design* (Baton Rouge: Louisiana State University Press, 2010), especially chapters 5–8. Michael Bess calls for more research on outer space in "Artificialization and Discontents," *Environmental History* 10, no. 1 (2005): 31–33.

11. For examples of such scholarship, see Kim McQuaid, "Selling the Space Age: NASA and Earth's Environment, 1958–1990," *Environment and History* 12 (2006): 128; and Henry Lambright, "NASA and the Environment: Science in a

Political Context," in *Societal Impact of Spaceflight*, ed. Steven Dick and Roger Launius (Washington, DC: NASA Office of External Relations, History Division, 2007), 313–330.

12. McQuaid, "Selling the Space Age," 128.

13. The natural advantages of Cape Canaveral as a spaceport are discussed in full in Charles D. Benson and William B. Faherty, *Gateway to the Moon: Building the Kennedy Space Center Launch Complex* (Gainesville, FL: University Press of Florida, 2001), 3, 49, 89.

14. Verne discusses Florida's geographic attributes for a space launch in "Chapter XI: Florida and Texas," in *From the Earth to the Moon* (New York: A. L. Burt, 1889).

15. "KSC Story—From Marshland to Spaceport," *Spaceport News* 7, no. 21 (10 October 1968): 2. *Spaceport News* is the official newsletter of the Kennedy Space Center. For additional descriptions of the Cape as a "vast wilderness," see "Counting Down with the Editor," *Spaceport News* 7, no. 15 (18 July 1968): 8.

16. For the use of these terms to describe Cape Canaveral before NASA arrived in the area, see "Rocket Pioneers Faced Rattlesnakes, Mosquitoes," *Spaceport News* 7, no. 21 (10 October 1968): 4; Benson and Faherty, *Gateway to the Moon*, 4; and Cliff Lethbridge, *Cape Canaveral: 500 Years of History, 50 Years of Rocketry* (Space Coast Cover Service, 2000), especially "Chapter 2: The Missile Range Takes Shape (1949–1958)," online at http://www.spaceline.org/capehistory/2a.html (accessed 14 April 2014).

17. For a brief history of Native Americans at the Cape, see Cliff Lethbridge, "The History of Cape Canaveral Chapter 1: Cape Canaveral before Rockets (B.C.–1948)," available online at www.spaceline.org/capehistory/1a.html (accessed 7 December 2016). On the early orange industry in the Cape Canaveral area, see Alfred Jackson Hanna and Kathryn Abbey Hanna, *Florida's Golden Sands* (Indianapolis: Bobbs-Merrill, 1950), 232–244.

18. "Rocket Pioneers Faced Rattlesnakes, Mosquitoes," 4.

19. On NASA's land purchases on the Cape, see Benson and Faherty, *Gateway to the Moon*, 93, 105–107.

20. On clearing land at Cape Canaveral and the construction of Complex 39, see Benson and Faherty, *Gateway to the Moon*, 105–107, 247. Similar activities began even earlier at the Cape when in June 1950 the U.S. military cleared six square kilometers of underbrush and built a concrete launchpad for missile testing. On this military history at the Cape prior to NASA, see Benson and Faherty, *Gateway to the Moon*, 3–4; and Lethbridge, *Cape Canaveral*, especially chapter 2.

21. "New Look at the Cape," *Time* 85, no. 13 (26 March 1965): 104–107.

22. General Electric's Ray Forbes is quoted in Bernard Weinraub, "Moon Shot Center Uneasy over Space Cutbacks," *New York Times*, 20 May 1969, 20.

23. Much has been written regarding the impact of Sputnik on American culture, yet perhaps the most engaging coverage still remains Tom Wolfe, *The Right Stuff* (New York: Farrar, Straus, and Giroux 1979), 52–57. For analysis of the American media's portrayal of Sputnik as signifying a technological crisis for the United States, and Americans' anxiety in response to such accounts, see Jack Lule, "Roots of the Space Race: Sputnik and the Language of U.S. News in 1957," *Journalism Quarterly* 68, no. 1/2 (Spring/Summer 1991): 76–86.

24. For an analysis of the technological display at the 1964 World's Fair, see Michael Smith, "Making Time: Representations of Technology at the 1964 World's Fair," in *The Power of Culture: Critical Essays in American History*, ed. Richard Wightman Fox and T. J. Jackson Lears (Chicago: University of Chicago Press, 1993), 223–244.

25. Michael Smith discusses the role of space technology in general at the fair, as well as NASA's Space Park in particular, in "Making Time," 229–236.

26. "U.S. Space Park, New York World's Fair 1964–1965," Official World's Fair Post Cards by Dexter, West Nyack, NY, circa 1964.

27. On the evolution of bus tours at the Kennedy Space Center, see "NASA Tours Marks Sixth Anniversary," *Spaceport News*, 27 July 1972, 3; and Sanders LaMont, "Moonport Visits Soaring into the Millions," *Florida Today*, 19 July 1970, 8F.

28. Edwin P. Hoyt, "A Visitor's Guide to Kennedy Space Center: Orbiting the Apollo, Gemini and Saturn V," *Baltimore Sun*, 11 May 1969, 19.

29. Ibid.

30. LaMont, "Moonport Visits." NASA also built a similar Rocket Garden outside its Marshall Space Flight Center in Huntsville, Alabama; the Johnson Space Center in Houston; and its Goddard Space Flight Center in Greenbelt, Maryland.

31. Hoyt, "Visitor's Guide to Kennedy Space Center."

32. Although postcards have been previously shunned as historical evidence because of questions regarding their photographic honesty, scholars have increasingly turned to them as historical texts that can be analyzed. One of the best recent uses of postcards as historical sources is Alison Eisenberg, *Downtown America: A History of the Place and the People Who Made It* (Chicago: University of Chicago Press, 2004), especially chapter 2. See also Steven Dotterer and Galen Cranz, "The Picture Postcard: Its Development and Role in American Urbanization," *Journal of American Culture* 5, no. 1 (Spring 1982): 46. I collected the vintage NASA postcards depicted in this book, and approximately seventy others, over a five-year period beginning in 2005. Although during the time period I am studying several companies produced these postcards, NASA retained some control over their production; the majority of these postcards note, on their address side, that the space agency supplied the photograph depicted on the front side. On the role of visual culture in promoting space

exploration, see William R. Macauley, "Crafting the Future: Envisioning Space Exploration in Post-War Britain," *History and Technology* 28, no. 3 (September 2012): 281–309.

33. This postcard of a NASA rocket being assembled inside the VAB was one of twelve "Jumbo Prints" that came in a set documenting the Kennedy Space Center bus tour. See "Print 1: Inside the V.A.B.," in "Kennedy Space Center, 12 Jumbo Prints," published and distributed by Promotions Original Products, Orlando, FL. All the prints included in this set depict space technology, including the exterior of the VAB with a Saturn V rocket being transported to its launchpad, the "crawler" that transports Apollo rockets from the VAB to the launchpad, and a Saturn V rocket poised for launch. The set of Jumbo prints is in the possession of the author.

34. The postcard of visitors touring a mockup of the lunar landing module in the Rocket Garden was another of the twelve Jumbo prints that came in a set documenting the Kennedy Space Center bus tour. See "Print 2: Lunar Module," in "Kennedy Space Center, 12 Jumbo Prints." For the postcard depicting the interior of the Kennedy Space Center's Space Museum, see "John F. Kennedy Space Center, N.A.S.A.," postcard produced by NASA TOURS, Conducted by TWA, P.O. Box 21222, Kennedy Space Center, FL, circa mid-1960s. For an example of a postcard portraying the exterior of the Kennedy Space Center's visitor center, see "Entrance to Visitors' Information Center, Kennedy Space Center, Florida, N.A.S.A.," NASA TOURS, Conducted by TWA, P.O. Box 21222, Kennedy Space Center, FL, circa mid-1960s. All these postcards are in the possession of the author.

35. Diamond describes what he calls the "space circus on TV" in his book *The Rise and Fall of the Space Age* (New York: Doubleday, 1964), 95–98.

36. For a similar argument, see Michael Smith, "Selling the Moon: The U.S. Manned Space Program and the Triumph of Commodity Scientism," in *The Culture of Consumption: Critical Essays in American History, 1880–1980*, ed. Richard Wightman Fox and T. J. Jackson Lears (New York: Pantheon, 1983), 177.

37. "The Bright Side of the Moon Mission . . . ," *Washington Post*, 20 July 1969, 40.

38. Erma Bombeck, "At Wit's End: Awe-Inspiring Visit to the Land of the Saturns," *Evening Independent* (St. Petersburg, FL), 14 March 1968, 12A.

39. "Gemini Crew Makes More Records, Photos," *Evening Independent* (St. Petersburg, FL), 14 November 1966, 1.

40. For critical coverage of space "junk" left in Earth's orbit, see "Getting Crowded in Space, *Philadelphia Inquirer*, 20 May 1969, article from Apollo 10 clipping file, National Air and Space Museum Library, National Air and Space Museum, Washington, DC, 28; and "Apollo Rocket Becomes Junk in 9 Minutes," *Chicago Tribune*, 22 December 1968, A2.

41. Rose DeWolf, "Astronauts Are Neat in Their Cabin But Will Be Litterbugs on the Moon," (Philadelphia) *Evening Bulletin*, 17 July 1969, 1.

42. The beer-can comment appeared in ibid. On Apollo 11 being ticketed for littering, see William McCann, "Litter Is Remnant of Man's First Moon Visit," *Plain Dealer* (Cleveland, OH), 21 July 1969, article from Apollo 11 clipping file, National Air and Space Museum Library, National Air and Space Museum, Washington, DC, 208. For similar examples, see "Lunar Litter," *Newsweek*, 28 July 1969, 10; and "One Giant Step—One Giant Leap," *Reader's Digest*, October 1969, 292.

43. Douglas Borgstedt, "You'd Think They'd at Least Pick Up!," (Philadelphia) *Sunday Bulletin*, 27 July 1969. For a similar cartoon portraying aliens on the moon complaining about Apollo 11 litterbugs, see also Bill Mauldin, "First They'll Bring Us an Atmosphere, Then They'll Pollute It," *Daily Reporter* (Dover, OH), 25 July 1969, 4.

44. Scott Long, "Mankind's Magnificent Obsession . . . ," *Minneapolis Tribune*, 23 July 1969.

45. "Moon Shot," *The Screw* 1, no. 10 (1969): n.p., online at http://asp6new.alex anderstreet.com/sixt/sixt.object.details.aspx?id=1004100623&view=fulltextview (accessed 1 May 2014).

46. "Sky-High Pollution," *Newsweek*, 4 February 1963, 48.

47. For an example of environmentalist opposition to the SST, including that of the Sierra Club, the Wilderness Society, and the Conservation Foundation, see Bruce L. Welch, "SST: Coming Threat to Wilderness," *National Park Magazine*, March 1968, 9–11. For a general overview of this opposition by environmentalists, see also Erik Conway, *High-Speed Dreams: NASA and the Technopolitics of Supersonic Transportation, 1945–1999* (Baltimore, MD: Johns Hopkins University Press, 2005), 127–132; and Joshua Howe, *Behind the Curve: Science and the Politics of Global Warming* (Seattle: University of Washington Press, 2014), 44–50.

48. Ronald Kotulak, "Warns of Freon in Atmosphere," *Chicago Tribune*, 3 April 1974, 5. For additional examples of such media coverage, see also "Booms and Fumes," *Christian Science Monitor*, 15 November 1967, 22; and Jim Bishop, "Jim Bishop Reporter," *Terre Haute (IN) Tribune*, 19 March 1971, 4.

49. See "The Moon Age," a special section in *Newsweek*, 7 July 1969, 40–68, especially John G. Mitchell, "Good Earth?," 57–59.

50. Victor Cohn, "Space Program Adopts New Look," *Boston Globe*, 12 April 1970, 33.

51. Gene Payne, "He Says He's from the Moon," *Baltimore Sun*, 27 July 1969.

52. L. D. Warren, "Let's Take a Few More Deep Breaths," (Philadelphia) *Evening Bulletin*, 22 July 1969. Warren also implicates the United States government by placing the American flag front and center as well.

53. Stuart Udall, *The Quiet Crisis* (New York: Holt, Rinehart and Winston, 1963), vii.

54. Ralph Nader, *Unsafe at Any Speed: The Designed-In Dangers of the American Automobile* (New York: Grossman, 1965), 147, as quoted in Rome, *Genius of Earth Day*. On the role of these technological disasters in bringing together environmentalists during the late 1960s, see Kirkpatrick Sale, *The Green Revolution: The American Environmental Movement, 1962–1992* (New York: Hill and Wang, 1993), 18–19.

55. On Earth Day events being focused on fighting pollution, see Rome, *Genius of Earth Day*, especially chapter 3.

56. James Reston, "Fiery Run, VA: Something to Muse about on a Summer's Day," *New York Times*, 12 August 1962, E10. For similar complaints lodged the following year by high-ranking scientists, see James Reston, "The Man on the Moon and the Men on the Dole," *New York Times*, 5 April 1963, 35; and Albert Eisele, "Nobel Winners Criticize Moon Project," *Washington Post*, 6 May 1963, A3.

57. "Sniping at NASA on Space," *Business Week*, 11 May 1963, 92.

58. P. H. A., "Manned Lunar Landing," *Science* 140, no. 3564 (19 April 1963): 267.

59. Everly Driscoll, "A Last Fling for Science on the Moon," *Science News* 102, no. 17 (21 October 1972): 268.

60. The trials and tribulations of NASA's life sciences efforts are detailed in Joan Vernikos, "Life Sciences in Space," chapter 3 in *Exploring the Unknown: Selected Documents in the History of the U.S. Civil Space Program, Volume VI: Space and Earth Science*, ed. John M. Logsdon (Washington, DC: NASA History Series, 2004), 267–303, especially 273–277; and Donald A. Beattie, *Taking Science to the Moon: Lunar Experiments and the Apollo Program* (Baltimore, MD: Johns Hopkins University Press, 2001), 78–106.

61. For a history of the Early Apollo Scientific Experiment Package (EASEP) and the Apollo Lunar Surface Experiments Package (ASLEP), see Beattie, *Taking Science to the Moon*, 125–145. On the decision to reduce the lunar experiments package for the Apollo 11 mission, see ibid., 137–138.

62. "Disenchantment with Apollo," *Science News* 96, no. 6 (9 August 1969): 112–113. For additional discussions of scientists quitting NASA for similar reasons, see also Richard Lewis, "Requiem for the Scientist-Astronauts," *Bulletin of the Atomic Scientists* 27, no. 5 (May 1971): 17–18.

63. "Before We Start to Mars," *Washington Post*, 8 August 1969, A22.

64. For a description of these events, see Brian O'Leary, "Rebellion among the Astronauts," *Ladies' Home Journal* 87, no. 3 (March 1970): 143–146. Another scientist-astronaut, Dr. F. Curtis Michel, resigned for similar reasons. On Michel's resignation, see Jonathan Eberhart, "Science Gets a Chance," *Bulletin of the Atomic Scientists* 96, no. 16 (18 October 1969): 355.

65. Allan H. Brown, "The Post-Apollo Era—Decisions Facing NASA," *Bulletin of the Atomic Scientists* 23, no. 4 (April 1967): 11–16.

66. For additional examples of such articles published in the *Bulletin*, see Jonathan Eberhart, "Science Gets a Chance," *Bulletin of the Atomic Scientists* 96, no. 16 (18 October 1969): 355–356; Eugene Rabinowitch, "Reflections on Apollo 8," *Bulletin of the Atomic Scientists* 25, no. 3 (March 1969): 2–3, 12; Richard S. Lewis, "End of Apollo: End of an Era," *Bulletin of the Atomic Scientists* 27, no. 1 (January 1971): 26–28; and Richard Lewis, "End of Apollo: The Ambiguous Epic," *Bulletin of the Atomic Scientists* 28, no. 10 (December 1972): 39–44. For an earlier version of this critique, see also Homer E. Newell, "NASA and Space," *Bulletin of the Atomic Scientists* 17, issue 5/6 (May 1961): 222–229. For an example of *Science* magazine picking up this critique, see Jonathan Eberhart, "NASA's Post-Moon Pitch: Research," *Science* 91, no. 3 (21 January 1967): 60–61. The *Bulletin* called for a congressional investigation of NASA regarding its aversion to developing scientific experiments for its missions in Lewis, "Requiem for the Scientist-Astronauts," 18.

67. Rachel Carson, *Silent Spring* (Boston: Houghton Mifflin, 1962), 6. On Carson's comparison of chemical pesticides and nuclear fallout, see Ralph Lutts, "Chemical Fallout: Rachel Carson's Silent Spring, Radioactive Fallout, and the Environmental Movement," *Environmental Review* 9, no. 3 (Autumn 1985): 210–225.

68. For an insightful examination of Barry Commoner's central role among this new generation of politically active postwar scientists, see Egan, *Barry Commoner*, 9–11, 15–46.

69. There is a rich literature on this new generation of scientists who were politically active in the postwar environmental movement. See especially Donald Flemming, "Roots of the New Conservation Movement," in *Perspectives in American History* 6 (New York: Cambridge University Press, 1972), 7–91; Michael Egan, "Shamans of the Spring: Environmentalism and the New Jeremiad," in *New World Coming: The Sixties and the Shaping of Global Consciousness*, ed. Karen Dubinsky (Toronto: Between the Lines, 2009), 286–303; and Rome, *Genius of Earth Day*, 20–29.

70. Lewis, "Requiem for the Scientist-Astronauts," 18. Italics added.

71. Lewis, "End of Apollo," 40, 43. Italics added.

72. For similar arguments, see also ibid., 26–28.

73. Similar divisions within the environmental movement have existed since the Progressive Era. For an analysis of "amateur" versus "expert" activists within the environmentalist tradition, see Stephen Fox, *John Muir and His Legacy: The American Conservation Movement* (Boston: Little, Brown, 1981).

74. Louis Harris, "Space Program Losing Support," *Washington Post*, 31 July 1967, A2.

75. Louis Harris, "Public Prefers Cutbacks in Space," *Philadelphia Inquirer*, 18 February 1969, 20. The Harris Survey reported that while the space program was the leading candidate for cutbacks among the American public,

programs to eliminate and control air and water pollution were the most popular.

76. For historical data on NASA's total budget for these years in both real and in 2008 inflation-adjusted dollars, see National Aeronautics and Space Council, *Aeronautics and Space Report of the President: Fiscal Year 2008 Activities* (Washington, DC: U.S. Government Printing Office, 2008), "Appendix D-1A: Space Activities of the U.S. Government, Historical Table of Budget Authority (in Millions of Real-Year Dollars)," 146, and "Appendix D-1B: Space Activities of the U.S. Government, Historical Table of Budget Authority (in Millions of Inflation-Adjusted FY 2008 Dollars)," 147.

77. Cohn, "Space Program Adopts New Look."

78. Personal Note #27, 18 July 1970, "Fiscal Year 1972, Budget and Programmatic Discussion," 12–14, Box 70, George Low Papers, Institute Archives and Special Collections, Rensselaer Polytechnic Institute, Troy, NY.

79. Personal Note #97, 7 July 1973, George Low Papers, Institute Archives and Special Collections, Rensselaer Polytechnic Institute, Troy, NY, 6.

80. John Donnelly, "Letter to James Fletcher," 30 January 1973, Folder 004157, NASA Historical Reference Collection, NASA Headquarters, Washington, DC, 1–4. The space agency rejected similar arguments made by William Pickering of the Jet Propulsion Laboratory back in 1960. On Pickering's failure to convince NASA to embrace earthly applications, see McQuaid, "Selling the Space Age," 133. McQuaid argues that NASA ultimately failed to truly engage the environmental movement.

81. On NASA's role in expanding the Merritt Island National Wildlife Refuge from 1969 through 1972, see Department of the Interior's Fish and Wildlife Service, "Wildlife Refuge at Space Center Being Expanded," news release, 11 November 1969, Box 1, LOC: 15A.6, Folder: 8-Accession, #88-50, NASA Department of Interior Agreement, Merritt Island National Wildlife Refuge, Kennedy Space Center Library Archives, Kennedy Space Center, FL; "Wildlife Refuge Expansion Pact Signing Friday," *Spaceport News*, 1 June 1972, Spaceport News Publications, Folders 28–29, Box: 2, LOC: 19D.2, Kennedy Space Center Library Archives, Kennedy Space Center, FL, 2. For a history of NASA's role in creating the wildlife refuge back in 1963, see Benson and Faherty, *Gateway to the Moon*, 107.

82. Debus is quoted in "Refuge's Growth Paralleled KSC's," *Spaceport News*, 4 October 1973, Spaceport News Publications, Folders 30–31, Box: 2, LOC: 19D.2, Kennedy Space Center Library Archives, Kennedy Space Center, FL, 5. For additional statements by Debus praising the Merritt Island Wildlife Refuge, see also "140,000-Acre Refuge on KSC Provides Wildlife Protection," *Spaceport News*, 29 June 1972, Spaceport News Publications, Folders 29–30, Box: 2, LOC: 19D.2, Kennedy Space Center Library Archives, Kennedy Space Center, FL, 5, 12.

83. Department of the Interior Stewart Udall to NASA Administrator James Webb, 5 April 1967, Box: 1, LOC: 15A.6, Folder: #2-National Seashore Park, Kennedy Space Center Library Archives, Kennedy Space Center, FL.

84. On NASA's reasons for opposing the creation of the Cape Canaveral National Seashore, see Dick Young, "Seashore Park Plan to Be Sifted," *Orlando Sentinel*, 28 May 1968, newspaper clipping located in Community Relations Collection, 214-Internal Correspondence, Box: 21, LOC: 18C.3, Kennedy Space Center Library Archives, Kennedy Space Center, FL; and "Seashore Park Area Surveyed," *Evening Star* (Tallahassee, FL), 30 April 1968, newspaper clipping located in Community Relations Collection, 214-Internal Correspondence, Box: 21, LOC: 18C.3, Kennedy Space Center Library Archives, Kennedy Space Center, FL.

85. George E. Mueller and Frank A. Bogart, "Establishment of the Florida National Seashore," memorandum prepared 4 February 1966, Community Relations Collection, 215-Internal Correspondence, Box: 21, LOC: 18C.3, Kennedy Space Center Library Archives, Kennedy Space Center, FL. While Mueller wrote this memorandum in 1966, it illustrates how environmental pressure on the space agency during the late 1960s influenced NASA administrators' decision to move forward with the creation of the Canaveral National Seashore. On NASA's cooperation in creating the Canaveral National Seashore, see "Management Instruction: NASA-Interior Agreement Use of Property at Kennedy Space Center, as Part of Canaveral National Seashore," Box: 1, LOC: 15A.6, Folder 8: Accession #88-50, NASA-Department of Interior Agreements, Kennedy Space Center Library Archive, Kennedy Space Center, FL.

86. For a description of the final agreement between NASA and the National Park Service for Canaveral National Seashore, see "Canaveral National Seashore: NASA, Department of Interior Sign Agreement," *Spaceport News*, 17 April 1975, Spaceport News Publications, 1975, Folders 34–35, Box: 2, LOC: 19D.2, Kennedy Space Center Library Archives, Kennedy Space Center, FL, 1; and "Canaveral Seashore Includes Some of KSC," *Spaceport News*, 23 January 1975, , Spaceport News Publications, 1975, Folders 34–35, Box: 2, LOC: 19D.2, Kennedy Space Center Library Archives, Kennedy Space Center, FL, 7.

87. D. G. Lawrence, "Shore Park for Florida Seen by '69," *Orlando Sentinel*, 24 December 1968, 1, 2A.

88. For coverage of NASA's cooperation with counting and banding birds on the Merritt Island National Wildlife Refuge, see "161 Bird Species Identified during Refuge Bird Count," *Spaceport News*, 28 December 1972, , Spaceport News Publications, 1972, Folders 28–29, Box: 2, LOC: 19D.2, Kennedy Space Center Library Archives, Kennedy Space Center, FL, 3; and "1,000 Ducks Caught for 'Operation Bird Banding,'" *Spaceport News*, 11 February 1965, Spaceport News Publications, 1965, Folders 11–14, Box: 1, LOC: 19D.1, Kennedy Space Center Library Archives,

Kennedy Space Center, FL, 3. On similar efforts regarding brown pelican re-introduction and alligator removal and relocation, see "Pelicans Moved to New Home," *Spaceport News*, 23 July 1976, Spaceport News Publications, 1976, Folders 36–37, Box: 2, LOC: 19D.2, Kennedy Space Center Library Archives, Kennedy Space Center, FL, 8; and "Kasey Hears Call of the Wild," *Spaceport News*, 4 October 1973, Spaceport News Publications, 1973, Folders 30–31, Box: 2, LOC: 19D.2, Kennedy Space Center Library Archives, Kennedy Space Center, FL, 7. For examples of correspondence between NASA administrators and the local Audubon Society regarding the protection of bald eagles at the Merritt Island National Wildlife Refuge, see Vice President of the Indian River Audubon Society Bald Eagle Research Committee Lon E. Ellis to Kennedy Space Center Chief of Mission Support Operations Harold Collins, 31 July 1969, Community Relations Collection, Folder: 211-Requests (Wildlife Refuge), Box: 21, LOC: 18C.3, Kennedy Space Center Library Archives, Kennedy Space Center, FL. The comment on the dangers posed by DDT to eagles can be found in "Counting Down with the Editor," *Spaceport News*, 23 May 1968, Spaceport News Publications, 1968, Folders 20–21, Box: 1, LOC: 19D.1, Kennedy Space Center Library Archives, Kennedy Space Center, FL, 8.

89. On these three programs, see "Tiny Tagged Turtles Trot toward Tide," *Spaceport News*, 28 June 1973, Spaceport News Publications, 1973, Folders 30–31, Box: 2, LOC: 19D2, Kennedy Space Center Library and Archives, Kennedy Space Center, FL, 4; "Endangered Young Green Turtles Released on Space Center Beaches," *Spaceport News*, 13 June 1974, Spaceport News Publications, 1974, Folders 32–33, Box: 2, LOC: 19D.2, Kennedy Space Center Library and Archives, Kennedy Space Center, FL, 8; "31 Species of Fish Found on Spaceport," *Spaceport News*, 24 January 1974, Spaceport News Publications, 1974, Folders 32–33, Box: 2, LOC: 19D2, Kennedy Space Center Library and Archives, Kennedy Space Center, FL, 7; and "Manatees from KSC Waters to Be Wired for Sound," *Spaceport News*, 18 March 1976, Spaceport News Publications, 1976, Folders 36–37, Box: 2, LOC: 19D2, Kennedy Space Center Library and Archives, Kennedy Space Center, FL, 3.

90. On NASA's involvement in dune-barrier stabilization at the Cape, see "Playalinda Dunes Get Overhaul during Skylab Preparations," *Spaceport News*, 14 June 1973, Spaceport News Publications, 1973, Folders 30–31, Box: 2, LOC: 19D2, Kennedy Space Center Library and Archives, Kennedy Space Center, FL, 5; and "History, Geology Combine in Mosquito Lagoon Study," *Spaceport News*, 9 August 1973, Spaceport News Publications, 1973, Folders 30–31, Box: 2, LOC: 19D2, Kennedy Space Center Library and Archives, Kennedy Space Center, FL, 6.

91. "Wildlife Refuge Manager Finds It's the Good Life," *Spaceport News*, 1 January 1970, Spaceport News Publications, 1970, Folders 24–25, Box: 2, LOC: 19D.2, Kennedy Space Center Library Archives, Kennedy Space Center, FL, 6.

92. On the construction of the Merritt Island National Wildlife Refuge visitor center on KSC land, see "Wildlife Refuge Gets New Headquarters on KSC Land," *Spaceport News*, 19 November 1970, Spaceport News Publications, 1970, Folders 24–25, Box: 2, LOC: 19D.2, Kennedy Space Center Library Archives, Kennedy Space Center, FL, 7. NASA promoted these nature trails in "40 Youths Work in Summer Program at Refuge," 6 August 1976, Spaceport News Publications, 1976, Folders 36–37, Box: 2, LOC: 19C.2, Kennedy Space Center Library Archives, Kennedy Space Center, FL, 4. On NASA's role in developing "conservation-oriented" campgrounds in the refuge, and offering campers a "package deal" that included a tour of KSC, see "For Boy Scouts, Happiness Is—A Weekend Spaceport Campout," *Spaceport News*, 16 March 1967, Spaceport News Publications, 1967, Folders 18–19, Box: 1, LOC: 19D.1, Kennedy Space Center Library Archives, Kennedy Space Center, FL, 8; and "Draft: Plan for Operation of the Dummitt Cove Campground," Community Relations Collection, Folders 204—Camp Sites, KSC, Dummitt Cove Campground, Box: 21, LOC: 18C.3, Kennedy Space Center Library Archives, Kennedy Space Center, FL. For coverage of NASA's involvement in the public education program at the Merritt Island National Wildlife Refuge, see "Wildlife in Action Course Scheduled by MINWR, BCC," *Spaceport News*, 22 March 1973, Spaceport News Publications, 1973, Folders 30–31, Box: 2, LOC: 19D.2, Kennedy Space Center Library Archives, Kennedy Space Center, FL, 6. The screening of wildlife films produced by both NASA and the National Wildlife Service at the refuge's visitor center is described in "Wildlife Refuge Set's Open House," *Spaceport News*, 21 September 1972, Spaceport News Publications, 1972, Folders 28–29, Box: 2, LOC: 19D.2, Kennedy Space Center Library Archives, Kennedy Space Center, FL, 4.

93. "Wildlife Drive to Be Dedicated May 28," news release, Merritt Island National Wildlife Refuge, U.S. Department of the Interior, 26 May 1976, Folder 3: Wildlife Refuge, Box: 1, LOC: 15A.6, Kennedy Space Center Archive and Library, Kennedy Space Center, FL.

94. "Refuge Visits High during 1975," *Spaceport News*, 5 February 1976, Spaceport News Publications, 1975, Folders 36–37, Box: 2, LOC: 19D.2, Kennedy Space Center Library Archives, Kennedy Space Center, FL, 7.

95. Regarding this Apollo 11 postcard, see "John F. Kennedy Space Center, N.A.S.A.," produced by Dexter Press, Inc., West Nyack, NY, for NASA Tours, Conducted by TWA, P.O. Box 21222, Kennedy Space Center, FL. The front of this postcard mistakenly identified the spaceship as "Apollo 2 Saturn 5," but the address side correctly identifies the photograph as of Apollo 11 (the postcard producer most probably mistook the number "11" for a Roman numeral "II" or 2). For similar examples, see the following. For Apollo 8, see "Apollo 8—Lunar Vehicle," produced by Florida Natural Color, Inc., 190 N.E. 186th Terr., Miami, FL; and for Apollo 16, see "John F. Kennedy Space Center, N.A.S.A.," NASA Tours, Conducted

by TWA, P.O. Box 21222, Kennedy Space Center, FL. The address sides of these postcards indicate that NASA supplied the color photographs for the other side. These postcards are in the possession of the author.

96. For this caption, see the postcard "John F. Kennedy Space Center, N.A.S.A.," in the possession of the author. Another postcard for Apollo 11 includes this side caption: "The Apollo 11 Saturn V space vehicle passes scenic Florida landscape along Crawlerway during rollout from the Vehicle Assembly Building." For this postcard, see "Kennedy Space Center, FLA," NASA Tours, Conducted by TWA, P.O. Box 21222, Kennedy Space Center, FL, in the possession of the author.

97. "NASA-Interior Pact Expands Refuge," *Spaceport News*, 15 June 1972, Spaceport News Publications, Folders 28–29, Box: 2, LOC: 19D.2, Kennedy Space Center Library Archives, Kennedy Space Center, FL, 1.

98. This report was published as National Aeronautic and Space Administration, *Remote Measurement of Pollution*, SP-285 (Washington, DC, 1971), 1, 5. For a description of this conference, see Ellis E. Remsberg, "Remote Measurement of Pollution—A 40-Year Langley Retrospective: Part II—Aerosols and Clouds," NASA/TM-2012-217578, NASA, Langley Research Center, Hampton, VA.

99. This argument is made in Erik Conway, *Atmospheric Science at NASA: A History* (Baltimore, MD: Johns Hopkins University Press, 2008), 129.

100. On NASA's development and the use of its TIROS, NIMBUS, and ATS satellites, see Richard Benjamin Leshner, "The Evolution of the NASA Earth Observing System: A Case Study in Policy and Project Formulation" (Ph.D. dissertation, George Washington University, 2007), 21–22; Conway, *Atmospheric Science at NASA*, 27–38, 39–63; and Pamela Mack, *Viewing the Earth: The Social Construction of the Landsat Satellite System* (Cambridge, MA: MIT Press, 1990), 19–20.

101. This point is made most clearly in Mack, *Viewing the Earth*, 53.

102. On this political move by Udall and the Department of the Interior, see ibid., 60–64; and Leshner, "Evolution of the NASA Earth Observing System," 30.

103. Evert Clark, "Sensing Satellite to Study Earth Resources," *New York Times*, 21 September 1966, 1.

104. On NASA receiving negative media coverage for months after Udall's announcement, and such coverage forcing NASA administrators to speed up the development of Landsat, see Mack, *Viewing the Earth*, 64.

105. "Researchers Say Satellites Can Monitor Pollution," *Evening Capital* (Annapolis, MD), 3 March 1970, 2; Associated Press, "Orbs to Spy on Polluters," *Huntsville (AL) Times*, 3 March 1970, clipping file, Folder 5754: Earth Resources Satellite, 1970, NASA Historical Materials, NASA Headquarters Archive, Washington, DC.

106. "Researchers Say Satellites Can Monitor Pollution," 2; Associated Press, "Orbs to Spy on Polluters."

107. On the history of Seasat, see Alan Buis, "Seafaring Satellite Sets 25 Year Trend," National Aeronautics and Space Administration, online at http://www .nasa.gov/vision/earth/lookingatearth/Seasat_25.html (accessed 4 June 2014).

108. NASA, *Remote Measurement of Pollution*, 6.

109. Conway, *Atmospheric Science at NASA*, 141.

110. "New Division Chief Sees Vital Role for Earth Resources Program," *Roundup* (newspaper of the Johnson Space Center) 9, no. 25 (9 October 1970): 1, Record Number 206558, Record Number 206594, Date: 10/09/1970, Johnson Space Center Collection, University of Houston, Clear Lake, Clear Lake, TX.

111. For a discussion of the role of these NASA scientists in initially raising the shuttle exhaust ozone problem, see Conway, *Atmospheric Science at NASA*, 347 (footnote 40); and Lydia Dotto and Harold Schiff, *The Ozone War* (Garden City, NY: Doubleday, 1978), 127.

112. AA/Manager, Sciences, Applications, and ASTP Office to MD/Director, Life Science Office, memorandum prepared by William H. Rock on 27 December 1974, Record Number 162171, Report Number AA-STA-1, Location: Box 010-36, Johnson Space Center Collection, University of Houston, Clear Lake, Clear Lake, TX.

113. Frank von Hippel and Robert H. Williams, "Solar Technologies," *Bulletin of the Atomic Scientists* 31, no. 9 (November 1975): 29. Regarding the claim that NASA leadership worried that environmental opposition to the space shuttle, on the grounds that its exhaust would deplete the ozone layer, could seriously erode public support for the space program, see W. Henry Lambright, *The Case of Ozone Depletion*, Monograph in Aerospace History No. 38, NASA SP-2005-4538 (Washington, DC: NASA, 2005), 5. Lambright also claims that the national media and environmentalists seized on these claims and called for action from Congress.

114. On these efforts by Fletcher, see Conway, *Atmospheric Science at NASA*, 137–140; and Leshner, "Evolution of the NASA Earth Observing System, 55, 58. Here Conway also discusses the lobbying done by NASA within Congress and at the National Science Foundation to make NASA the lead agency in these ozone efforts.

115. This speech by Fletcher was later published in pamphlet form as National Aeronautics and Space Administration, *Spaceship Earth: A Look Ahead to a Better Life* (Washington, DC: U.S. Department of Health, Education, and Welfare, 1973), 28. Here I am countering Kim McQuaid's dismissal of this statement by Fletcher as a shallow public relations campaign. See McQuaid, "Selling the Space Age," 129.

116. On such congressional rewards for NASA, see Lambright, *Case of Ozone Depletion*, 8–9. For another analysis of NASA's role in the ozone crisis, see Karen T. Litfin, *Ozone Discourses: Science and Politics in Global Environmental Cooperation* (New York: Columbia University Press, 1994), 63.

117. On the 1977 amendment to the Clean Air Act instructing NASA to report biennially to Congress regarding the ozone crisis, see W. Henry Lambright, "NASA, Ozone, and Policy-Relevant Science," *Research Policy* 24 (1995): 750.

118. On the pollution instruments designed into Nimbus 7, see the Earth Observing Portal webpage, "Nimbus-7," online at https://directory.eoportal.org/web/eoportal/satellite-missions/n/nimbus-7 (accessed 10 June 2014); National Aeronautics and Space Administration webpage, "Total Ozone Mapping Spectrometer-Earth Probe (TOMS-EP)," online at http://eospso.gsfc.nasa.gov/missions/total-ozone-mapping-spectrometer-earth-probe (accessed 10 June 2014); and Conway, *Atmospheric Science at NASA*, 142–144.

119. "Nimbus-7," online at https://directory.eoportal.org/web/eoportal/satellite-missions/n/nimbus-7 (accessed 10 June 2014).

120. "Jacques Cousteau," 31 May 1975, Personal Note #145, 3 Box 66, George Low Papers, Institute Archives and Special Collections, Rensselaer Polytechnic Institute, Troy, NY, 3.

121. On the NASA experiment sparking the idea for *Oasis in Space*, see Peter Wood, "The Cousteau Team Charts a More Serious Course," *New York Times*, 20 February 1977, 105.

122. "Personal Notes: Visit to Calypso, 26–30 December 1974," esp. 8–9, Box 66, Low / RPI, as quoted in McQuaid, "Selling the Space Age," 149.

123. Ralph Shapiro, *Nimbus Program History: Earth-Resources Research Satellite Program*, NASA Goddard Space Flight Center, NP-2004-10-672-FSFC, 2004, online at http://atmospheres.gsfc.nasa.gov/uploads/files/Nimbus_History.pdf (accessed 12 June 2014), 28.

124. During the late 1970s and 1980s NASA developed dozens of space technologies that scientifically studied various natural environments on Earth. Rather than covering all of these technologies, here I have focused my analysis instead on the two most important, Landsat and Nimbus-7, for the study of Earth's ecology. For a thorough examination of these other technologies regarding atmospheric science, see Conway, *Atmospheric Science at NASA*; regarding oceanography, see Ian S. Robinson, *Measuring the Oceans from Space: The Principles and Methods of Satellite Oceanography* (New York: Springer-Praxis, 2004); and regarding land-based natural resources, see Mack, *Viewing the Earth*.

125. "NASA Modifying Saturn-Apollo Equipment, Disposing of Some," news release, mailed 16 December 1976, 31, Folder OA-250417-01, Apollo Project, NASA Releases 1970s, NASM Archives Division, Technical Reference Files, Space History Series, National Air and Space Museum Library, National Air and Space Museum, Washington, DC.

126. David Brower, quoted in John McPhee, *Encounters with the Archdruid* (New York: Farrar, Straus and Giroux, 1971), 80.

127. Scholars have similarly analyzed the photograph. See Goldberg, *Power of Photography*, 52–57; Cosgrove, "Contested Global Visions," 270–294; Sachs, "Ambiguous Modern Icon," 170–175; and Garb, "Perspective or Escape?," 264–278.

128. As the editor of the *Whole Earth Catalog*, Brand discusses this image in Stewart Brand, ed., *The Last Whole Earth Catalog* (Menlo Park, CA: Portola Institute and Random House, 1971), 1. The best histories of Stewart Brand and his *Whole Earth Catalog* are Andrew G. Kirk, *Counterculture Green: The Whole Earth Catalog and American Environmentalism* (Lawrence: University of Kansas Press, 2007); and Fred Turner, *From Counterculture to Cyberculture: Stewart Brand, the Whole Earth Network, and the Rise of Digital Utopianism* (Chicago: University of Chicago Press, 2006).

129. As quoted in Ron Schick and Julia Van Haaften, *The View from Space: American Astronaut Photography, 1962–1972* (New York: C. N. Potter, 1988), 8, 11.

130. On the cultural importance of *Earthrise*, see Poole, *Earthrise*.

131. NASA released to the public numerous examples of photographs of Earth from space that predated *Whole Earth* in 1972. For examples from the Mercury mission, see "Sunset on Earth Seen from Space, Photographed by Scott Carpenter," *Life* 52, no. 23 (8 June 1962): 24–28. For examples from the lunar orbiter, see "Lunar View of a Socked-In Earth," *Life* 61, no. 11 (9 September 1966): 347. For examples from Apollo 8, see John Noble Wilford, "Astronauts Give Television Show on Way to Moon," *New York Times*, 23 December 1968, 1; and the cover of "The Moon Age," a special issue of *Newsweek* (7 July 1969).

132. "View from Apollo 8 Shows Most of Western Hemisphere with South America Near the Center," *Washington Post*, 31 December 1968, 1; "We Look Brilliant and Splendid," Associated Press photograph appearing in the *Spartanburg (SC) Herald*, 24 December 1968, 1; and "Apollo Astronaut's View from Earth," Associated Press photograph appearing in the *Spokesmen-Review* (Spokane, WA), 30 December 1968, 16.

133. For examples of corporate advertisements using the image of Earth taken from NASA's ATS satellite in 1967, see "The View from Outer Space," Ashland Oil and Refining Company advertisement, *Fortune* 80, no. 2 (1 August 1969): back cover. For examples of *Earthrise* used to promote a company's role in developing space technology, see "Peace on Earth Seems So Simple from 251,000 Miles Away," McDonnell Douglas advertisement, *New York Times*, 10 October 1973, 43; "Leading in Communications towards a Bright Tomorrow," AT&T advertisement, *Fortune* 79, no. 6 (15 May 1969): 271; and "The Programming Skills Man Used to Conquer Space Now Also Help Control His Roadways in the Sky," IBM advertisement, *Aviation Week*, 13 November 1972, 43.

134. On the sale of these Earth photographs to the general public, see NASA, "Lunar Surface Photos Available," news release no. 69:83Ja, 18 August 1969, Folder:

OA-250416-01, Apollo Project, NASA Releases, 1969, Space History Series, Technical Reference Files, Archives Division, National Air and Space Museum, Washington, DC. I personally own a View-Master reel that depicts Earth from space. See "NASA's Apollo Project Moon Landing 1969," View-Master reel number B 6632, Sawyer's, Inc., Portland, OR.

135. For a discussion of the ATS-3 Earth photograph, see Poole, *Earthrise*, 85–87. On the Apollo 8 image, see ibid., 21–22.

136. Few newspapers initially linked these images of Earth from space to the environmental crisis of the late 1960s. For exceptions see "Another Giant Step," *Christian Science Monitor*, 28–30 December 1968, 14; and "Beneficent Nature" (letter to the editor), *New York Times*, 7 January 1969, 40.

137. Archibald MacLeish, "A Reflection: Riders on Earth Together, Brothers in Eternal Cold," *New York Times*, 25 December 1968, 1.

138. For instance, *National Geographic* quoted MacLeish's essay accompanied by a color centerfold reproduction of *Earthrise*. See "A Most Fantastic Voyage," *National Geographic* 135, no. 5 (May 1969): 593–631. For examples of local newspapers reproducing the essay, see "To See the Earth as It Truly Is . . . ," *Morning Record* (Meriden, CT), 31 December 1968, 6; "Now, Men May Be Brothers Riding Earth," *Miami News*, 26 December 1968, 1; and "Man Has New Idea of Earth," *Spokesman-Review* (Spokane, WA), 24 December 1968, 11.

139. Denis Cosgrove makes a similar argument about *Earthrise*, and MacLeish's essay, representing the idea of a unified world in Cosgrove, "Contested Global Visions," 283. That *Earthrise* promoted the idea of a shared Earth was also evident in media coverage of MacLeish's essay. For examples see "Apollo 8 Gives Man a New Idea of Himself," *Kansas City Star*, 26 December 1968; and "Seeing Earth as a Whole," *Sunday Denver Post*, 29 December 1968.

140. The organizers of the first Earth Day, a nonprofit group called Environmental Action, never archived promotional materials for the first Earth Day, and the Earth Day Network, which currently organizes the annual event, is also unaware of any archived promotional material from the first Earth Day in 1970. To trace Earth Day's visual culture I therefore relied on historical photo galleries documenting the events of Earth Day 1970, collected by the following organizations: *National Geographic*; the Boston public radio station WGBH and its American Experience history series; the archive at the University of Wisconsin at Green Bay; WWLP, an NBC-affiliated television station in Western Massachusetts; and the *Wall Street Journal*. I also conducted an extensive online image search of posters, flyers, and pamphlets from Earth Day 1970, as well as of media photographs capturing the events of the day in an effort to further identify visual trends in the event's iconography. This search resulted in more than one hundred images, only six of which depicted the Earth in any form. For access to these Earth Day 1970 photo galleries, see "PHOTOS: The First Earth Day—Bell-Bottoms and Gas

Masks," *National Geographic Daily News*, online at http://news.nationalgeographic
.com/news/2009/04/photogalleries/first-earth-day-1970-pictures/(accessed 10
July 2014); "Photo Gallery: Earth Day across America," WGBH Boston Public
Radio, online at http://www.pbs.org/wgbh/americanexperience/features/photo
-gallery/earthdays/ (accessed 10 July 2014); "Earth Day Memories: A Photo Gallery
of 1970s Activism at Eco U," University of Wisconsin at Green Bay, online at
http://news.uwgb.edu/multimedia/photos/04/03/earth-day-memories/ (accessed 10
July 2014); "First Earth Day in 1970," WWLP public television, online at http://
interactives.wwlp.com/photomojo/gallery/11922/225372/first-earth-day-in-1970
/gas-masks-magnolia-blossoms/(accessed 10 July 2014); and "Earth Day over the
Years," *Wall Street Journal*, online at http://www.wsj.com/articles/SB10001424052
70230342550457735147273973222 (accessed 14 September 2015).

141. On the prominence of gas-mask iconography during Earth Day 1970, see
Finis Dunaway, "Gas Masks, Pogo, and the Ecological Indian: Earth Day and the
Visual Politics of American Environmentalism," *American Quarterly* 60, no. 1
(March 2008): 67–99.

142. Robert Leydenfrost, "Earth Day, April 22, 1970," poster for original Earth
Day with photograph by Don Brewster, 1970, online at http://collections.vam
.ac.uk/item/O100756/earth-day-poster-leydenfrost-robert/ (accessed 23 August
2016).

143. To access the visual culture of Earth Day 1980 I undertook similar research
as explained in note 140. The only Earth Day 1980 materials depicting the planet
that I could locate included a poster from the city of Denver with a drawing of
several hands holding up a globe, a poster from New York City depicting a drawn
heart-shaped globe, and an Earth Day poster that depicted several individuals
stretching out a net to catch a falling Earth. I was also able to find a photograph of
an individual wearing a papier-mâché Earth for a head.

144. For examples of *Whole Earth* appearing on Earth Day 1990 flags, see the
iconic photograph of the celebration on the Mall in Washington, DC, online
at http://www.pollutionissues.com/Co-Ea/Earth-Day.html (accessed 14 September 2015). For examples of *Whole Earth*–like images appearing on Earth Day
1990 posters, see http://monsantoblog.com/2011/04/22/thinking-about-dirt/
(accessed 14 September 2015). On Jim Hansen's public service announcement
with Kermit the Frog singing "Bein' Green" from within a *Whole Earth*–like set,
see http://muppet.wikia.com/wiki/Earth_Day_1990 (accessed 14 September 2015).

145. Adam Rome discusses this logo, but not its deeper history associated with
Whole Earth, in his book on Earth Day, *Genius of Earth Day*, 277.

146. On the IGY's impact on the scientific collection of global data through
orbiting satellites and computer modeling, see Paul N. Edwards, "Representing
the Global Atmosphere: Computer Models, Data, and Knowledge about Climate
Change," in Miller and Edwards, *Changing the Atmosphere*, 31–65; Paul N.

Edwards, *A Vast Machine: Computer Models, Climate Data, and the Politics of Global Warming* (Cambridge, MA: MIT Press, 2010), 202–207; Conway, *Atmospheric Science at NASA*, 16–26; and Roger Launius, "Toward the Poles: A Historiography of Scientific Exploration during the International Polar Years and the International Geophysical Year," in *Globalizing Polar Science: Reconsidering the International Polar and Geophysical Years*, ed. Roger Launius, James Rodger Fleming, and David H. Devorkin (New York: Palgrave Macmillan, 2010), 47–81.

147. For an explanation of the incompleteness of global satellite data sets, see Edwards, "Representing the Global Atmosphere," 31–65.

148. On NASA's role in GARP, see Conway, *Atmospheric Science at NASA*, 65–93.

149. On the local nature of early ozone research, see Sebastian Vincent Grevsmuhl, "The Creation of Global Imaginaries: The Antarctic Ozone Hole and the Isoline Tradition in the Atmospheric Sciences," in *Image Politics of Climate Change: Visualizations, Imaginations, Documentations*, ed. Brigit Schneider and Thomas Nocke (New York: Columbia University Press, 2014), 36. On the local context of early global warming research, see Howe, *Behind the Curve*, 41.

150. The BAS team published its findings in J. C. Faran, B. G. Gardiner, and J. D. Shanklin, "Large Losses of Total Ozone in Antarctica Reveal Seasonal ClO_x/NO_x Interaction," *Nature* 315 (16 May 1985): 207–210. Charles Keeling promoted his research regarding rising levels of CO_2 around the Mauna Loa Observatory in numerous publications during this period. See especially C. D. Keeling, "The Concentration and Isotopic Abundances of Carbon Dioxide in the Atmosphere," *Tellus* 12 (1960): 200–203; and Charles D. Keeling, "The Influence of Mauna Loa Observatory on the Development of Atmospheric CO_2 Research," in *Mauna Loa Observatory: A 20th Anniversary Report*, ed. John Miller (Washington, DC: National Oceanic and Atmospheric Administration, 1978), 35–54.

151. NASA scientists were understandably alarmed because their own satellites, which had not been calibrated finely enough to identify the slight increases recorded by the BAS team, missed the ozone hole. On this embarrassing situation within NASA, see Grevsmuhl, "Creation of Global Imaginaries," 34.

152. On James Hansen's efforts at the Goddard Institute for Space Studies to model climate change, see Conway, *Atmospheric Science at NASA*, 199–206; J. Hansen et al., "Climate Impact on Increasing Carbon Dioxide," *Science* 213, no. 4511 (28 August 1981): 957–966; and J. Hansen et al., "Climate Sensitivity: Analysis of Feedback Mechanisms," in *Climate Processes and Climate Sensitivity, Geophysical Monograph* 29 (Washington, DC: American Geophysical Union, 1984), 130–163. Hansen was one of the first climate scientists to raise sustained alarm over global warming, and he is today considered one of the most important voices in the climate-change debate.

153. *Remote Measurement of Pollution*, cover.

154. *Inadvertent Climate Modification: Report of the Study of Man's Impact on Climate* (SMIC) (Cambridge, MA: Massachusetts Institute of Technology, 1971), cover; Conella H. Meadows, Dennis L. Meadows, Jorgen Randers, and William W. Behrens III, *Limits to Growth* (New York: Universe, 1972), first and second edition covers; James Lovelock, "More on Gaia and the End of Gaia," *Coevolution Quarterly* 31 (Fall 1981): 36; and James Lovelock, *Gaia: A New Look at Life on Earth* (London: Oxford University Press, 1987), cover.

155. For this ozone depletion graph, see Faran, Gardiner, and Shanklin, "Large Losses of Total Ozone," 208.

156. On this approach by NASA, see Grevsmuhl, "Creation of Global Imaginaries," 37–47. On the first presentation of this image at a scientific conference, see NASA, "Discovering the Ozone Hole: Q&A with Pawan Bhartia," online at http://www.nasa.gov/topics/earth/features/bhartia-qa.html (accessed 24 August 2016).

157. Walter Sullivan, "Low Ozone Level Found above Antarctica," *New York Times*, 7 November 1985, B21.

158. For a general description of NASA's Scientific Visualization Studio, see "Scientific Visualization Studio," online at https://svs.gsfc.nasa.gov/(accessed 25 July 2016); and for an interview with Dr. Horace Mitchell, the director of NASA's Scientific Visualization Studio, see Lauren Hockenson, "How NASA Makes Scientific Data Beautiful," Mashable, online at http://mashable.com/2012/08/28/nasa -svs/#GNcFUNDH2Eqz (accessed 25 July 2016).

159. NASA oceanographer Gene Carl Feldman as quoted in Michael S. Rosenwald, "Telling NASA's Tales with Hollywood's Tools; Space Center Uses Pixar's Palette to Artfully Explain Scientific Data," *Washington Post*, 21 August 2006, D01.

160. On the importance of telling stories within the practice of scientific visualization, and the important role of NASA's Scientific Visualization Studio within this practice, see Kwan-Liu Ma and Isaac Liao, Jennifer Frazier, Helwig Hauser, and Helen-Nicole Kostis, "Scientific Storytelling Using Visualization," *Computer Graphics and Applications, IEEE* 32, no. 1 (January–February 2012): 12–19.

161. Rosenwald, "Telling NASA's Tales."

162. This process was described to the author by Dr. Horace Mitchell, the director of the Scientific Visualization Studio, in a phone call interview conducted on 25 July 2016.

163. NASA / Goddard Space Flight Center Scientific Visualization Studio, "Global Temperature Anomalies from January 2016," available in color online at https://svs.gsfc.nasa.gov/4438 (accessed 24 August 2016).

164. This story is recounted by Brand in Brand, "First Whole Earth Photograph," 188. Italics added.

165. The historian of science Sheila Jasanoff has also argued that *Whole Earth* and similar images of the planet from space *became* culturally powerful *gradually*, as their social meaning cohered during the final decades of the twentieth century. I have tried to push past Jasanoff's reliance on her rather fuzzy category of "culture," to pinpoint exactly what types of cultures caused this shift. I have also argued against her notion that a new global consciousness arising during the 1980s *made Whole Earth* an environmental symbol. See Jasanoff, "Image and Imagination," 309–337.

166. Six months before *Silent Spring* was published, Murray Bookchin published *Our Synthetic Environment*, which similarly warned of the dangers of synthetic pesticides. See Peter McCord, "Divergences on the Left: The Environmentalisms of Rachel Carson and Murray Bookchin," *Left History* 13, no. 1 (Spring / Summer 2008): 14–34.

167. Brand, "First Whole Earth Photograph," 188.

4. Heavenly Bodies

1. For a description of this landing, see Robert Lindsey, "Shuttle Returns, Diverted to Land on the West Coast," *New York Times*, 25 June 1983, 1. On Sally Ride rejecting a bouquet of flowers from NASA officials during STS-7's "homecoming ceremony," see Associated Press, "No White Roses for a Crew Lady," *Washington Post*, 26 June 1983, A2.

2. Lindsey, "Shuttle Returns." For similar public comments regarding Sally Ride's desire to be considered an astronaut rather than a female astronaut, see also Associated Press, "No White Roses for a Crew Lady," A2.

3. Sally Ride used this phrase often when describing herself to the national media. See United Press International, "Sally Ride: I'm Just One of the Guys," *Salina (KS) Journal*, 19 June 1983, 6.

4. On this intense public relations tour by Ride, see Harry F. Rosenthal, "Woman Astronaut Caught Up in Celebrity Whirl," *Morning Journal* (Daytona Beach, FL), 25 July 1983, 7A. On the conference for women in space, see "Women in Space Confab Planned," *Morning Journal* (Daytona Beach, FL), 25 July 1983, 7A.

5. United Press International, "Female Astronaut Says Her Flight Helped Other Women," *Durant (OK) Daily Democrat*, 24 July 1983, 13.

6. Here and in the rest of this chapter I focus on liberal feminists, and I understand that second-wave feminism was an extremely diffuse movement that encompassed other types of feminism as well. I also understand that while NOW was a major political player within the feminist movement, it does not represent second-wave feminism as a whole. On the diversity within second-wave feminism, see the essays in Sarah Gamble, ed., *The Routledge Companion to Feminism*

and Postfeminism (London: Routledge, 2002); Jane Gerhard, *Desiring Revolution: Second-Wave Feminism and the Rewriting of American Sexual Thought* (New York: Columbia University Press, 2001); Maylei Blackwell, *Chicana Power! Contested Histories of Feminism in the Chicano Movement* (Austin: University of Texas Press, 2011); and Lisa Duggan and Nan D. Hunter, *Sex Wars: Sexual Dissent and Political Culture* (London: Routledge, 2006). Regarding socialist feminism, see the essays in Karen V. Hansen and Ilene J. Philipson, eds., *Women, Class, and the Feminist Imagination: A Socialist-Feminist Reader* (Philadelphia: Temple University Press, 1989). On the decline of second-wave feminism in the early 1980s, see the works cited earlier in this note, as well as Dorchen Leidholdt and Janice Raymond, eds., *The Sexual Liberals and the Attack on Feminism* (New York: Pergamon, 1990); and, on a more popular level, Susan Faludi, *Backlash: The Undeclared War against American Women* (New York: Crown, 1991).

7. Associated Press, "Ride Qualified," *St. Joseph (MO) News-Press*, 23 April 1982, 11B. For similar statements by Crippen regarding the skills of female astronauts, see also Associated Press, "A Texan Speaks Out on Female Astronauts," *Spokesman-Review* (Spokane, WA), 18 June 1983, 2.

8. Lindsey, "Shuttle Returns." Italics added.

9. Ibid.

10. While the symbolic meaning of these flowers has shifted over time, within Western culture they have consistently denoted "distinction" and "beauty." See "The Carnation Flower: Its Meanings & Symbolism," *Flower Meaning*, available at http://www.flowermeaning.com/carnation-flower-meaning/ (accessed 28 October 2016); and "The Rose Flower: Its Meanings and Symbolism, *Flower Meaning*, available at http://www.flowermeaning.com/rose-flower-meaning/ (accessed 28 October 2016).

11. On Ride refusing to be televised on the *Challenger* downlink performing these tasks, see Encyclopedia Astronautica, s.v., "Ride," http://www.astronautix .com/r/ride.html (accessed 31 October 2013).

12. The links between women's bodies, women's health, and the feminist movement are most obvious in the history of the Boston Women's Health Book Collective, which published *Our Bodies, Ourselves* in 1971. For an overall discussion of the role of the female body within the feminist movement, see also Sara Evans, "Sons, Daughters, and Patriarchy: Gender and the 1968 Generation," *American Historical Review* 114, no. 2 (April 2009): 340–345. On the history of connections between feminism and the birth control pill, see Elaine Tyler May, *America and the Pill: A History of Promise, Peril, and Liberation* (New York: Basic Books, 2011); and Elizabeth Siegel Watkins, *On the Pill: A Social History of Oral Contraceptives, 1950–1970* (Baltimore, MD: Johns Hopkins University Press, 2001).

13. News articles about the *Challenger*'s bathroom became ubiquitous prior to Ride's flight. For several examples, see Robert Alexander, "'No Big Deal' Says First U.S. Woman Astronaut," *Daytona Beach (FL) Sunday News-Journal*, 29 May 1983, 2F; Olive Talley, "Women in Space: Historic Ride," *Telegraph-Herald* (Dubuque, IA), 14 June 1983, 9; and Ellen Goodman, "Ride, Sally Ride, Ride!," *News and Courier* (Charleston, SC), 24 June 1983, 10A. For a scholarly examination of the gendered politics of bathrooms, see Julia Edwards and Linda McKie, "Women's Public Toilets: A Serious Issue for the Body Politic," in *Embodied Practices: Feminist Perspectives on the Body*, ed. Kathy Davis (London: SAGE, 1997), 135–149.

14. Associated Press, "Female Astronaut Feels Added Pressure," *Observer-Reporter* (Washington, PA), 30 April 1982, A8.

15. Although several scholars have examined the struggle of women to become astronauts, few place this struggle within the broader context of the postwar feminist movement. See, especially, Leslie Haynesworth and David Toomey, *Amelia Earhart's Daughters: The Wild and Glorious Story of American Women Aviators from World War II to the Dawn of the Space Age* (New York: William Morrow, 1998); Stephanie Nolen, *Promised the Moon: The Untold Story of the First Women in the Space Race* (New York: Four Walls Eight Windows, 2002); Martha Ackmann, *The Mercury 13: The Untold Story of Thirteen American Women and the Dream of Space Flight* (New York: Random House, 2003); and Margaret Weitekamp, *Right Stuff, Wrong Sex: America's First Women in Space Program* (Baltimore, MD: Johns Hopkins University Press, 2004). The two books that best link the histories of women astronauts and the women's movement are Bettyann Holtzmann Kevles, *Almost Heaven: The Story of Women in Space* (New York: Basic Books, 2003); and Amy Foster, *Integrating Women into the Astronaut Corps: Politics and Logistics at NASA, 1972–2004* (Baltimore, MD: Johns Hopkins University Press, 2011).

16. The field of gender and technology studies has produced a rich literature around the belief that gender and technology are mutually constitutive. For an overview of the field of gender and technology studies and this literature, see Judy Wajcman, "Reflections on Gender and Technology Studies: In What State Is the Art?," *Social Studies of Science* 30, no. 3 (June 2000): 447–464. On the role of technology in shaping gender, see also Donald MacKenzie and Judy Wajcman, "Introductory Essay," and Cynthia Cockburn, "Caught in the Wheels: The High Cost of Being a Female Cog in the Male Machinery of Engineering," both in Donald MacKenzie and Judy Wajcman, eds., *The Social Shaping of Technology: How the Refrigerator Got Its Hum* (Philadelphia: Open University Press, 1985), 2–25 and 126–134, respectively. See also Donna Haraway, "A Manifesto for Cyborgs: Science, Technology, and Socialist Feminism in the 1980s," *Socialist Review* 15, no. 2 (1985): 65–108; and Donna Haraway, *Modest_Witness@Second_Millennium. FemaleMan@Meets_OncoMouseTM: Feminism and Technoscience* (New York: Routledge, 1997). In arguing that gender is performative, Judith Butler also sug-

gests that certain technologies can shape gender identities, see *Gender Trouble* (New York: Routledge, 1990). Butler's theory on gender as performative is used to analyze the Apollo program in Daniel Sage, "Giant Leaps and Forgotten Steps: NASA and the Performance of Gender," *Sociological Review* 57, issue supplement (15 May 2009): 146–163.

17. The body has become a category of analysis of increasing importance in the field of environmental history. For examples of this work, see Richard White, "'Are You an Environmentalist or Do You Work for a Living?': Work and Nature," in *Uncommon Ground: Rethinking the Human Place in Nature*, ed. William Cronon (New York: W. W. Norton, 1996), 171–185; Linda Nash, *Inescapable Ecologies: A History of Environment, Disease, and Knowledge* (Berkeley: University of California Press, 2007); Conevery Valencius, *The Health of the Country: How American Settlers Understood Themselves and Their Land* (New York: Basic Books, 2002); and Neil M. Maher, "A New Deal Body Politic: Landscape, Labor, and the Civilian Conservation Corps," *Environmental History* 7, no. 3 (July 2002): 435–461. For an insightful example of the importance of historicizing bodies within their natural landscapes, see Gregg Mitman, *Breathing Spaces: How Allergies Shape Our Lives and Landscapes* (New Haven, CT: Yale University Press, 2008). For a more theoretical discussion by nonenvironmental historians of cultural inscription regarding astronaut bodies, both male and female, see Monica J. Casper and Lisa Jean Moore, "Inscribing Bodies, Inscribing the Future: Gender, Sex, and Reproduction in Outer Space," *Sociological Perspectives* 38, no. 2 (Summer 1995): 311–333, especially 316–319.

18. "First Woman in Space: Valentina Vladimirovna Tereshkova," *New York Times*, 17 June 1963, 8.

19. "She Orbits over the Sex Barrier," *Life* 54, no. 26 (28 June 1963): 28.

20. On these tests by Soviet scientists and coverage by TASS, see Henry Tanner, "Bykovsky Nears a Flight Record," *New York Times*, 18 June 1963, 1; and "Soviet Women," *Houston Post*, 17 June 1963, 6.

21. For examples of Soviet media praising Tereshkova's physical fitness, see "First Woman in Space."

22. This image of Tereshkova as the ultimate proletarian in the Soviet media is analyzed in Roshanana P. Sylvester, "She Orbits over the Sex Barrier: Soviet Girls and the Tereshkova Moment," in *Into the Cosmos: Space Exploration and Soviet Culture*, ed. James T. Andrews and Asif A. Siddiqi (Pittsburgh: University of Pittsburgh Press, 2011), 198.

23. *Uchitel'skaya gazeta*, 23 June 1963, 2, as quoted in Sue Bridger, "The Cold War and the Cosmos: Valentina Tereshkova and the First Woman's Space Flight," in *Women in the Khrushchev Era*, ed. Melanie Ilic, Susan E. Reid, and Lynne Attwood (New York: Palgrave Macmillan, 2004), 231. For photographs of Tereshkova holding an enormous bouquet of flowers, see "Nikita Khrushchev, Valentina

Tereshkova, Yuri Gagarin and Valery Bykovsky in Red Square, 1963," online at http://www.diomedia.com/stock-photo-nikita-khrushchev-valentina-tereshkova -yuri-gagarin-and-valery-bykovsky-in-red-square-1963-image20261032.html (accessed 18 May 2016).

24. Dorothy McCardle, "New Cultural Attaché Says: USSR Training Women by Dozens for Space," *Washington Post*, 20 June 1963, B3. For a similar reference to the equality of Soviet women to Soviet men, see also *Moscow News*, 23 June 1963, as quoted in Sylvester, "She Orbits over the Sex Barrier," 205.

25. Such news reports are discussed in Clare Boothe Luce, "But Some People Simply Never Get the Message," *Life*, 28 June 1963, 31. For firsthand accounts, see "Muscovite Women Are Proud," *New York Herald Tribune* (European Edition, Paris), 17 June 1963, 1.

26. Luce, "But Some People."

27. Friedan is quoted in Sue Solet, "The Reasons Why Soviet Sent a Woman into Space," *New York Herald Tribune*, 17 June 1963, 1.

28. Associated Press, "U.S. Names Seven Men Pioneer Space Fliers," *Los Angeles Times*, 10 April 1959, 1. For an additional example of newspapers reporting on the Mercury Seven's bodies during this press conference, see John Norris, "7 Vie to Be First Man in Space," *Washington Post*, 10 April 1959, A1.

29. On the physical tests endured by the Mercury astronaut candidates at the Lovelace Clinic, see Mae Mills Link, *Space Medicine in Project Mercury* (Washington, DC: NASA Scientific and Technical Information Division, 1965), 48. See also Joseph L. Myler, "First 7 Spacemen Selected," *Washington Post*, 7 April 1959, A1; and James MacNees, "U.S. Identifies Seven Selected for Trips into Space," *Baltimore Sun*, 10 April 1959, 1.

30. MacNees, "U.S. Identifies Seven."

31. Glennan made these remarks at a dinner hosted by the Institute of the Aeronautical Sciences. On Glennan's remarks, see Richard Witkin, "110 Selected as Potential Pilots for Nation's First Space Flight," *New York Times*, 28 January 1959, 1.

32. As quoted in John Dille, "Those Seven—An Introduction," in *We Seven* (New York: Simon and Schuster, 1962), 5–22.

33. Ronald Thompson, "Astronauts Aren't Supermen; Just Stay in Top Shape," *Capital Times* (Madison, WI), 6 September 1967, 18. On John Glenn's obsession with running two miles every day, see "The Astronauts Exclusive Report: Part VII, We're Going Places No One Has Ever Travelled in a Craft No One's Flown," *Life* 50, no. 4 (27 January 1961): 46.

34. Ronald Thompson, "Astronauts Don't Have to Be Supermen," *Independent* (Long Beach, CA), 8 September 1967, A6.

35. On the bodily monitoring of astronauts, see Charles Berry, "Space Medicine in Perspective: A Critical Review of the Manned Space Program," *Journal of*

the American Medical Association 201, no. 4 (24 July 1967): 232–233; "Heart Study of Astronauts," *Science News Letter*, 31 March 1962, 199; and "Brain Monitor Operates inside Astronaut Helmet," *Science News Letter*, 31 August 1963, 133.

36. "Space Doctor for the Astronauts," *Ebony* 17 (April 1962): 35.

37. Scott Carpenter, "Astronauts' Training in Weightlessness Part II: 'The Eerie World of Zero G,'" *Life* 48, no. 11 (21 March 1960): 54. To maintain the health of its astronauts, NASA also developed special food. On the safety and nutritional characteristics of astronaut food for the Mercury, Gemini, and Apollo projects, see Jennifer Ross-Nazzal, "'From Farm to Fork': How Space Food Standards Impacted the Food Industry and Changed Food Safety Standards," in *Societal Impact of Spaceflight*, ed. Steven J. Dick and Roger D. Launius (Washington, DC: NASA History Division, 2007), 219–236.

38. Loudon Wainwright, *The Great American Magazine: An Inside History of "Life"* (New York: Knopf, 1986), 263, as quoted in James Kauffman, *Selling Outer Space: Kennedy, the Media, and Funding for Project Apollo, 1961–1963* (Tuscaloosa: University of Alabama Press, 1994), 73.

39. "The Chosen Three for First Space Ride," *Life* 50, no. 9 (3 March 1961): 26; "Put Them High on the List of Men Who Count," *Life* 62, no. 5 (3 February 1967): 23; "The Freshmen in the Crew: Two Self-Made Men Who Finally Have the Jobs They Want," *Life* 62, no. 20 (19 May 1967): 37. For critical histories of NASA's deal with *Life* magazine, see Kauffman, *Selling Outer Space*, 68–92; Robert Sherrod, "The Selling of the Astronauts," *Columbia Journalism Review* (May–June 1973): 16–25; and Edwin Diamond, *The Rise and Fall of the Space Age* (New York: Doubleday, 1964), 79–98.

40. Reporter Warren Burkett wrote a six-part series in the *Houston Chronicle* about the Mercury Seven's physical examinations, titled "Astronaut: The Magnificent Male." See especially part three, Warren Burkett, "Brains as Well as Bodies Have to Rate with the Best," *Houston Chronicle*, 6 November 1962, 1.

41. Norman Mailer, "The Psychology of Astronauts," *Life* 67, no. 20 (14 November 1969): 60.

42. "The Long Road Home," *Washington Daily News*, 22 July 1969. For a historical analysis of astronaut bodies being used to portray the American space program as masculine, see Amy Foster, "Sex in Space: The Politics and Logistics of Sexually Integrating NASA's Astronaut Corps" (Ph.D. dissertation, Auburn University, 2005), 86.

43. On astronauts as pioneers, see "The Astronauts—Ready to Make History," *Life* 47, no. 11 (14 September 1959): 26; and "Space: An American Necessity," *Life* 47, no. 22 (30 November 1959): 36. On exploring space as America's destiny, see "With All the Cost and Risk, Why Go?," *Life* 52, no. 17 (17 April 1962): 85. On space being compared to the U.S. West, see "Space: An American Necessity," *Life* 47, no. 22 (30 November 1959): 36. On this linkage between space exploration and

settlement of the U.S. West, see De Witt Kilgore, *Astrofuturism: Science, Race, and Visions of Utopia in Space* (Philadelphia: University of Pennsylvania Press, 2003), especially the introduction; and Kauffman, *Selling Outer Space*, 78–79.

44. "World Will Be Ruled from Skies Above," *Life* 54, no. 20 (17 May 1963): 4.

45. "What's It to Us?," *Newsweek*, 7 July 1969, 64. For additional coverage of the "whiteness" of NASA's astronaut corps, see Lynn Spigel, "White Flight," in *The Revolution Wasn't Televised: Sixties Television and Social Conflict*, ed. Lynn Spigel and Michael Curtin (New York: Routledge, 1997), 47–71; "What's It to Us?"; "Lily-White NASA," *Chicago Daily Defender*, 29 July 1969, 13; and Norman Mailer, *Of a Fire on the Moon* (Boston: Little, Brown, 1970), 125. Mailer's book was serialized in *Life* magazine in 1969 and 1970.

46. "Beneath the Smiles: Hard Work," *Life* 53, no. 8 (24 August 1962): 23.

47. On the Soviet cosmonaut-selection process, its emphasis on physical fitness, and the medical monitoring of cosmonaut bodies from space, see Asif Siddiqi, *Challenge to Apollo: The Soviet Union and the Space Race, 1945–1974* (Washington, DC: NASA History Division, Office of Policy and Plans, 2000), 243–262, 293. On the cosmonauts' physical training regimen, see also "Always Aim to Be First: The Yuri Gagarin Story . . . Using Words from the Cosmonaut, His Family, and Colleagues," Novosti Information Agency and TASS, 1991, as translated and published in *Quest: The History of Spaceflight Quarterly* 13, no. 1 (2006): 56.

48. "Always Aim to Be First," as translated and published in *Quest: The History of Spaceflight Quarterly* 13, no. 1 (2006): 62, 58. For a similar argument regarding Soviet cosmonauts being represented as ordinary, everyday men, see also Iina Kohonen, "The Heroic and the Ordinary: Photographic Representations of Soviet Cosmonauts in the Early 1960s," in *Soviet Space Culture: Cosmic Enthusiasm in Socialist Societies*, ed. Eva Maurer, Julia Richers, Monica Ruthers, and Carmen Scheide (New York: Palgrave Macmillan, 2011), 110–112.

49. On the Communist Party's demand that the first cosmonaut come from a common working-class background, see Siddiqi, *Challenge to Apollo*, 262. For a description of Gagarin as the quintessential Soviet proletarian, see Reuters, "Soviet Astronaut 'Introduced' to World," *Christian Science Monitor*, 12 April 1961, 3.

50. John F. Kennedy, "The Soft American," *Sports Illustrated* 13 (26 December 1960): 16, as quoted in Robert L. Griswold, "The 'Flabby American,' the Body, and the Cold War," in *A Shared Experience: Men, Women, and the History of Gender*, ed. Laura McCall and Donald Yacovone (New York: New York University Press, 1998), 324.

51. "The Sixties: The Decade of Man in Space," *Newsweek*, 14 December 1959, 34.

52. This masculinity crisis was also related to the rise of counterculture. For an examination of hippies rebelling against their "father's masculinity" during

the late 1960s, see Sara Evans, "Sons, Daughters, and Patriarchy: Gender and the 1968 Generation," *American Historical Review* 114, no. 2 (April 2009): 332–335. See also Timothy Hodgdon, *Manhood in the Age of Aquarius: Masculinity in Two Countercultural Communities* (New York: Columbia University Press, 2009), 40; and Susan Faludi, *Stiffed: The Betrayal of the American Man* (New York: William Morrow, 2000), 452–463.

53. Roulhac Hamilton, "'Squares' Inherit the Moon," *Columbus (OH) Dispatch*, 27 July 1969, as reprinted in "'Squares' Inherit the Moon," 115th Cong., *Congressional Record*, 11 August 1969, 23610.

54. Austin C. Wehrwein, "Orbiting Potato May Be Life Clue: Biologist Believes It Might Show How Body Ticks," *New York Times*, 31 October 1965, 72. Space historian Roger Launius encourages historians to explore more fully the role of the extreme environment of space in his essay "Forum: Writing the History of Space's Extreme Environment," *Environmental History* 15 (July 2010): 526–532.

55. The best history of the science conducted in association with the launching of animals into space is Colin Burgess and Chris Dubbs, *Animals in Space: From Research Rockets to the Space Shuttle* (Chichester, UK: Springer, 2007).

56. Neil Armstrong, "Flying the Mission in Make Believe," essay in a longer article titled "Weird Astronaut Machines," *Life* 57, no. 13 (25 September 1964): 138.

57. Warren Kornberg, "Space: Unjustified Risk," *Los Angeles Times*, 30 November 1969, F1.

58. William E. Lawrence, "Man on the Moon: New U.S. Goal Requires Solution of Some Difficult Problems," *New York Times*, 28 May 1961, E7.

59. Albert Rosenfeld, "How Will Man Suit Up for Space? Freezing, Frying, Bubbling to Death," *Life* 58, no. 15 (16 April 1965): 56. *Life* magazine reporter John Dille similarly described space as an unknown, hostile environment in his introduction to *We Seven, by the Astronauts Themselves* (New York: Simon and Schuster, 1962), 4. For similar coverage, see also Albert Q Maisel, "The Fantastic Hazards of Landing on the Moon," *Reader's Digest*, September 1963, 106.

60. There is an increasingly rich literature bridging the fields of environmental and medical history that argues for the examination of human bodies within their natural landscapes. For an introduction to this literature, see Gregg Mitman, Michelle Murphy, and Christopher Sellers, eds., *Landscapes of Exposure: Knowledge and Illness in Modern Environments*, Osiris 19 (Chicago: University of Chicago Press, 2004), especially Gregg Mitman, Michelle Murphy, and Christopher Sellers, "Introduction: A Cloud over History."

61. Berry, "Space Medicine in Perspective," 232–241. For additional contemporary descriptions of the possible dangers of space travel, see Harold M. Schmeck Jr., "New Hazards for Men in Space," *New York Times*, 20 November 1966, A9; and Associated Press, "U.S. Space Doctor Asks New Studies," *New York Times*, 9 June 1969, 13.

62. On the dangers of meteorites in space, see "How Soon to the Moon?," *Grumman Horizons* (newsletter of the Grumman Aircraft Engineering Corporation) 4, no. 2 (1964): 8, located in the National Air and Space Museum Archive Division, Technical Reference File, Space History Series, Folder: 0A-250025-01 "Apollo Project, Articles, 1964.

63. Maisel, "Fantastic Hazards," 103–104.

64. A minority of biomedical scientists proposed altering the human body for space travel instead of building protective spacesuits. See especially Manfred E. Clynes and Nathan S. Kline, "Cyborgs and Space," *Astronautics*, September 1960, 27, 76.

65. "Apollo News Reference: Crew Personal Equipment," news media information brochure by Grumman Aerospace Corporation, located at National Air and Space Museum, Archives Division, Technical Reference Files, Space History Series, Folder: OA-250520-01, "Apollo Project, Press Kits, Grumman," National Air and Space Museum, Washington, DC.

66. Here I focus my analysis on the extravehicular version of the A7L suit worn by lunar module astronauts, such as Armstrong and Aldrin. On the history of NASA's Apollo spacesuit, see Douglas N. Lantry, "Man in Machine: Apollo-Era Suits as Artifacts of Technology and Culture," *Winterthur Portfolio* 30, no. 4 (Winter 1995): 203–230; Nicholas de Monchaux, *Spacesuit: Fashioning Apollo* (Cambridge, MA: MIT Press, 2011), 55–66; Lillian D. Kozloski, *U.S. Space Gear: Outfitting the Astronauts* (Washington, DC: Smithsonian Institution, 1994), 1–40; and Matthew H. Hersch, "High Fashion: The Women's Undergarment Industry and the Foundations of American Spaceflight," *Fashion Theory* 13, no. 3 (2009): 348–351. NASA's Wernher von Braun discussed this history in his article "Space Suits—From Pressurized Prison to Mini-Spacecraft," *Popular Science* 201, no. 1 (July 1972): 62–64, 121.

67. Douglas N. Lantry, "Dress for Egress: The Smithsonian National Air and Space Museum's Apollo Spacesuit Collection," *Journal of Design History* 14, no. 4 (2001): 345.

68. This term is used in official NASA publications, as well as in NASA-approved publications by its contractors, to describe the portable life support system. See "Apollo News Reference: Crew Personal Equipment," 4–5; and Kozloski, *U.S. Space Gear*, 89.

69. On the Apollo spacesuit's Pressure Garment Assembly, see "Apollo News Reference: Crew Personal Equipment," 1–2; Kozloski, *U.S. Space Gear*, 87; and Lantry, "Man in Machine," 212–213.

70. Associated Press, "Moonsuits Cost Sum of $300,000," *News and Courier* (Charleston, SC), 21 July 1969, 7A.

71. On the control of "climate" within NASA spacesuits, see Don Bane, "Air Conditioned Comfort Will Surround Astronauts," (Los Angeles) *Herald-Examiner*,

10 July 1969, A-6; Wernher von Braun, "What an Astronaut Will Wear on the Moon," *Popular Science* 186, no. 4 (April 1965): 210; "Apollo News Reference: Crew Personal Equipment," 3–5; and Kozloski, *U.S. Space Gear*, 89.

72. On the Apollo spacesuit's Liquid Cooling Garment, see Burnham M. Lewis, "Earthman in Space," located at National Air and Space Museum Archive Division, Technical Reference File, Space History Series, Folder-0A2500250-01, "Apollo Project Articles, 1964," National Air and Space Museum, Washington, DC; "Apollo News Reference: Crew Personal Equipment," 1.

73. "Lunar Suit Has It's Own Electricity, Water, Oxygen, Fan, Refrigerator, Etc.," *New York Times*, 21 July 1969, 14.

74. On the Apollo spacesuit guarding against micrometeorites, see "Lunar Suit"; and Kozloski, *U.S. Space Gear*, 88. On NASA's Lunar Extravehicular Visor Assembly, see Kozloski, *U.S. Space Gear*, 87.

75. Bane, "Air Conditioned Comfort."

76. Walter Schirra, "A Suit Tailor-Made for Space," *Life* 49, no. 5 (1 August 1960): 36.

77. This International Latex Corporation chart for Apollo spacesuit depicts measurements 30–46. Image courtesy of ILC Dover, LP. For another example of these measuring charts see Lantry, "Dress for Egress," 350.

78. Along with being extremely skilled craftswomen, the seamstresses at the International Latex Corporation were also encouraged to collaborate with the company's engineers regarding improvements in the construction of the Apollo spacesuits. On these seamstresses, see de Monchaux, *Spacesuit*, 208–226; and Kozloski, *U.S. Space Gear*, 80–81. On the manufacture of the early Mercury suits, which were equally personalized, see Hersch, "High Fashion," 358; and Kozloski, *U.S. Space Gear*, 44–46.

79. Jim Lovell to the Girls of Frederica, as quoted in Lantry, "Dress for Egress," 350.

80. This press release is "For Release: June 26, 1964," Photo No. 64-Apollo-104 64, Record Group 255-G, National Archives Still Picture and Photograph Collection, National Archives and Records Administration, College Park, MD. This NASA press release is also discussed in de Monchaux, *Spacesuit*, 198–199.

81. NASA distributed this press release to the media on June 26, 1964. For this photograph, see NASA Image 64-Apollo-104, Record Group 255-G, Still Picture and Photograph Collection, National Archives at College Park, College Park, MD. The press release and image are also discussed in de Monchaux, *Spacesuit*, 234.

82. United Press International, "Spacesuits Will Restrict Astronauts on the Moon," *European Stars and Stripes*, 14 September 1964. For a similar medieval description, see also Associated Press, "Glenn, Like Man with Comfortable Pair of Shoes, Has Favorite Suit," *Register* (Danville, VA), 9 February 1962, 6-B. On

space-suited astronauts representing modern-day medieval knights, see also Roger Launius, "Heroes in a Vacuum: The Apollo Astronaut as Cultural Icon," *Florida Historical Quarterly* 87, no. 2 (Fall 2008): 202; and Rachel Quinn Fischhoff, "Locating the Astronaut Body in Space" (Ph.D. dissertation, Wesleyan University, 2008), 16.

83. Neil Armstrong, "Flying the Mission in Make Believe," part of a longer article titled "Weird Astronaut Machines," *Life* 57, no. 13 (25 September 1964): 138.

84. While space historians and historians of technology have examined how simulators helped to familiarize astronauts with NASA's space technology during the Mercury, Gemini, and Apollo eras, historians, including environmental historians, have refrained from examining the role of environmental simulators in these same missions. On the history of NASA's simulators, see James E. Tomayko, "Computers in Spaceflight: The NASA Experience" (Washington, DC: NASA History Office, Scientific and Technical Information Division, 1988), online at http://history.nasa.gov/computers/Compspace.html (accessed 5 December 2013); and de Monchaux, *Spacesuit*, especially chapter 12, 163–180.

85. On environmental simulations during the post–World War II era, see Matthew Farish, "Creating Cold War Climates: The Laboratories of American Globalism," in *Environmental Histories of the Cold War*, ed. J. R. McNeill and Corinna R. Unger (New York: Cambridge University Press, 2010), 51–83; and Matthew Farish, "The Lab and the Land: Overcoming the Arctic in Cold War Alaska," *Isis* 104, no. 1 (2013): 1–29. Also see Farish, *The Contours of America's Cold War* (Minneapolis: University of Minnesota Press, 2010).

86. On the amount of time that Mercury, Gemini, and Apollo crewmen spent training on simulators, see Tomayko, "Computers in Spaceflight"; and de Monchaux, *Spacesuit*, 176. *Reader's Digest* explained to its readers that astronauts had to "learn to manipulate their bodies in a drastically different environment." See "1969: The Year of the Moon," condensed from Ira Wolfert, "Empire," *Denver Post*, 13 April 1969.

87. Armstrong, "Flying the Mission in Make Believe." On NASA's far-flung network of simulations, see Bernard Kovit, "Space on Earth," *Grumman Horizons* 4, no. 2 (1964): 22–25, located at National Air and Space Museum Archive Division, Technical Reference File, Space History Series, Folder-0A2500250-01, "Apollo Project Articles, 1964."

88. On NASA's use of the U.S. Navy's Johnsville centrifuge, see C. H. Woodling, Stanley Faber, John J. Van Bockel, Charles C. Olasky, Wayne K. Williams, John L. C. Mire, and James R. Homer, "Apollo Experience Report—Simulation of Manned Space Flight for Crew Training," NASA TN D-7112 (Washington, DC: National Aeronautics and Space Administration, 1973), 7, 31–32, 47; and Witkin, "110 Selected as Potential Pilots," 2.

89. On the Manned Spacecraft Center / Johnson Space Center centrifuge, see Woodling et al., "Apollo Experience Report," 51. For the quote on the cen-

trifuge in *Time* magazine, see Mark Wolverton, "The G Machine," *Air and Space* (May 2007), online at http://www.airspacemag.com/history-of-flight/the_g _machine.html (accessed 4 December 2013).

90. Michael Collins, *Carrying the Fire: An Astronaut's Journey* (New York: Farrar, Straus and Giroux, 1974), 133.

91. Scott Carpenter discusses his experiences with SEALAB and how they influenced his recommendation to NASA for underwater astronaut training in Scott Carpenter, interviewed by Michelle Kelly, 30 March 1998, "Oral History Transcript," Johnson Space Center Oral History Project, NASA, 17–18.

92. "Astronauts Turn Aquanauts as NASA Shakes Up the Lagging Moon Program," *Life* 62, no. 20 (19 May 1967): 35.

93. On NASA's Water Immersion Facility, see Martin D. Robbins, John A. Kelley, and Linda Elliott, "Mission-Oriented R & D and the Advancement of Technology: The Impact of NASA Contributions, Final Report, Volume II, Contract NSR 06-004-063," prepared for NASA by the Industrial Economic Division, Denver Research Institute, University of Denver, May 1972, located under call number T174.3.R62 at the Hagley Museum and Library, Wilmington, DE. For a more popular description of the facility, see "Ugly, Unearthly, Bug: We'll Land It on the Moon," *Life* 57, no. 14 (2 October 1964): 99. On this and other underwater astronaut training by NASA, see Valerie Olson, "American Extreme: An Ethnography of Astronautical Visions and Ecologies" (Ph.D. dissertation, Rice University, 2010), especially chapter 1.

94. NASA's lunar orbiter program was responsible for gathering data and images for possible Apollo landing sites. Between 1966 and 1967 the program's unmanned satellites mapped 99 percent of the moon's surface to assess Apollo landing sites. On the history of the lunar orbiter, see Bruce K. Byers, *Destination Moon: A History of the Lunar Orbiter Program* (Washington, DC: NASA History Office, 1977), online at http://www.hq.nasa.gov/office/pao/History/TM-3487/top .htm (accessed 28 December 2013).

95. On the NASA's Lunar Module Mission Simulator, see Woodling et al., "Apollo Experience Report," 14–16, 24–27; "Lunar Module Mission Simulator," clipping from publication by the Link Group of General Precision Systems, Inc., located at National Air and Space Museum (NASM) Archives Division, Technical Reference Files, Space History Series, Folder: OA-251200-01 "Apollo Lunar Module (LEM, LM), NASM, Washington, D.C."; "Apollo Mission Simulator," clipping from publication by the Link Group of General Precision Systems Inc., located at National Air and Space Museum (NASM) Archives Division, Technical Reference Files, Space History Series, Simulations, Folder: OA-250250-04 "Apollo Project, General Systems, NASM, Washington, D.C."; David Mindell, *Digital Apollo: Human and Machine in Spaceflight* (Cambridge, MA: MIT Press, 2008), 209–210; and Tomayko, "Computers in Spaceflight."

96. On the lunar surface models of the Lunar Module Simulator, see Woodling et al., "Apollo Experience Report," 27, 50.

97. For Alan Shepard's comment, see "Landing at Fra Mauro," Apollo 14 Lunar Surface Journal, Corrected Transcript and Commentary, compiled by Eric M. Jones, 1995, last revised 18 December 2001, online at http://www.history.nasa.gov /alsj/a14/a14.landing.html (accessed 9 December 2013). For more popular coverage, see "NASA Simulator Shows Apollo 14 Descent toward Landing Site in Fra Mauro Region," *Aviation Week and Space Technology*, 18 January 1971, 40–41.

98. For a description of the USGS's efforts at identifying these moonlike landscapes on Earth, and then altering them with explosives to make them even more "moonlike," see Donald A. Beattie, *Taking Science to the Moon: Lunar Experiments and the Apollo Program* (Baltimore, MD: Johns Hopkins University Press, 2001), 159–165, 182. Beattie was a NASA engineer who worked with the USGS on training astronauts. For media coverage of such trips, see Associated Press, "Astronaut Will Make Simulated Moon Tests," *Pittsburgh Post-Gazette*, 23 August 1964, 26.

99. Colonel David R. Scott, "Apollo 15: Three Views of the Moon," *Reader's Digest*, November 1971, 73, as condensed from Colonel David R. Scott, "Finding the Golden Easter Egg," *New York Times*, 13 August 1971, 1.

100. Beattie, *Taking Science to the Moon*, 186.

101. The creation of this fake moonscape at Sunset Crater is described in Beattie, *Taking Science to the Moon*, 182. For the photograph of the detonation of these craters, see "Construction of Cinder Lake Crater Field #2 on 27 July 1968; (a) aerial view of firing of sequence one of 354 craters during construction of Cinder Lake Crater Field # 2; USGS photo 768227-3 USGS Open-File Report 2005-1190, Figure 056a.," USGS, 1968, online at https://www.sciencebase.gov/catalog/item /51dda4bae4b0f72b4471e347 (accessed 22 March 2016).

102. Beattie, *Taking Science to the Moon*, 182.

103. Interestingly, as better data on the lunar surface of this area became available, NASA planned to alter the simulated landscape in Houston accordingly. On this simulated landscape at the Johnson Space Center, see Bernard Kovit, "Space on Earth," *Grumman Horizons* 5, no. 2 (1964), located at National Air and Space Museum Archive Division, Technical Reference File, Space History Series, Folder-0A2500250-01, "Apollo Project Articles, 1964."

104. Neil Armstrong, "Flying the Mission in Make Believe," essay in a longer article titled "Weird Astronaut Machines," *Life* 57, no. 13 (25 September 1964): 139.

105. Apollo astronauts spent between one-third and one-half of their overall training time practicing in simulators, often fully suited. On the amount of time spent by NASA's astronauts inside simulators, see Woodling et al., "Apollo Experience Report," 2.

106. Armstrong made this statement at the press conference held on 12 August 1969. For an edited transcript of this press conference, see *The First Lunar Landing: As Told by the Astronauts*, Office of Public Affairs, NASA, circa 1989, 6–7.

107. Robert Gannon, "With the Apollo Astronauts: Training for the Moon," *Popular Science* 199, no. 1 (July 1971): 53.

108. Colonel James B. Irwin, "Apollo 15: Three Views of the Moon," *Reader's Digest*, November 1971, 73, as condensed from James Irwin, "I Will Lift Up Mine Eyes," *New York Times*, 14 August 1971, 1.

109. For this photograph, see *Life* 47, no. 12 (21 September 1959): cover.

110. Ron Laytner and Donald Mclachlan, "Ride, Sally Ride: Her Place Is Space," *Chicago Tribune*, 24 April 1983, H1.

111. This decision to seek astronauts from America's test pilot ranks was made by President Eisenhower. On these requirements, see Weitekamp, *Right Stuff, Wrong Sex*, 42; Foster, "Sex in Space," 17, 79, and Joseph D. Atkinson Jr. and Jay M. Shafritz, *The Real Stuff: A History of NASA's Astronaut Recruitment Program* (New York: Praeger, 1985), 10.

112. "No Ladies in Orbit," *Saturday Evening Post* 235, no. 30 (August 1962): 86.

113. Richard Witkin, "Simulated Trips Part of Program: Space Travelers Will Begin to Master New Techniques Sometime Next Year," *New York Times*, 3 August 1962, 6.

114. Congress, House, Committee on Science and Astronautics, *Qualifications for Astronauts: Hearings before the Special Subcommittee on the Selection of Astronauts of the Committee on Science and Astronauts*, 87th Cong., 2nd sess., 17–18 July 1962 (no. 9): 67.

115. "Space Women Expensive," *Science News Letter*, 4 August 1962, 70.

116. "The Craftier Sex Is Cleared for Space," *Los Angeles Times*, 28 August 1960, E5. For a similarly sexist article regarding women as unfit for space exploration, see "A Lesson from Levitt," *Time*, 18 July 1969, 116.

117. Guernsey LePelley, "Okay, Dear, Come On . . . Children Are Hungry . . . Something's Wrong with the Water . . . ," *Christian Science Monitor*, 26 July 1969.

118. Bill Mauldin, ". . . And I Can't Get the Car Fixed . . . ," *Chicago Sun-Times*, 22 July 1969, 35.

119. Italics added. For newspaper coverage of von Braun making this remark numerous times, see Associated Press, "News Brief: Lint Men Told Rocketry Pays," *Thomasville (GA) Times-Enterprise*, 18 May 1962, 1; and "So They Say," *Reno (NV) Evening Gazette*, 29 June 1962, 4.

120. "A Mrs. in the Missile?," *Los Angeles Times*, 7 September 1958, 12. For an interesting examination of this sort of reproductive rhetoric during the space race, see Lisa Ruth Rand, "The Case for Female Astronauts: Reproducing Americans in

the Final Frontier," *Appendix: A New Journal of Narrative and Experimental History* 2, no. 3 (July 2014), online at http://theappendix.net/issues/2014/7/the-case
-for-female-astronauts-reproducing-americans-in-the-final-frontier (accessed 2
August 2016).

121. Associated Press, "Women's Place in Outer Space: Is NASA for Coeds?,"
Washington Post, 17 March 1968, H28.

122. This was the charge of Representative James Fulton, who argued in Congress that women were being kept out of space by NASA and the all-male astronaut corps because of a "protective attitude." On this statement, see "Space Women
Expensive," *Science News Letter*, 4 August 1962, 70.

123. For examples of this, see Harold M. Schmeck Jr., "Experts See Long-Term
Value in Flight of Woman Astronaut," *New York Times*, 17 June 1963, 8; and Louis
Lasagna, "Why Not 'Astronauttes' Also?," *New York Times*, 21 October 1962, 268.

124. Isaac Asimov, "No Space for Women?," *Ladies' Home Journal* 88, no. 3
(March 1971): 204.

125. Arthur J. Snider, "The Progress of Medicine: Space May Be a Man's World,"
Science Digest 57, no. 3 (March 1965): 23–24.

126. Carl R. Konkel and William G. Holder, "Sputnik to Mutnik to—Picnic!,"
Space World, vol. 1-12-108 (December 1972): 48. For similar coverage on female
astronauts' menstrual cycles, see also Lionel Tiger, "Male Dominance? Yes, Alas.
A Sexist Plot? No," *New York Times*, 25 October 1970, S18.

127. "No Women in Space," *Science News Letter*, 18 October 1960, 230.

128. Associated Press, "Women Seen for Space Shuttles," *Los Angeles Times*, 18
December 1972, A23.

129. Barbara Rowes, "Housewives in Space," *Omni*, June 1982, 66.

130. Jack Anderson, "Would-Be Astronauts: Legion of Angry Women," *Parade*, 19 November 1967, 7. For an informative essay on discrimination against
women in the provision of bathroom facilities in general, see Julia Edwards and
Linda McKie, "Women's Public Toilets: A Serious Issue for the Body Politic," in
Davis, *Embodied Practices*, 135–149.

131. Jack Smith, "Craftier Sex Is Cleared for Space," *Los Angeles Times*, 28 August 1960, E5.

132. Jim Lange, "Women in Space, Too?," *Daily Oklahoman*, 19 July 1962.

133. "Space Women Expensive," *Science News Letter*, 4 August 1962, 70.

134. Ironically, Flickinger later became a proponent of biologically testing
women for space travel, and he worked with Lovelace during the Women in Space
Program in Albuquerque, New Mexico. This statement was made in "No Women
in Space," 230.

135. Anderson, "Would-Be Astronauts," 7.

136. Richard P. Powers, "Senator's Wife Protests Ban: Jane Hart Is Eager for
Space," *Lawrence* (KS) *Journal-World*, 22 May 1963, 10.

137. In an internal memo written immediately after his meeting with Cobb and Hart, Johnson wrote "Let's stop this now!" in reference to women in space. On this memo, see Weitekamp, *Right Stuff, Wrong Sex*, 137.

138. Jane B. Hart, oral history interview, edited transcribed tape recording, Mackinac City, MI, 7 October 1997, 4–5, 20, as quoted in Weitekamp, *Right Stuff, Wrong Sex*, 135.

139. For analysis of the Lovelace Clinic's tests on female astronauts in 1961, see Weitekamp, *Right Stuff, Wrong Sex*, 71–76, 91–117, especially 110–111 on Lovelace being denied access to pressure suits and space simulators.

140. The Pentagon response was reported in United Press International, "Space Age Research," *Oxnard (CA) Press-Courier*, 8 April 1963, 14.

141. On these political maneuvering, see Weitekamp, *Right Stuff, Wrong Sex*, 140.

142. Congress, *Qualifications for Astronauts*, 58. Glenn's comment before Congress was covered extensively by the media. For an example, see Associated Press, "U.S. Aims High in New Space Efforts," *News and Courier* (Charleston, SC), 19 July 1962, 1; and "No Ladies in Orbit," 86.

143. Marie Smith, "Space Training Lack Bars Women," *Washington Post, Times Herald*, 21 March 1962, C1. There is a long history of political activism by women's clubs. See, in particular, Jennifer Price, *Flight Maps: Adventures with Nature in Modern America* (New York: Basic Books, 2000), especially chapter 2, "When Women Were Women, Men Were Men, and Birds Were Hats," 57–111.

144. Marie Smith, "NASA to Use Women in Space," *Washington Post, Times Herald*, 23 October 1963, D1.

145. "For Equal Rights: BPW Buzzes NASA, State," *Washington Post, Times Herald*, 20 July 1963, A9.

146. This phone call is described in Pamela Freni, *Space for Women: A History of Women with the Right Stuff* (Santa Ana, CA: Seven Locks, 2002), 124. Hart's role as a founding board member of NOW is also mentioned in Weitekamp, *Right Stuff, Wrong Sex*, 176.

147. On NOW filing discrimination charges against NASA, see Associated Press, "Women Astronauts, Justices of Supreme Court Are Urged," *Toledo (OH) Blade*, 8 December 1969, 3.

148. For a description of this event, and an explanation of the letters written by NOW members in response to Conrad's comment, see Vermillion, South Dakota NOW chapter leader Catherine Dunlap to NOW President Wilma Scott Heide, 19 September 1973, Manuscript Collection 495: Wilma Scott Heide, 1921–1985, Box 20, Folder 5: "NASA [Correspondence, Notes, Etc., 1973]," Schlesinger Library, Radcliffe Institute, Harvard University, Cambridge, MA.

149. Tampa Chapter NOW President Von Drury Kerik to Mr. James Lovell, 22 October 1973, Manuscript Collection 495: Wilma Scott Heide, 1921–1985, Box 20,

Folder 5: "NASA [Correspondence, Notes, Etc., 1973]," Schlesinger Library, Radcliffe Institute, Harvard University, Cambridge, MA. Similar mock award ceremonies were held by other local NOW chapters. For an example of New York City's NOW chapter holding annual "Keep Her in Her Place" awards, see Laurie Johnston, "'Women Power' Protests 'Male Domination' of Wall St.," *New York Times*, 24 August 1973, 39.

150. Lovell's invitation to the NOW "Barefoot and Pregnant Ball" because of this statement can be found in "The Decline and Fall of Lunar Royalty," *Texas Monthly*, January 1974, 38. Lovell's statement was covered widely in the press. See, for instance, Nancy Q. Keefe, "Liberation Comes in Subtle Ways," *Berkshire (MA) Eagle*, 3 November 1972, 8; "People Etc.," *Waterloo (IA) Sunday Courier*, 15 July 1972, 3; and "Gazette News," *Ms.*, September 1973, 21.

151. Associated Press, "Women Astronauts," 3.

152. "Tentative Outline—Meeting with Fletcher—NASA—11/29/73," Wilma Scott Heide, 1921–1985, Box 20, Folder 5: "NASA [Correspondence, Notes, Etc., 1973]," Schlesinger Library, Radcliffe Institute, Harvard University, Cambridge, MA.

153. These data on professional women hired at NASA come from Kim McQuaid, "Race, Gender, and Space Exploration: A Chapter in the Social History of the Space Age," *Journal of American Studies* 41 (2007): 405–454, 410. McQuaid's essay is also a thorough history of NASA's EEO Program.

154. "Official Statement of National Organization for Women (NOW), January 24, 1974," Manuscript Collection 496: National Organization for Women, Records 1959–2002 (Inclusive), 1966–1998 (Bulk), Carton 44, Folder 5: NASA [Correspondence, Notes, Clippings, Press Releases, Etc.], 1973–1974, Schlesinger Library, Radcliffe Institute, Harvard University, Cambridge, MA.

155. This reference to "Space City U.S.A." and the 1964 Civil Rights Act appears in what seems to be a first draft of the press release. See "NOW Calls for Resignation of NASA Administrator," press release by the National Public Information Office of the National Organization for Women, copy located at Manuscript Collection 496: National Organization for Women, Records 1959–2002 (Inclusive), 1966–1998 (Bulk), Carton 44, Folder 5: NASA [Correspondence, Notes, Clippings, Press Releases, Etc.], 1973–1974, Schlesinger Library, Radcliffe Institute, Harvard University, Cambridge, MA.

156. Ibid.

157. Chase Untermeyer, "Women's Unit Asks Impeachment: NASA Head Also Assailed," *Washington Post*, 29 May 1974, A4.

158. On NOW's petition to Bella Abzug regarding demands that Fletcher resign from NASA, see NOW Coordinator to Launch NASA into Feminist Space (NOW Compliance Task Force) to Honorable Bella Abzug, U.S. House of Representatives, 26 June 1974, Manuscript Collection 496: National Organization for

Women, Records 1959–2002 (Inclusive), 1966–1998 (Bulk), Carton 44, Folder 5: NASA [Correspondence, Notes, Clippings, Press Release, Etc.], 1973–1974, Schlesinger Library, Radcliffe Institute, Harvard University, Cambridge, MA.

159. Joan McCullough, "13 Who Were Left Behind," *Ms.* 11, no. 3 (September 1973): 41–45. Several magazines published shorter articles on these thirteen female pilots during their tests at the Lovelace Clinic, but none had the public impact of the *Ms.* article. This was due, in part, to the tone of these earlier essays, which never examined the gender discrimination at the heart of the NASA astronaut-selection program. For examples of earlier coverage of the Lovelace women astronaut examinations, see "Damp Prelude to Space," *Life* 49, no. 17 (24 October 1960): 81; "A Lady Proves She's Fit for Space Flight," *Life* 49, no. 9 (29 August 1960): 72–74; and "The Lady Wants to Orbit," *Look* 24, no. 3 (1960): 112–119.

160. This protest is described in a letter from Catherine Dunlap to NOW President Ms. Wilma Scott Heide, 19 September 1973, Manuscript Collection 495: Wilma Scott Heide, 1921–1985, Box 20, Folder 5: "NASA [Correspondence, Notes, Etc., 1973]," Schlesinger Library, Radcliffe Institute, Harvard University, Cambridge, MA.

161. On the Radcliffe symposium, see "Space for Women: Perspectives on Careers in Science," *Space World*, vol. 0-3-171 (March 1978): 4–21. For a description of the Mills College teach-in symposium, see Kevles, *Almost Heaven*, 41–42.

162. This surreptitious protest was described at length in "The Decline and Fall of Lunar Royalty," *Texas Monthly*, January 1974, 38.

163. A copy of the fake ballot, along with several cartoons, including the ones discussed here, can be found at Manuscript Collection 496: National Organization for Women, Records 1959–2002 (Inclusive), 1966–1998 (Bulk), Carton 44, Folder 5: NASA [Correspondence, Notes, Clippings, Press Releases, Etc.], 1973–1974, Schlesinger Library, Radcliffe Institute, Harvard University, Cambridge, MA.

164. On NOW coordinating ZAPs to protest against NASA's all-male astronaut program, see Judy Flander, "NOW Banks Too Hard on One Zap," *Washington Star-News*, 27 August 1974, C1–2; and Barbara Bright-Sagnier, "NOW Protesters Stage 3 Rallies," *Washington Post*, 27 August 1974, A10. For other examples of NOW using ZAPs to fight gender discrimination, see Johnston, "'Women Power' Protests," 39.

165. Louis Harris, "Harris Survey: Landing on Moon Opposed by Most," *Milwaukee (WI) Sentinel*, 17 February 1969, 7.

166. Another primary motivator for these hearings was the firing, by NASA, of Ruth Bates Harris, an African American woman who had worked in NASA's EEO Program. On this case, see McQuaid, "Race, Gender, and Space Exploration."

167. *Review of NASA's Equal Employment Opportunity Program*, Hearing before the Committee on Aeronautical and Space Sciences, 93rd Cong. (Senate), 2nd Sess., 98–99 (24 January 1974) (written statement of Gayla Salinas, National

Compliance Task Force, and Ann Scott, Vice President for Legislation, National Organization for Women). For a deeper history of NASA's early, flawed EEO Program, see Foster, *Integrating Women*, 71–73.

168. Austin Scott, "Senators Eye NASA EEO Goals," *Washington Post*, 25 January 1974, A5.

169. Some high-level NASA administrators wanted to keep these studies on flight nurses secret. See Brian B. King, "Next Space Shuttle Candidate Could Be of the Fairer Sex," *Reading (PA) Eagle*, 18 October 1973, 25. Officials at NASA also went out of their way to explain to the public that these tests in no way obligated the space agency to accept female candidates in its astronaut corps. On this, see King, "Next Space Shuttle Candidate," 25.

170. "Space for Women," *Time*, 5 November 1973, 19.

171. The internal body thermometer is mentioned in "Nurses Finish Weeks of Medical Testing," *Roundup* (newsletter of the Johnson Space Center), 23 November 1973, 2. On these tests in 1973 by NASA, see also "Ladies on the Pad?," *Time*, 22 October 1973, 79; Associated Press, "NASA Begins Studies to Learn Suitability of Women in Space," *New York Times*, 22 September 1973, 61; and, especially, David J. Shayler and Ian A. Moule, *Women in Space: Following Valentina* (New York: Springer, 2005), 148–151.

172. King, "Next Space Shuttle Candidate," 25.

173. "Space for Women," 19. See also Associated Press, "NASA Begins Studies," 61.

174. For a description of the simulations used by NASA during these 1973 tests, see "Space for Women," 19; and Shayler and Moule, *Women in Space*, 148–151.

175. The NASA spokesperson, Donald L. Zylstra, was quoted in King, "Next Space Shuttle Candidate," 25.

176. For a thorough and often humorous description of NASA's "Housewives in Space" study, see Barbara Rowes, "Housewives in Space," *Omni*, June 1982, 64–67, 128.

177. "Space Suits Smaller for Women," 29 December 1978, Women in Space: 1978–90 file, NASA Headquarters, as quoted in Foster, *Integrating Women*, 115. Foster here includes an interesting and much more extensive discussion of development by NASA of spacesuits for women during the shuttle era.

178. Los Angeles Times Service, "Women Pass Space Tests with High Flying Colors," *Milwaukee (WI) Journal*, 24 October 1973, 6.

179. For a description of the medical results of both tests, see Shayler and Moule, *Women in Space*, 149.

180. These results are discussed in Mary Connors, Albert A. Harrison, and Faren R. Akins, *Living Aloft: Human Requirements for Extended Spaceflight* (Honolulu: University Press of the Pacific, 2005), 28–30.

181. NASA's Dr. Sandler is quoted in Rowes, "Housewives in Space," 128.

182. "NASA to Recruit Space Shuttle Astronauts," NASA News Release, No. 76-122, 8 July 1976, NASA History Office.

183. United Press International, "Astronauts Wanted; Women, Minorities Are Urged to Apply," *New York Times*, 8 July 1976, 23.

184. On NASA's relationship with Nichelle Nichols, see Shayler and Moule, *Women in Space*, 152–156. For discussion of the NASA symposium on women in space-related professions, including astronauts, see "Federal Women's Program at JSC Attracts Unprecedented Crowds," *Roundup* 17, no. 8 (28 April 1978): 1; and "Women in Aerospace," *Roundup* 19, no. 5 (7 March 1980): 1. For an example of NASA targeting women's magazines regarding its call for female astronaut candidates, see "Coming Soon: Women in Space," *McCall's* 104, no. 7 (April 1977): 56. Sally Ride mentions NASA targeting women's science organizations on college campuses in "Gloria Steinem Interview with Sally Ride," transcript located in Sophia Smith Collection, College Archives, Smith College, Northampton, MA.

185. Sara Sanborn, "Sally Ride, Astronaut the World Is Watching," *Ms.*, January 1983, 47.

186. Richard D. Lyons, "35 Chosen Astronaut Candidates; Six Are Women and Three Blacks," *New York Times*, 17 January 1978, 14.

187. Sanborn, "Sally Ride," 88.

188. Vera Glaser, "Senator's Wife Delivers Verbal Slap at Astronaut," *Youngstown (OH) Vindicator*, 25 January 1969, 12. While such language was used during the early 1960s in the debate over women astronauts, it was used less frequently and went largely unnoticed, suggesting that while these two ideologies existed side-by-side during this period, their respective cultural significance waxed and waned. For examples from the early 1960s of references to women's biological differences being beneficial for spaceflight, see Drew Pearson, "U.S. Lady Pilots Mad at NASA Chief," *Nevada Daily Mail*, 20 June 1963, 2; and "Jerrie Cobb Grounded," *Milwaukee (WI) Sentinel*, 18 June 1963, 5.

189. There is a rich literature on this split within the feminist movement between those calling for equality with men versus those highlighting differences between men and women. For analyses of the role of the human body within this feminist split, see, especially, Ann Snitow, "A Gender Diary," in *Conflicts in Feminism*, ed. Marianne Hirsch and Evelyn Fox Keller (New York: Routledge, 1990), 9–43; Kathy Davis, "Embody-ing Theory: Beyond Modernist and Postmodernist Readings of the Body," in Davis, *Embodied Practices*, 1–23; and Joan Scott, "Deconstructing Equality-Versus-Difference: Or, the Uses of Poststructuralist Theory for Feminism," *Feminist Studies* 14, no. 1 (Spring 1988): 33–50. For a more general examination of this debate regarding equality and difference within second-wave feminism, see also Alison M. Jaggar, "Sexual Difference and Sexual Equality," in *Theoretical Perspectives on Sexual Differences*, ed. Deborah L. Rhode (New Haven, CT: Yale University Press, 1992); Josephine Donovan, *Feminist Theory* (New York:

Frederick Ungar, 1985); Hester Eisenstein, *Contemporary Feminist Thought* (Boston: G. K. Hall, 1980); Hester Eisenstein and Alice Jardine, eds., *The Future of Difference* (Boston: G. K. Hall, 1980); Zillah R. Eisenstein, *Feminism and Sexual Equality: Crisis in Liberal America* (New York: Monthly Review Press, 1984); Juliet Mitchell, *Women's Estate* (New York: Pantheon, 1971); and Juliet Mitchell and Ann Oakley, eds., *What Is Feminism?* (New York: Pantheon, 1986).

190. I borrow the term "equality feminists" from Snitow, "Gender Diary," 24–29. In her essay Snitow traces the origin of this "equality" camp back to Mary Wollstonecraft's *A Vindication of the Rights of Woman*. Kathy Davis also places Simone de Beauvoir in this group, as well as Judith Butler, who provides the most radical refusal of biological difference. See Davis, "Embody-ing Theory," 8–9.

191. Carol Gilligan is perhaps best known for making this the argument regarding women's uniqueness. See Carol Gilligan, *In a Different Voice: Psychological Theory and Women's Development* (Cambridge, MA: Harvard University Press, 1982). Gilligan's book launched its own debate within feminism regarding equality versus difference. On this debate, see Linda K. Kerber, Catherine G. Green, Eleanor E. Maccoby, Zella Luria, Carol B. Stack, and Carol Gilligan, "On *In a Different Voice*: An Interdisciplinary Forum," *Signs* 11, no. 2 (Winter 1986): 304–333.

192. For a helpful discussion of these "difference feminists," see Snitow, "Gender Diary," 24; and Scott, "Deconstructing Equality," 38–39. Kathy Davis places French feminist scholars Hélène Cixous, Luce Irigaray, and Julia Kristeva in this camp. See Davis, "Embody-ing Theory," 8–9.

193. As reported in Glaser, "Senator's Wife Delivers Verbal Slap," 12.

194. There is a rich literature on the women's health movement and the Boston Women's Health Book Collective. See, especially, Kathy Davis, *The Making of Our Bodies, Ourselves: How Feminism Travels Across Borders* (Durham, NC: Duke University Press, 2007; and Wendy Kline, "The Making of *Our Bodies, Ourselves*: Rethinking Women's Health and Second-Wave Feminism," in *Feminist Coalitions: Historical Perspectives on Second-Wave Feminism in the United States*, ed. Stephanie Gilmore (Chicago: University of Illinois Press, 2008).

195. McCullough, "13 Who Were Left Behind," 41–45.

196. Rosenthal, "Woman Astronaut Caught Up in Celebrity Whirl," 7A. Although STS-7's four male crewmembers accompanied Ride during this schedule, most of the events, according to this reporter, were aimed at celebrating America's first woman astronaut.

197. Ken Alexander, "Flight of the Space Shuttle Challenger as Seen by the Media," *San Francisco Examiner*, 20 June 1983, B2.

198. This statement by Ride's mother was covered extensively by the press. See Joyce Wadler, "The Feminist at 50: A Glittery Birthday Bash for Gloria Steinem at 50," *Washington Post*, 24 May 1984, D1.

199. Scott, "Deconstructing Equality," 48.

200. United Press International, "Feminists Hail Historic Flight into Space," *Salina (KS) Journal*, 19 June 1983, 6.

5. The New Right's Stuff

1. Kunen describes his experiences covering the Apollo 8 launch in James Simon Kunen, "The Great Rocketship," in *US: Paperback Magazine*, no. 1 (New York: Bantam, 1969), 14, 16. Although Kunen could not remember his attire exactly, he surmised that he was most likely wearing blue jeans and "a necklace of small beads." James Kunen, email message to author, 4 August 2015.

2. William F. Buckley Jr., "Can Men Make Miracles?," from his syndicated column titled "On the Right," *Victoria (TX) Advocate*, 30 December 1968, 2A.

3. Kunen, "Great Rocketship," 15–16, 18.

4. I describe Kunen as a member of the counterculture because others at the Apollo 8 launch identified him as such. At the time Kunen was best known for his book *The Strawberry Statement: Notes of a College Revolutionary*, which chronicled his experiences as a student during the Columbia University uprising from 1966 to 1968.

5. The idea of opposing "the Establishment" became a key concept during the 1960s, especially among the counterculture. On the use of the phrase during this era, see the writings of Howard Zinn, especially *A People's History of the United States: 1492–Present* (New York: Harper Perennial Modern Classics, 2005), 541–562.

6. This point is made repeatedly by scholars of the 1960s. See especially Marilyn B. Young, "Foreword," and Beth Bailey, "Sex as a Weapon: Underground Comix and the Paradox of Liberation," in *Imagine Nation: The American Counterculture of the 1960s and '70s*, ed. Peter Braunstein and Michael William Doyle (New York: Routledge, 2002).

7. Because members of the counterculture often called themselves "hippies," I use the term as well. My view of the 1960s counterculture as an intentioned social movement has been influenced by theoretical literature that posits distinctions between undisciplined mobs on the one hand and crowds that function with some semblance of intent on the other. For a brief review of this literature, see Christian Borch, *The Politics of Crowds: An Alternative History of Sociology* (New York: Cambridge University Press, 2012), 11–12. Here I am drawn to the ideas of what Borch calls the "Lefebvre–Rudé–Thompson–Hobsbawm tradition."

8. Nixon's speech was reported by Ron Yogman in the *Evening Independent* (St. Petersburg, FL), 28 September 1968, 1.

9. Richard Nixon, "Address to the Nation on the War in Vietnam," 3 November 1969, *American Presidency Project*, online at http://www.presidency.ucsb.edu/ws/index.php?pid=2303 (accessed 9 August 2016). On the composition of

Nixon's silent majority and its geography across the Southern Sunbelt, see Matthew D. Lassiter, *The Silent Majority: Suburban Politics in the Sunbelt South* (Princeton, NJ: Princeton University Press, 2006), especially the introduction. As Lassiter explains, this courtship by Nixon was part of the Republican Party's "Southern strategy."

10. Although sociologists and journalists had used the term "New Right" in the 1950s, it was first used as self-identification in 1962 by the conservative student activist group Young Americans for Freedom. It was Buckley's *National Review*, however, that popularized the term "New Right" to describe an emergent conservative movement.

11. For examples of this scholarship, see Roger Launius and Howard Mc-Curdy, "Epilogue: Beyond NASA Exceptionalism," in *Spaceflight and the Myth of Presidential Leadership*, ed. Roger Launius and Howard McCurdy (Urbana: University of Illinois Press, 1997), 221–244; and Andrew J. Butrica, "The 'Right' Stuff: The Reagan Revolution and the U.S. Space Program," in *Remembering the Space Age: Proceedings of the 50th Anniversary Conference*, ed. Steven J. Dick (Washington, DC: NASA History Division, 2008), 121–134.

12. The best account of Apollo's impact on American culture, and the one that comes closest to engaging the social and political movements of the 1960s era, is Matthew D. Tribbe, *No Requiem for the Space Age: The Apollo Moon Landings and American Culture* (New York: Oxford University Press, 2014).

13. United Press International (hereafter UPI), "'Triumph of the Squares': Future Outlined by NASA Leader," *Spokane (WA) Daily Chronicle*, 27 December 1968, 15.

14. Thomas O. Paine, "Squareland, Potland and Space," 7 June 1970, Worcester Polytechnic Institute commencement address, Box 11, Folder 420, UA009 University Subject Files, Curation, Preservation, and Archives, Gordon Library, Worcester Polytechnic Institute, 2.

15. Ayn Rand, "Apollo and Dionysus," speech given on 9 November 1969 at Suffolk University's Ford Hall Forum, and reprinted in Ayn Rand, *Return of the Primitive: The Anti-Industrial Revolution* (New York: Penguin, 1999), 100, 109.

16. "The Moon and 'Middle America,'" *Time*, 1 August 1969, 11.

17. Ernest B. Furgurson, "Sons of the Forgotten," *Sun* (Baltimore, MD), 22 July 1969, A10. Furgurson is referring to Nixon's speech of 8 August 1968, when he accepted the presidential nomination at the Republican National Convention. On the rise of the silent majority, see Lassiter, *Silent Majority*. For newspaper coverage using similar language, see Roulhac Hamilton, "'Squares' Inherit the Moon," *Columbus (OH) Dispatch*, 27 July 1969, as reprinted in "'Squares' Inherit the Moon," 115th Cong., *Congressional Record*, 11 August 1969, 23610.

18. Senator Barry Goldwater, "A Realistic Space Program for America," *Science and Mechanics* (June 1964): 44.

19. See Edwin McDowell, "Can Man Turn His Back on Space?," *Wall Street Journal*, 21 April 1972, 8; "NASA Meets Small Business," *Business Week*, 17 February 1962, 140; *The Space Industry: America's Newest Giant*, by the editors of *Fortune* magazine (Englewood Cliffs, NJ: Prentice Hall, 1962), especially chapter 6, titled "Hitching the Economy to the Infinite," by Gilbert Burck.

20. Orr Kelly, "Washington Close-Up: Dollars for Space Getting Scarcer," *Washington Star*, 27 May 1969, clipping file on newspaper coverage of Apollo 10 located at the National Air and Space Museum, Washington, DC. For other examples of such language, see "The Moon Landing and the ABM," *Chicago Tribune*, 24 July 1969, N20; and Henry MacArthur, "The Golden Goose," *Lodi (CA) News-Sentinel*, 13 November 1963, 14. For examples of NASA's critics, such as Lewis Mumford, who also associated the aerospace industry with the military-industrial complex, see Nick Thimesch, "Astronauts Revived Our Spirits," *Cleveland (OH) Plain Dealer*, 25 July 1969, found on page 337 of clipping file on newspaper coverage of Apollo 11 located at the National Air and Space Museum, Washington, DC.

21. Ayn Rand, "Apollo 11," *Objectivist* 8, no. 9 (September 1969): 8.

22. Thomas O'Toole, "$24 Billion—20,000 Contractors: Companies Wage Industrial Effort for Apollo Equivalent to Fighting Small War," *Washington Post* (Business and Financial section), 13 July 1969, 1; James Webb, "Impact of NASA Program on Industry," an address at the 68th Annual Congress of American Industry, New York City, 6 December 1963, Hagley Museum and Library, Wilmington, DE, 6; and John Dille, "The Revolution Isn't Coming—It Is Already Here," *Life* 57, no. 13 (25 September 1964): 94–103.

23. "Apollo Footnotes," *Birmingham News*, 22 July 1969, found on page 221 of clipping file on newspaper coverage of Apollo 11 located at the National Air and Space Museum, Washington, DC.

24. Buckley, "Can Men Make Miracles?"

25. Rand, "Apollo 11," 1.

26. Rand, "Apollo and Dionysus," 101, 109.

27. The historian David Nye has termed such thinking the "technological sublime." For his analysis of Apollo 11's aesthetic as what he calls the "dynamic sublime," see David Nye, *American Technological Sublime* (Cambridge, MA: MIT Press, 1994), 225–256.

28. Rand, "Apollo 11," 6. Italics in the original.

29. Paine, "Squareland, Potland and Space," 5, 8.

30. Webb, "Impact of NASA Program on Industry," 1.

31. Rand, "Apollo 11," 2.

32. President Richard Nixon, "Second Annual Report to the Congress on United States Foreign Policy," 25 February 1971, reprinted in *Public Papers of the Presidents of the United States: Richard Nixon, Containing the Public Messages,*

Speeches and Statements of the President, 1971 (Washington, DC: U.S. Government Printing Office, 1972), 339.

33. Wernher von Braun, "For Space Buffs—National Space Institute—You Can Join," *Popular Science* (May 1976): 73.

34. John F. Kennedy, "Address of Senator John F. Kennedy Accepting the Democratic Party Nomination for the Presidency of the United States, Memorial Coliseum, Los Angeles," 15 July 1960, online at http://www.presidency.ucsb.edu /ws/?pid=25966 (accessed 30 July 2015).

35. On Kennedy's New Frontier in space and America's western frontier, see Patricia Nelson Limerick, "The Adventures of the Frontier in the Twentieth Century," in *The Frontier in American Culture*, ed. James R. Grossman (Berkeley: University of California Press, 1994), 67–102; Susan K. Opt, "American Frontier Myth and the Flight of Apollo 13: From News Events to Feature Film," *Film and History* 26, nos. 1–4 (1996): 40–51; and Karl Leib, "Manifest Destiny in Space: Ideology, International Competition, and U.S. Policy," *Maxwell Review* 6, no. 1 (Spring 1998): 28–37.

36. John F. Kennedy, "Address at Rice University in Houston on the Nation's Space Effort," 12 September 1962, online at http://www.presidency.ucsb.edu/ws /index.php?pid=8862 (accessed 30 July 2015).

37. Kennedy, "Address of Senator John F. Kennedy Accepting the Democratic Party Nomination."

38. Richard Nixon, "Fourth Annual Report to the Congress on United States Foreign Policy," 3 May 1973, Public Papers of the Presidents, online at http://www .presidency.ucsb.edu/ws/index.php?pid=3832 (accessed 12 August 2015).

39. Richard Nixon, "Statement About the Space Program," 19 December 1972, online at: http://www.presidency.ucsb.edu/ws/index.php?pid=3718 (accessed 29 October 2016).

40. Mayo Mohs, "God, Man and Apollo," *Time*, 1 January 1973, 51.

41. Rand, "Apollo 11," 15.

42. E. A. Kendall, "Moon Flags—The Other Side," letter to the editor, *Washington Post*, 20 June 1969, A26. For examples of similar sentiment, see also John C. Snow, "Moon Flags—The Other Side," letter to the editor, *Washington Post*, 20 June 1969, A26.

43. For background on NASA's lunar flags and legislation regarding them, see Anne M. Platoff, "Where No Flag Has Gone Before: Political and Technical Aspects of Placing a Flag on the Moon," NASA Contractor Report 188251, online at http://www.jsc.nasa.gov/history/flag/flag.htm (accessed 24 August 2015).

44. For a brief biography of Charles Brooks, including Nixon's praise of him, see Michael Cavna, "RIP, Charles Brooks Sr.: Editorial Cartoonist / Editor Dies at 90," *Washington Post*, 1 October 2011, online at http://www.washingtonpost.com

/blogs/comic-riffs/post/rip-charles-brooks-sr-editorial-cartoonisteditor-dies-at
-90/2011/10/01/gIQAZMvzBL_blog.html (accessed 24 August 2015).

45. Charles Brooks, "Flag-Raising," *Birmingham (AL) News*, 22 July 1969, 16.
For additional examples of cartoons on Apollo referencing America's old frontier,
see C. P. Houston (Clyde Peterson), "To the Planets or Bust," *Houston Chronicle*,
22 July 1969, 16; and York, "Remember, When They Land Don't Show Them How
to Plant Corn or Build Wigwams," *Evening Star* (Washington, DC), 3 July 1969.

46. Kunen, "Great Rocketship," 14, 15–16, 18.

47. On Mailer serving as a literary bridge between the Beats and the counterculture, see Peter Braunstein, "Forever Young: Insurgent Youth and the Sixties
Culture of Rejuvenation," in Braunstein and Doyle, *Imagine Nation*, 256.

48. Norman Mailer, *Of a Fire on the Moon* (Boston: Little, Brown, 1969), 16,
17, 227.

49. Abbie Hoffman, *Woodstock Nation: A Talk-Rock Album* (New York:
Random House, 1969), 7–8.

50. Ibid., 41.

51. R. Crumb, "R. Crumb on Assignment for the *CoEvolution Quarterly* Goes
to the . . . Space Day Symposium (Or Whatever the Hell It Was Called)," *CoEvolution Quarterly*, Fall 1977, 48–51. The *CoEvolution Quarterly* was the successor to
Stewart Brand's *Whole Earth Catalog* and was read widely by the counterculture
during the 1970s.

52. For additional critiques by the counterculture of the Apollo mission on
economic and military grounds, see Kunen, "Great Rocketship," 19; Victor Chen,
"Majority of Youth Laud Missions," *Boston Globe*, 18 July 1969, 3; and Mailer, *Of a
Fire on the Moon*, 65.

53. Paine, "Squareland, Potland and Space," 6.

54. Chip Lord, "Space Colonies Are Tail Fins," *CoEvolution Quarterly*, Spring
1976, 19.

55. Jefferson Airplane, "Have You Seen the Saucers," from the album *Mexico*
(RCA, 1970).

56. Paul Kantner and Jefferson Starship, *Blows against the Empire* (RCA
Victor, 1970). For a similar analysis of Jefferson Airplane's music and its relationship to the space race, see Tribbe, *No Requiem for the Space Age*, 146–149.

57. David Bowie, "Space Oddity," Mercury Records (1969).

58. Wendell Berry, *CoEvolution Quarterly*, Spring 1976, 8–9.

59. See Gary Snyder, "Space Colonies," *CoEvolution Quarterly*, Spring 1976,
45; and Anne Waldman, "Space Colonies," *CoEvolution Quarterly*, Spring 1976, 47.

60. United Press International, "Hippies Play Astronaut Buying Moon Off
Indian," NASA Current News, Prepared by [NASA] Public Information Division, Office of Public Affairs, July and August 1969, Apollo 11 Special Part 2,

Handwritten Page Number 162, NASM Archives File: OA-253815-02- Apollo 11 Flight, Articles, NASA News Compilations, Archives Department, National Air and Space Museum, Smithsonian Institution, Washington, DC. On the importance of guerrilla theater to the politics of the counterculture during the 1960s, see Michael William Doyle, "Staging the Revolution: Guerrilla Theater as a Countercultural Practice, 1965–68," in Braunstein and Doyle, *Imagine Nation*, 71–98.

61. Hoffman, *Woodstock Nation*, 42–43.

62. Ibid., 43.

63. "Checklist for ISA (Inner Space Astronauts)," *San Francisco Oracle* 1, no. 3 (November 1966): 6.

64. Hoffman, *Woodstock Nation*, 13.

65. "Year of the Commune," *Newsweek*, 18 August 1969, 89.

66. Woodstock Music and Art Fair advertisement, *Village Voice*, 14 August 1969, 21.

67. "Year of the Commune," 89.

68. Hoffman, *Woodstock Nation*, 43. The number of hippies living on communes during the 1960s was most probably even higher, approaching one million individuals living on tens of thousands of communes. On these statistics, see Timothy Miller, *The 60's Communes: Hippies and Beyond* (Syracuse, NY: Syracuse University Press, 1999), xviii–xx.

69. Timothy Leary, *Flashbacks: An Autobiography* (Los Angeles: J. P. Tarcher, 1983), 190. Thomas Wolfe describes Leary shunning Kesey's Merry Pranksters at Millwood in *The Electric Kool-Aid Acid Test* (New York: Farrar, Straus and Giroux, 1968), 107.

70. For analyses of these communes' relationship to the natural environment, especially in rural areas, see Andrew G. Kirk, *Counterculture Green: The Whole Earth Catalog and American Environmentalism* (Lawrence: University of Kansas Press, 2007), 43–73; and Andrew Kirk, "'Machines of Loving Grace': Alternative Technology, Environment, and the Counterculture," in Braunstein and Doyle, *Imagine Nation*, 353–378. On the more general history of these rural communes, see Timothy Miller, "The Sixties-Era Communes," in Braunstein and Doyle, *Imagine Nation*, 327–351. According to Miller the number of communes during the 1960s era most probably reached into the tens of thousands. See Miller, *60's Communes*, xviii–xx.

71. "Year of the Commune," 89.

72. This language was used by Langdon Winner to describe the audience of the *Whole Earth Catalog*. See Langdon Winner, "Building the Better Mousetrap: Appropriate Technology as a Social Movement," in *Appropriate Technology and Social Values—A Critical Appraisal*, ed. Franklin A. Long and Alexandra Oleson (Pensacola, FL: Ballinger, 1980), 31.

73. Inside the cover of the inaugural edition of his *Whole Earth Catalog* Brand lamented technology created by "government" and "big business" and praised instead tools that were more "intimate" and "personal." See "Purpose," in *Whole Earth Catalog: Access to Tools* (Menlo Park, CA: Portola Institute, 1968), 2. For Brand and his readers, E. F. Schumacher's *Small Is Beautiful: A Study of Economics as If People Mattered* was equally appealing.

74. See "Space Structures," in *Whole Earth Catalog*, 13.

75. *Original Whole Earth Catalog, Special 30th Anniversary Issue*, ed. Peter Warshall and Stewart Brand (Portola, CA: Whole Earth, 1998), 3, as quoted in Kirk, "'Machines of Loving Grace,'" 363. On Brand visiting communes during the mid- to late 1960s, see Fred Turner, *From Counterculture to Cyberculture: Stewart Brand, the Whole Earth Network, and the Rise of Digital Utopianism* (Chicago: Chicago University Press, 2006), 82.

76. John Todd often referenced the Odums' experiments for NASA with bio-regenerative life support systems when explaining the New Alchemists' arks. See John Todd, "Space Colonies," *CoEvolution Quarterly*, Spring 1976, 21.

77. For a general description of the New Alchemists "arks," see "The New Al-chemists," *Time* 105, no. 11 (17 March 1975): 100; "New Alchemy Institute: Search for an Alternative Agriculture," *Science, News Series* 187, no. 4178 (28 Feb-ruary 1975): 727–729; and "New Alchemy's Ark: A Proposed Solar Heated and Wind Powered Greenhouse and Aquaculture Complex Adapted to Northern Cli-mates," *Journal of the New Alchemists* 2 (1974): 35–43. For discussion of the New Alchemists "solar ponds," see John Todd, "The New Alchemists," *CoEvolution Quarterly*, Spring 1976, 58–60.

78. On Brand and Rusty Schweikart visiting the New Alchemists on Cape Cod, see Nancy Jack Todd, *A Safe and Sustainable World: The Promise of Ecolog-ical Design* (Washington, DC: Island Press, 2006), 64.

79. Walter P. Stewart, "The Ark: Prince Edward Island's Spaceship to the Future," *Atlantic Advocate* 67, no. 1 (September 1976): 44.

80. On the politics of these home technologies, see Kirk, "'Machines of Loving Grace,'" 363.

81. Kunen, "Great Rocketship," 13.

82. Kim McQuaid makes this argument about Paine blaming hippies in Kim McQuaid, "Selling the Space Age: NASA and Earth's Environment, 1958–1990," *Environment and History* 12, no. 2 (February 2006): 136.

83. Paine, "Squareland, Potland and Space," 13.

84. See NASA's press release on this collaboration, "NASA, Science Foundation Sign Agreement on Energy Studies," NASA News Release No. 74-302, 18 November 1974, located in NASA Headquarter Archives, NASA Historical Collection, Folder 9825: Energy Solar (1974), National Aeronautics and Space Administration,

Washington, DC. Much of the research of this committee focused on the unsuccessful development of solar satellites that could beam solar energy back to Earth.
For a detailed description of this failure, see Jeff Womack, "Pipe Dreams for Powering Paradise: Solar Power Satellites and the Energy Crisis," in *American Energy
Policy in the 1970s*, ed. Robert Lifset (Norman: University of Oklahoma Press,
2014), 205–220.

85. Richard Nixon, "Special Message to the Congress on Energy Resources," 4
June 1971, Public Papers of the Presidents, online at http://www.presidency.ucsb
.edu/ws/index.php?pid=3038 (accessed 29 October 2015).

86. For examples of Nixon proposing nuclear and solar power, see Richard
Nixon, "Special Message to the Congress on Energy Policy," 18 April 1973, Public
Papers of the Presidents, online at http://www.presidency.ucsb.edu/ws/?pid=3817
(accessed 29 October 2015).

87. Richard Nixon, "Address to the Nation about Policies to Deal with the
Energy Shortages," 7 November 1973, Public Papers of the Presidents, online at
http://www.presidency.ucsb.edu/ws/?pid=4034 (accessed 29 October 2015).

88. On NASA's involvement in Nixon's Project Independence, see Joseph G.
Mason, "Solution Is Still Sought for Petroleum Scarcity," *Sarasota (FL) Journal*, 14
October 1974, 4B.

89. Thomas Sweeney, "If Verne Could Look at NASA," *National Review* 21,
no. 19 (20 May 1969): 490. Although Sweeney wrote this article before Nixon announced Project Independence, it and the other commentary from conservatives
capture their support of the possibility of solar-powered space colonies.

90. "Can the Moon Be of Any Earthly Use?," *Time*, 18 July 1969, 23; "A Question of Space," editorial, *Wall Street Journal*, 27 May 1969, 22.

91. Von Braun mentions the role of solar power in all three *Colliers* special
issues. See Wernher von Braun, "Crossing the Last Frontier," in "Man Will Conquer Space Soon," special issue, *Colliers* (22 March 1952): 72; Wernher von Braun,
"The Journey," in "Man on the Moon," special issue, *Colliers* (18 October 1953):
52–59; and Fred L. Whipple and Wernher von Braun, "The Exploration," in "Man
on the Moon: The Exploration," special issue, *Colliers* (25 October 1952): 46.

92. See "Man in Space," an episode of *Disneyland*, which originally aired on
ABC on 9 March 1955. The solar-power technology described in this episode seems
to have derived from a 1952 sketch by von Braun in preparation for the *Colliers*
articles. On this similarity, see John Tierney, "Von Braun's Spaceship," written 14
October 2008 and online at http://tierneylab.blogs.nytimes.com/2008/10/14/von
-brauns-spaceship/?_r=0 (accessed 4 November 2015).

93. Gerard O'Neill, *The High Frontier: Human Colonies in Space*, as quoted in
Patrick McCray, *The Visioneers: How a Group of Elite Scientists Pursued Space Colonies, Nanotechnologies, and a Limitless Future* (Princeton, NJ: Princeton University Press, 2013), 84. O'Neill published his calculations regarding the efficiency of

NASA's satellite solar-power stations in Gerard K. O'Neill, "Space Colonies and Energy Supply to the Earth," *Science* 190, no. 4218 (5 December 1975): 943–947.

94. Gerard O'Neill, "The Colonization of Space," *Physics Today* 27, no. 9 (September 1974), online at http://www.askmar.com/Massdrivers/1974-9%20Space%20 Colonization.pdf (accessed 4 November 2015), 6.

95. For accounts of O'Neill's relationship with NASA, see De Witt Douglas Kilgore, *Astrofuturism: Science, Race, and Visions of Utopia in Space* (Philadelphia: University of Pennsylvania Press, 2003), 161; and McCray, *Visioneers*, 47, 53–59, 84–87. For an example of NASA incorporating O'Neill's space-colony concepts into its own promotional materials, see "Johnson Space Center, Houston, Texas: 41 Full Color Pictures of the 'Nerve-Center" of America's Aerospace Development" (Houston: Astrocard, 1978), especially the materials on "Space Solar Power Concept" and "Space Colonies."

96. Walter Sullivan, "Proposal for Human Colonies in Space Is Hailed by Scientists as Feasible Now," *New York Times*, 13 May 1974, A1; Gerard O'Neill, *The High Frontier: Human Colonies in Space* (New York: William Morrow, 1977).

97. The artist for this image was Rick Guidice, *Stanford Torus Cutaway View*, 1975, courtesy of NASA, online at http://www.nss.org/settlement/space /stanfordtorus.htm (accessed 4 November 2015). The short film now appears on YouTube under the name *Space Colonization Stanford Torus Type Station 1970s NASA Video*, online at https://www.youtube.com/watch?v=UcO_BjXfhhc (accessed 4 November 2015).

98. Allen M. Steele, *Clarke County, Space* (New York: Ace, 1990), 91. Scholars have also compared O'Neill's space colonies to suburban communities. See especially Kilgore, *Astrofuturism*, 157, 171–172.

99. "Priorities: Romance and Reality," *Ledger-Star* (Norfolk, VA), 22 July 1969.

100. "NASA—Natch—Studies Sun Heat," *La Crosse (WI) Tribune*, 23 June 1974, 4. See also "Solar Energy for Heating and Cooling of Buildings," NASA, Office of Aeronautics and Space Technology (March 1974), publication located at NASA Headquarter Archives, NASA Historical Collection, Folder 9825: Energy Solar, National Aeronautics and Space Administration, Washington, DC.

101. On NASA's solar heating and cooling program, see "Solar Energy for Heating and Cooling of Buildings"; "MSFC Lead Center for NASA's Solar Heating-Cooling Effort," NASA News Release No. 74-175, 24 September 1974, NASA Headquarter Archives, NASA Historical Collection, Folder 9825: Energy Solar (1974), National Aeronautics and Space Administration, Washington, DC; and "Solar Energy Collectors Supply Plenty of Heat," NASA News Release No. 74-234, 2 December 1974, NASA Headquarter Archives, NASA Historical Collection, Folder 9825: Energy Solar (1974), National Aeronautics and Space Administration, Washington, DC.

102. NASA publications also referred to Tech House as NASA's Technology Utilization House. See "Your Home, Now: The 'Tech House,'" *Spinoff* (1976): 31. NASA began publishing *Spinoff*, a glossy magazine promoting to the public its technology transfer program, in 1976.

103. For an example of coverage in a local newspaper, see Frank Macomber, "The House That NASA Built," *Rome (GA) News-Tribune*, 19 February 1978, 2E; in a national magazine, see Hans Fantel, "Our Year in NASA's Far-Out House," *Popular Mechanics*, June 1979, 77–79, 126–127. The author received a copy of the short film, titled *The House That NASA Built* (1978), from archivists at the NASA Headquarters Archives in Washington, DC. This film is also available in a collection of three short films titled *Eagle Has Landed: The Flight of Apollo 11*; *Apollo 11 for All Mankind*; and *The House That NASA Built*, Manned Spacecraft Center, National Aeronautics and Space Administration, Washington, DC.

104. Fantel, "Our Year in NASA's Far-Out House," 77–79, 126–127. On the Tech House's economic and energy savings, see "NASA Technology Utilization House: A 'House for the Future Is Ready Today,'" *NASA Tech Brief* (Winter 1976), i, National Aeronautics and Space Administration, online at http://crgis.ndc.nasa .gov/crgis/images/4/4b/1976_NASA_Tech_Briefs.pdf (accessed 9 November 2015); and Ira H. A. Abbott, Kenneth A. Hopping, and Warren D. Hypes, "An Evaluation of the NASA Tech House Including Live-In Results," vol. 1, NASA Technical Paper 1564 (November 1979), 28, 31, National Aeronautics and Space Administration, Washington, DC.

105. Fantel, "Our Year in NASA's Far-Out House," 79.

106. This line was repeated over and over again, in both NASA publicity on Tech House and in media coverage of the experiment. See "The House That NASA Built"; and Macomber, "House That NASA Built."

107. For an architectural drawing of Tech House's interior, see "Floor Plan of the Technology Utilization House," online at http://crgis.ndc.nasa.gov/historic /File:Floorplan.jpg (accessed 16 September 2016). For a photograph of the Tech House exterior, see *NASA Tech House* (Washington, DC: NASA, 1977), cover, online at http://crgis.ndc.nasa.gov/crgis/images/1/11/NasaTechHouse.pdf (accessed 24 August 2016).

108. Fantel, "Our Year in NASA's Far-Out House," 78. Earlier efforts to solarize suburban homes are discussed by Adam Rome, *The Bulldozer in the Countryside: Suburban Sprawl and the Rise of American Environmentalism* (New York: Cambridge University Press, 2001), 45–87.

109. The Swains' experiences living in Tech House, including Elaine Swain's activities cooking, cleaning, and washing dishes, were analyzed for NASA by Gregory M. Frech in "Lifestyle Impacts of Residence in the Technology Utilization House," NASA Contractor Report 158963, NASA Contract NAS1-14193-Task 38 (September 1978), National Aeronautics and Space Administration, Wash-

ington, DC. The photograph of the Swains in their kitchen appears in "Tech House," *Spinoff* (1978): 46–47, which can also be found online at http://crgis.ndc .nasa.gov/crgis/images/e/ed/1978-01_Family_in_Tech_House.pdf (accessed 11 November 2015). This photo of the New Alchemists appeared in Nancy Jack Todd, "New Alchemy's Ark: A Proposed Solar Heated and Wind Powered Greenhouse and Aquaculture Complex Adapted to Northern Climates," *Journal of the New Alchemists* 2 (1974), 43. For other examples of photographs of their members, see Nancy Jack Todd, "New Alchemy: Creation Myth and Ongoing Saga," *Journal of the New Alchemists* 6 (1980): 9, 13, 14.

110. NASA's do-it-yourself handbook for the general public is mentioned in Graham Gross, "You Can Build NASA's Low-Cost Solar Heating System," *Popular Science* 212, no. 2 (February 1978): 106. For NASA's guide for contractors, see National Aeronautics and Space Administration, *Contractor's and Engineer's Guide to Space Heating through Solar Technology* (Washington, DC: U.S. Government Printing Office, 1976).

111. "A New Look for Greenbelt," *Spinoff* (1978): 102–105.

112. "Solar Energy Demonstrations," *Spinoff* (1979): 49.

113. *The House That NASA Built* (Valencia Entertainment International, 1978), located at the John F. Kennedy Space Center.

114. Frech, "Lifestyle Impacts of Residence," 3.

115. Peter Calthorpe and Susan Benson, "Beyond Solar Suburbia," *RAIN* (August/September 1979): 12–14. This article also appeared as "In Perspective—The Solar Shadow: A Discussion of Issues Eclipsed," *Progressive Architecture* 4, no. 79 (April 1979): 45–46, 50. Calthorpe invited the New Alchemist John Todd to a weeklong workshop, with thirty other leading innovators, to discuss ecology and community design. Calthorpe and his associate Sim Van der Ryn published the participants' contributions, including Todd's, in the now-classic *Sustainable Communities: A New Design Synthesis for Cities, Suburbs and Towns* (San Francisco: Sierra Club, 1986).

116. On the failure of NASA's solar-power satellites, see Jeff Womack, "Pipe Dreams for Powering Paradise: Solar Power Satellites and the Energy Crisis," in Lifset, *American Energy Policy in the 1970s*, 203–220.

117. "Space Crescent Transforms Gulf Area: The Impact of NASA's $40-Billion Program," *Business Week*, 24 March 1962, 72. There are numerous articles on the rise of NASA's "space crescent." For a brief, but informative, description of the "space crescent," see "Moon Dollars," *Newsweek*, 7 July 1969, 54.

118. Editors of *Fortune*, *The Space Industry: America's Newest Giant* (Englewood Cliffs, NJ: Prentice Hall, 1962), 83–88. On the economic geography of NASA contractors, see also O'Toole, "$24 Billion," F1.

119. "Space Crescent Transforms Gulf Area," 72.

120. There is a rich literature on the symbolic power of urban corporate skyscrapers before World War II. See especially Diana Balmori, "George B. Post: The

Process of Design and the New American Architectural Office (1868–1913)," *Journal of the Society of Architectural Historians* 46 (December 1987): 342–355; and Kenneth Turney Gibbs, *Business Architectural Imagery in America, 1870–1930* (Ann Arbor, MI: UMI Research Press, 1984). See Susan Wagner Schwartz, "The Suburban Corporate Headquarters in the United States: An Evolving Building Type" (Ph.D. thesis, Penn State University, 1995).

121. These push factors are covered most succinctly by Louise A. Mozingo, *Pastoral Capitalism: A History of Suburban Corporate Landscapes* (Cambridge, MA: MIT Press, 2011), 21–25. On nuclear attack, civic defense, and corporate geography, see also Jennifer Light, *From Warfare to Welfare: Defense Intellectuals and Urban Problems in Cold War America* (Baltimore, MD: Johns Hopkins University Press, 2005), especially chapter 1. "Duck and cover" refers to the practice, taught during the 1950s to schoolchildren across the nation, to seek protection, often under their desks, during a nuclear blast.

122. The scholarship on this geographic shift is extensive. See especially Mozingo, *Pastoral Capitalism*, especially 21–43; R. Keith Semple, "Recent Trends in the Spatial Concentration of Corporate Headquarters," *Economic Geography* 49, no. 4 (October 1973): 309–318; Sally Ward, "Trends in the Location of Corporate Headquarters, 1969–1989," *Urban Affairs Quarterly* 29, no. 3 (March 1994): 468–478; and Matthew Drennan, "Headquarters City: New York and the Corporate Complex," *New York Affairs* 5, no. 1 (1978): 72–81.

123. This point is made most forcefully by Mozingo, *Pastoral Capitalism*, 11.

124. On the suburban headquarters and research lab as corporate symbol, see Sheryl Williams and Mary Anne Moffitt, "Corporate Image as an Impression Formation Process: Prioritizing Personal, Organizational, and Environmental Audience Factors," *Journal of Public Relations Research* 9, no. 4 (1997): 237–258; and Roland Marchand and Michael Smith, "Corporate Science on Display," in *Scientific Authority and Twentieth-Century America*, ed. Ronald G. Walters (Baltimore, MD: Johns Hopkins University Press, 1997).

125. As one scholar has noted, "By the late 1950s the impact of the General Motors Technical Center had been absorbed into the world of suburban corporate architecture." See Schwartz, "Suburban Corporate Headquarters," 83.

126. On Saarinen's embrace of GM's car technology in his architectural design, see Scott Knowles and Stuart Leslie, " 'Industrial Versailles': Eero Saarinen's Corporate Campuses for GM, IBM, and AT&T," *Isis* 92 (2001): 9. See also Reinhold Martin, *The Organizational Complex: Architecture, Media, and Corporate Space* (Cambridge, MA: MIT Press, 2003), 142–156.

127. On Saarinen's incorporation of nature into his designs for GM's Technical Center, see Mozingo, *Pastoral Capitalism*, 76–77, 241; Knowles and Leslie, " 'Industrial Versailles,' " 9–10; and Peter G. Rowe, *Making a Middle Landscape* (Cambridge, MA: MIT Press, 1991), 155.

128. "G.M. Technical Center," *Architectural Forum* 91, no. 1 (July 1949): 71, as quoted in Mozingo, *Pastoral Capitalism*, 79.

129. "G.M. Nears Completion," *Architectural Forum* 101, no. 5 (November 1954): 109, as quoted in Mozingo, *Pastoral Capitalism*, 79.

130. General Motors Public Relations Staff, *Where Today Meets Tomorrow* (Detroit, MI: General Motors Corporation, 1956), cover; "General Motors Technical Center," postcard circa late 1950s or early 1960s (West Nyack, NY: Dexter Press), in the possession of the author.

131. "Architecture for the Future: GM Constructs a 'Versailles of Industry,'" *Life*, 21 May 1956, 102.

132. Dwight D. Eisenhower, "The Rich Reward Ahead," in *The Greatest Frontier: Remarks at the Dedication Program, General Motors Technical Center* (Detroit, MI: General Motors Corporation, 1956), 26–27, as quoted in Mozingo, *Pastoral Capitalism*, 83.

133. Dennis Doordan, "Design at CBS," *Design Issues* 6, no. 2 (Spring 1990): 12–13.

134. This term is used by Stuart W. Leslie, "Spaces for the Space Age: William Pereira's Aerospace Modernism," in *Blue Sky Metropolis: The Aerospace Century in Southern California*, ed. Peter Westwick (Berkeley, CA: Huntington-USC Institute on California and the West, 2012), 127.

135. Bernard Weinraub, "MCC-H, USA: Neverland of Moonlight and Robots," *New York Times*, 23 March 1969, 22.

136. For a description of the layout of the Manned Spacecraft Center, see "Fact Sheet #60, Manned Spacecraft Center Constructing Permanent Facilities at Clear Lake," October 1962, press release, NASA, Johnson Space Center, Brochures and Articles, 1962–1988, n.d., Box: "Cabinet 2, Drawer 1–3, NASA Centers, JSC," JSC History Collection, University of Houston, Clear Lake, TX. Material on the landscaping of what became the Johnson Space Center comes from my own examination of old photographs and postcards of the center, also housed at the JSC History Collection, University of Houston, Clear Lake, TX.

137. Weinraub, "MCC-H, USA," 22.

138. This postcard is titled "Manned Spacecraft Center, Houston, Texas: Building 2 Exterior," distributed by G. P. Slide Co., Inc., 22608 Sunset Blvd., Houston, TX, postcard in the possession of the author. This postcard states on the address side that the photograph used was taken by NASA.

139. E. Wallace Lane, *Dreams, Hopes, Realities: NASA's Goddard Space Flight Center, the First Forty Years* (Washington, DC: NASA History Office, 1999), 20, online at http://history.nasa.gov/SP-4312/ch2.htm (accessed 2 February 2016).

140. On the history of the design and development of these two facilities, see the following. For the Marshall Space Flight Center, see Helen T. Wells, Susan H. Whiteley, and Carrie E. Karegeannes, *Origins of NASA Names* (Washington, DC:

NASA History Series, 1976), 154, online at http://history.nasa.gov/SP-4402/ch6
.htm (accessed 2 February 2016). For the Mississippi Test Facility, see Mack R. Her-
ring, *Way Station to Space: A History of the John C. Stennis Space Center* (Washington, DC: NASA History Office, 1997), 49, online at http://history.nasa.gov/SP
-4310/ch4.htm (accessed 2 February 2016).

141. Regarding these postcards, see "An Artist's Drawing of a Panoramic View
of Goddard Space Flight Center," produced by CAPSCO, Inc., Washington, DC;
"Headquarters Complex, Marshall Space Flight Center, Huntsville, Alabama,"
produced by Mike Roberts, Berkeley, CA; and "Mississippi Test Operations,
National Aeronautics and Space Administration, Hancock, County," produced by
H. S. Crocker, San Francisco, CA. All three of these postcards are in the possession of the author.

142. J. B. Jackson, "The Popular Yard," *Places* 4, no. 3 (1987): 26, 30, as quoted
in Mozingo, *Pastoral Capitalism*, 43.

143. "Profits from Precision—Where the Money Goes," *Newsweek* (special series on "The Business of Space"), 8 October 1962, 40.

144. Rand, "Apollo 11," 1, 8.

145. "Apollo's Builders Start Closing the Lines," *Business Week*, 17 May 1969, 77.

146. There is a rich literature on the role of corporate architecture and corporate
parks in shaping corporate image. See especially Williams and Moffitt, "Corporate
Image as an Impression Formation Process," 246; Marchand and Smith, "Corpo-
rate Science on Display," 148–182; Mozingo, *Pastoral Capitalism*, 41–43; Gibbs,
Business Architectural Imagery in America, 169; Schwartz, "Suburban Corporate
Headquarters," 8.

147. Anne Thompson, "Cow Pasture to Space Center," *Free Lance Star* (Freder-
icksburg, VA), 31 December 1966, 8-A. On this region prior to NASA's arrival, see
also Kevin M. Brady, "NASA Launches Houston into Orbit: The Economic and
Social Impact of the Space Agency on Southeast Texas, 1961–1969," in *Societal
Impact of Spaceflight*, ed. Steven J. Dick and Roger D. Launius (Washington, DC:
NASA History Division, 2007), 452.

148. On this population increase, see "At the Manned Spacecraft Center: Six
Years—and Spin-Off," *Houston* 38, no. 8 (September 1967): 63. On population
growth in the Houston metropolitan area during this period, see John Ste-
phens and Brian Holly, "City System Behavior and Corporate Influence: The
Headquarters Location of US Industrial Firms, 1955–1975," *Urban Studies* 18
(1981): 289.

149. On the development of Clear Lake City, see Humble Way (Fall 1963), Vertical File, "T-Cities and Towns, Clear Lake City," Johnson Space Center History
Collection, University of Houston, Clear Lake, TX; and "Barren Prairie to 'Instant City,' *Houston*, November 1963, clipping from vertical file, "T-Cities and

Towns, Clear Lake City," Johnson Space Center History Collection, University of Houston, Clear Lake, TX.

150. On these subdivisions and aerospace firms moving into the Clear Lake area, see Thompson, "Cow Pasture to Space Center." On similar planned communities in the Johnson Space Center vicinity, see also Fairmont Park advertisement in *Houston*, August 1962, 30; and "Team Effort: A Place Where 13,000 Men Can Feel Like They Are Columbus," *Newsweek*, 7 July 1969, 55.

151. Mary Weigers, "Space Community: Emphasis on Excellence," *Washington Post*, 13 July 1969, 143.

152. On NASA fostering suburban sprawl in Huntsville during the early 1960s, see Andrew J. Dunar and Stephen P. Waring, *Power to Explore: A History of Marshall Space Flight Center, 1960–1990* (Washington, DC: NASA History Office, 1999), 126. On Huntsville's demographic, geographic, and economic growth during the 1960s, see also "Moon Dollars," 54; and Rudy Abramson, "Huntsville Riding Atop Boom 'Crescent,'" *Tuscaloosa (AL) News*, 5 June 1966, 10.

153. See Ronald Thompson, "Sleepy Hamlets Bow to Moon Rocket Project," *Sarasota (FL) Journal*, 11 February 1966, 22, which described "half a dozen housing projects" going up soon after NASA moved into the area.

154. "Port Malabar: Invest for Your Future in Florida's New Center of Industrial Growth . . . The Cape Canaveral Area!" *Life*, 27 October 1961, 7.

155. On the suburbanization of the Kennedy Space Center area, see Harry A. Green, "Urban Growth in the Nation's Spaceport," published 1 January 1964, available through NASA Technical Reports Server (NTRS), Document Source: CAS, Document ID: 19640017477, Report Number: NASA-CR-56204, NASA History Division, 1–5. On the number of lots sold by General Development Corps. in Port Malabar, see Theodor A. Ediger, "New Boom for Florida in Zoom to the Moon," *Florida Sentinel*, 18 March 1962, 7-E.

156. Greg Hise, *Magnetic Los Angeles: Planning the Twentieth-Century Metropolis* (Baltimore, MD: Johns Hopkins University Press, 1977). On this phenomenon, see also Richard Walker and Robert D. Lewis, "Beyond the Crabgrass Frontier: Industry and the Spread of North American Cities, 1850–1950," *Journal of Historical Geography* 27, no. 1 (2001): 3–19.

157. On the aerospace industry's impact on Southern California, see Peter J. Westwick, "The Jet Propulsion Laboratory and Southern California," in Dick and Launius, *Societal Impact of Spaceflight*, 468–479; D. J. Waldie, *Holy Land: A Suburban Memoir* (New York: W. W. Norton, 1996); and Allen J. Scott, "Interregional Subcontracting Patterns in the Aerospace Industry: The Southern California Nexus," *Economic Geography* 69, no. 2 (April 1993): 142–156.

158. Tom Leech, "When Aerospace Was King," *San Diego Magazine*, December 1997, 98, as quoted in Stuart W. Leslie, "Spaces for the Space Age: William

Pereira's Aerospace Modernism," in Westwick, *Blue Sky Metropolis*, 134. Much of my thinking here comes from Leslie's essay.

159. NASA and its contractors were not solely responsible for this suburban sprawl, but they added to it. For an analysis of the high-tech industry's role in also fostering suburbanization, especially along Route 128 outside Boston, see Margaret Pugh O'Mara, *Cities of Knowledge: Cold War Science and the Search for the Next Silicon Valley* (Princeton, NJ: Princeton University Press, 2005), 71–81. On this trend on Long Island with Grumman, see Christopher Sellers, "Unsettling Ground: Suburban Nature and Environmentalism in 20th Century America," unpublished manuscript, 72. Here Sellers notes that soon after aerospace firms including Grumman moved onto vacant land along Long Island's Hempstead Plains, suburban sprawl followed.

160. These signs are mentioned in "Humble Helps to Build a City," unidentified newspaper clipping located in Vertical File: "T-Cities-Town Clear Lake City Shores," Johnson Space Center History Collection, University of Houston, Clear Lake, TX, 4.

161. The best analysis of the environmental problems associated with suburbanization during this period is Rome, *Bulldozer in the Countryside*.

162. The Clear Lake region was called a "barren prairie" in "Barren Prairie to 'Instant City,'" *Houston*, November 1963, clipping in vertical file "T-Cities and Towns: Clear Lake City," Johnson Space Center History Collection, University of Houston, Clear Lake, TX; the "back country" orange groves of the Cape are discussed in Bob Hudson, "Mullet Wrapper," 27 April 1970, newspaper clipping located in Community Relations Collection, Memos and Correspondence, #79-53, Folders 161-NPS Nov. 14 Visit, Box: 16, LOC: 18B.4, Kennedy Space Center Library Archives, Kennedy Space Center; and the towns around the Mississippi Test Facility were called "sleepy hamlets" in Thompson, "Sleepy Hamlets Bow."

163. According to Adam Rome, the rate of soil erosion from home building in the postwar years often was as severe as from clear-cut logging or strip-mining. He also notes that as a result of infilling by developers between 1950 and 1970, almost one million acres of marshes, swamps, bogs, and coastal estuaries were destroyed. See Adam Rome, "Building on the Land: Towards an Environmental History of Residential Development in American Cities and Suburbs, 1970–1900," *Journal of Urban History* 20, no. 3 (May 1994): 416–417; and Rome, *Bulldozer in the Countryside*, 200–216.

164. "What Houston Won When NASA Came to Town," *Business Week*, 11 September 1965, 96.

165. The link between sprawl and Southern California's water problem is well documented. See Norris Hundley, *The Great Thirst: Californians and Water: A History* (Berkeley: University of California Press, 2001); Jared Orsi, *Hazardous Metropolis: Flooding and Urban Ecology in Los Angeles* (Berkeley: University of

California Press, 2004); and Sarah Elkind, *How Local Politics Shape Federal Policy: Business, Power, and the Environment in Twentieth-Century Los Angeles* (Chapel Hill: University of North Carolina Press, 2011). On water shortages in the Cape Canaveral area during its mid-1960s suburban growth spurt, see Annie Mary Hartsfield, Mary Alice Griffin, and Charles M. Grigg, eds., *Summary Report: NASA Impact on Brevard County* (Tallahassee: Florida State University Institute for Social Research, 1966), 69.

166. The subsidence and sewage problems caused by suburban sprawl on the outskirts of Houston, including the region around the Johnson Space Center, are discussed in Andrew Kirby and Karen Lynch, "A Ghost in the Growth Machine: The Aftermath of Rapid Population Growth in Houston," *Urban Studies* 24 (1987): 590–592.

167. Hartsfield et al., *Summary Report*, 70.

168. The link between energy-inefficient suburban homes and increased fossil fuel use is discussed by Rome, *Bulldozer in the Countryside*, 47, 223. On air pollution caused by suburban lawn mowing, see Theodore Steinberg, *Down to Earth: Nature's Role in American History* (New York: Oxford University Press, 2012), 219–221; and Theodore Steinberg, *American Green: The Obsessive Quest for the Perfect Lawn* (New York: W. W. Norton, 2007), 153.

169. On air pollution in Southern California caused by the aircraft industry, and by implication the aerospace industry that followed in its footsteps, see Wade Graham, "The Urban and Environmental Legacies of the Air Industry," in Westwick, *Blue Sky Metropolis*, 247, 252.

170. Kirby and Lynch, "Ghost in the Growth Machine," 591.

171. Malvina Reynolds, "Little Boxes" (Schroder Music Company, 1962). The song, which was actually inspired by a new suburban development in Northern California just south of San Francisco, became a hit for Pete Seeger in 1963 and something of a countercultural anthem against suburbia.

172. Weigers, "Space Community," 143, K5.

173. Southern California was hostile to unionization in part because its economy depended on the military, and strikes could easily be portrayed as a threat to national security. As the aerospace industry grew in the region, this political stance carried over. On the region as anti-union, see Anita Seth, "Los Angeles Aircraft Workers and the Consolidation of Cold War Politics," in Westwick, *Blue Sky Metropolis*, 79–99. For examples of NASA supporting federal intervention to halt strikes affecting its program, see "Kennedy Enters Lockheed Strike," *New York Times*, 29 November 1962, 1; and Harry Bernstein, "Machinists Accept Lockheed Pact without Provision for Union Shop," *Washington Post*, 28 January 1963, A1.

174. "Moon Dollars," 54. Cashin's wife, Joan, was quoted in Rosemarie Tyler Brooks, "Huntsville: Famed for Space, Shamed for Race," *Chicago Daily Defender*, 10 June 1965, 6.

175. On racial exclusion in the aerospace suburbs of Southern California, see Waldie, *Holy Land*, 161–162.

176. "Now Profit from Progress: Canaveral Acres," *Ebony*, May 1963, 10. For additional examples of racial segregation in the suburbs of Huntsville, see "Space City Faces School Segregation Showdown," *Chicago Defender*, 28 March 1963, 10; and United Press International, "Ala. Bigotry May Cause Space Research Loss," *Los Angeles Sentinel*, 5 November 1964, A11. For similar segregation near NASA's centers near New Orleans, see Jack Langguth, "Space Boom Stirs Louisiana Bayous: NASA Center Transforming Swamp Town into City," *New York Times*, 8 September 1963, 55; and Rudy Abramson, "Frontier Town Caught Up in Race to Moon: Space-Age Dixie (Part II)," *Tuscaloosa (AL) News*, 6 June 1966, 2.

177. For a discussion of exclusionary public housing in Cocoa Beach, see Edwin Diamond, *The Rise and Fall of the Space Age* (Garden City, NY: Doubleday, 1964), 73–74.

178. M. G. Lord, "Cold Warrior's Daughter," in Westwick, *Blue Sky Metropolis*, 45–47.

179. Weigers, "Space Community," K5.

180. On the role of security clearance and classified work within the aerospace industry fostering conservative politics, see Mihir Pandya, "The Vanishing Act: Stealth Airplanes and Cold War Southern California," in Westwick, *Blue Sky Metropolis*, 105–118; and Westwick, "Jet Propulsion Laboratory," 469. For a short history on conservative thought and policy regarding NASA during Ronald Reagan's presidency, see Butrica, "'Right' Stuff," 121–134.

181. "Team Effort," 56.

182. Weigers, "Space Community," K5. See also "Team Effort," 55.

183. Lisa McGirr, *Suburban Warriors: The Origins of the New American Right* (Princeton, NJ: Princeton University Press, 2001), 39.

184. For a discussion of the spatial component of this conservative turn, see Lassiter, *Silent Majority*, 9–11. There is a large and growing literature on the New Right and suburbanization. See, especially, Thomas Sugrue, *The Origins of the Urban Crisis: Race and Inequality in Postwar Detroit* (Princeton, NJ: Princeton University Press, 1996), 181–207; Robert O. Self, *American Babylon: Race and the Struggle for Postwar Oakland* (Princeton, NJ: Princeton University Press, 2003), 256–290; and Kevin M. Kruse, *White Flight: Atlanta and the Making of Modern Conservatism* (Princeton, NJ: Princeton University Press, 2007), 105–131. On the current state of the field regarding the history of conservatism, see "Conservatism: A Round Table," *Journal of American History* 98, no. 3 (December 2001), which includes several historiographical essays.

185. Weigers, "Space Community," K5.

186. Ibid.

187. While Kunen mentions this tour directly, Buckley alludes to it in his coverage when he reports on the five and a half million components that went into building Apollo 8. See Kunen, "Great Rocketship," 17–18; and Buckley, "Can Men Make Miracles?"

188. Buckley, "Can Men Make Miracles?"

189. Kunen, "Great Rocketship," 17–18.

190. Ibid.; Buckley, "Can Men Make Miracles?"

191. On the grassroots conservative movement in Atlanta, see Kruse, *White Flight*; on Southern California, see McGirr, *Suburban Warriors*. On the role of the built environment in forging these conservative suburbs on the fringes of Atlanta and Charlotte, North Carolina, see Lassiter, *Silent Majority*.

192. There is very little literature chronicling the role of the aerospace industry in fostering suburbanization across the South. On the general rise of the aerospace industry in the South, see Bruce Schulman, *From Cotton Belt to Sunbelt: Federal Policy, Economic Development, and the Transformation of the South 1939–1980* (Durham, NC: Duke University Press, 1994), especially chapter 6.

Conclusion

1. "Astronauts Find 'Hair' Offensive: Lovell and Swigert Walk Out after First Act of Musical," *New York Times*, 6 June 1970, 22. Marionetti was quoted the day after the incident occurred.

2. Ibid.

3. "Flag Respect: Florence News," *Herald-Journal* (Spartanburg, SC), 18 June 1970, 4.

4. Regina Maher, interview by author, New York, NY, 30 October 2016.

5. Ed Bowes and Tom Bowes, interview by author, New York, NY, 30 October 2016.

6. Daniel Rodgers, *Age of Fracture* (Cambridge, MA: Belknap Press of Harvard University Press, 2011), 3. Rodgers identifies the rise of Milton Friedman's laissez-faire economic thinking, as well as the emergence of identity politics associated with the civil rights and women's movements, as two major drivers of this civic disaggregation during the 1970s.

7. Archibald MacLeish, "A Reflection: Riders on Earth Together, Brothers in Eternal Cold," *New York Times*, 25 December 1968, 1.

8. The idea of a "flattened" Earth, or level playing field, in the era of globalization, has been promoted most popularly in Thomas Friedman, *The World Is Flat: A Brief History of the Twenty-First Century* (New York: Farrar, Straus, and Giroux, 2005). Scholars from a variety of disciplines have criticized the notion of a "flag" world for obscuring a wide range of inequalities and differences in power that

continue to exist. For example, see Susan Christopherson, Harry Garretsen, and Ron Martins, "The World Is Not Flat: Putting Globalization in Its Place," *Cambridge Journal of Regions, Economy and Society* 1, no. 3 (2008): 343–349.

9. Several scholars have examined such concerns regarding NASA's photographs of Earth from space. Robin Kelsey, "Reverse Shot: Earthrise and Blue Marble in the American Imagination," in *New Geographies 4: Scales of the Earth* (Cambridge, MA: Harvard University Press, 2011), 14; and Yaakov Jerome Garb, "Perspective or Escape? Ecofeminist Musings on Contemporary Earth Imagery," in *Reweaving the World: The Emergence of Ecofeminism*, ed. Irene Diamond and Gloria Feman Orenstein (San Francisco: Sierra Club, 1990), 269.

10. Associated Press, "Historic Mission Recalled," *Herald* (Clinton, IA), 14 December 1988, 23.

11. J. N. Cooper, "Prestige and Apollo 8" (letter to the editor), *New York Times*, 12 January 1969, E13.

12. For other examples of similar editorial cartoons, see Anonymous, "The Illusion," *Times-Picayune* (New Orleans, LA), May 26, 1969, n.p.; Jacob Burck, "Well— Here We Are, Back on Earth," (Chicago) *Sun-Times*, 25 July 1969, 43; and Bill Crawford, *Ocala* (FL) *Star-Banner*, 21 May 1970, 4A.

13. Franklin Morse, "Didn't I Promise You the Moon?," (Los Angeles) *Herald Examiner*, 20 May 1969, A-10.

14. My thinking here has been influenced by the concept of "social movement spillover," which argues that different social and political movements gain strength from borrowing political strategies and ideologies from one another. For an introduction to this idea, and an exploration of the movement-to-movement influence of the women's movement on the U.S. peace movement of the 1980s, see David S. Meyer and Nancy Whittier, "Social Movement Spillover," *Social Problems* 41, no. 2 (May 1994): 277–298.

15. Julian Scheer, "The 'Sunday of the Space Age,'" *Washington Post*, 8 December 1972, A26.

16. Paul Bunge, "Voice of Youth: Do We Really Need to Land on the Moon?" *Denver Post*, July 1969.

17. Stewart Brand, "The First Whole Earth Photograph," in *Earth's Answer: Explorations of Planetary Culture at the Lindisfarne Conferences*, ed. Michael Katz, William P. Marsh, and Gail Gordon Thompson (New York: Lindisfarne Books / Harper and Row, 1977), 188. Gloria Steinem is quoted in United Press International, "Feminists Hail Historic Flight into Space," *Salina* (KS) *Journal*, 19 June 1983, 6.

18. Abbie Hoffman, *Woodstock Nation: A Talk-Rock Album* (New York: Random House, 1969), 13.

19. Sally Ride, testimony within the *House Hearing Report on the Science, Space, and Technology Committee Hearing on NASA's Long Range Goals*, 22 July 1987, as

quoted in Jenna Minicucci, "Review of NASA's Mission to Planet Earth Program," American Geological Institute Government Affairs Program, 7 July 1997, online at www.agiweb.org/legis105/mtpehist.html (accessed 6 November 2016), 8.

20. George H. W. Bush announced Mission to Planet Earth during a speech on the twentieth anniversary of the Apollo 11 landing. For this speech, see George H. W. Bush, "Remarks on the 20th Anniversary of the Apollo 11 Moon Landing," 20 July 1989, online at http://www.presidency.ucsb.edu/ws/?pid=17321 (accessed on 6 November 2016).

21. On the Bush administration's announcement regarding the mission's proposed budget see, William K. Stevens, "NASA Plans a 'Mission to Planet Earth,'" *New York Times*, 25 July 1989, C1. On Mission to Planet Earth's dwindling budget, see Gary Taubes, "Earth Scientists Look NASA's Gift Horse in the Mouth," *Science: News Series* 259, no. 5097 (12 February 1993), 913–914.

22. Bush, "Remarks on the 20th Anniversary."

23. See, for example, Rand Simberg, "Time to Rethink NASA," *National Review*, 21 July 2014, online at http://www.nationalreview.com/article/383220/time-rethink -nasa-rand-simberg (accessed 6 November 2016); and Edward L. Hudgins, "Commentary: Time to Privatize NASA," Cato Institute, online at https://www.cato.org /publications/commentary/time-privatize-nasa (accessed 6 November 2016).

24. Peter Diamandis, "Space: The Final Frontier of Profit?," *Wall Street Journal*, 13 February 2010, online at http://www.wsj.com/articles/SB1000142405274870338 2904575059350409331536 (accessed 6 November 2016).

25. Andy Pasztor, "NASA Official Lays Out Vision of Privatized Space Stations," *Wall Street Journal*, 29 April 2015, online at http://www.wsj.com/articles/nasa -official-lays-out-vision-of-privatized-space-stations-1430363264 (accessed 6 November 2016).

26. A similar argument is made, in a very different context, in Lizabeth Cohen, "From Town Center to Shopping Center: The Reconfiguration of Community Marketplaces in Postwar America," *American Historical Review* 101, no. 4 (1996): 1050–1081; and in her book, *A Consumer's Republic: The Politics of Mass Consumption in Postwar America* (Cambridge, MA: Harvard University Press, 2003).

27. Lyndon B. Johnson, *Vantage Point: Perspectives of the Presidency, 1963–1969* (New York: Holt, Rinehart and Winston, 1971), 285.

Acknowledgments

Like a successful Apollo launch, this book could not have gotten off the ground without the support of a far-flung network of hardworking, supportive individuals and institutions. Initial funding for the project came from the Smithsonian Institution's National Air and Space Museum, where I spent a year as the Verville Fellow, as well as from NASA's Science Mission Directorate, which provided me with a History of the Scientific Exploration of Earth and Space Research Award. I was also fortunate to spend two summers at the Library of Congress as a Kluge Fellow, a shorter stint conducting research at the Hagley Museum and Library, and an exceptionally productive year researching and writing at Harvard University's Charles Warren Center for Studies in American History. All of these experiences would not have been possible without the support of my home institution, the New Jersey Institute of Technology (NJIT), which generously allowed me to take time away from campus to conduct this research.

The intellectual spark for *Apollo in the Age of Aquarius* came from a throwaway comment made during a fast break while playing pickup basketball, and I'd like to thank a fellow New York University history graduate student, Scott Messinger, for that, and also our teammate Andrew Darien for nurturing this idea in my mind ever since. More recently, my colleagues in the Federated History Department at NJIT and Rutgers University, Newark, have been a source of both professional and personal support, especially Richard Sher, Stephen Pemberton, Alison Lefkovitz, Liz Petrick, Kyle Riismandel, Scott Kent, Lisa Nocks, Maureen O'Rourke, and Jessica Witte. I would also like to thank my research assistants, John Holoduek and Holly

Stanton, for helping me track down permission rights to the many images that appear in this book, as well as my current dean, Kevin Belfield, and my former dean and current provost at NJIT, Fadi Deek, for their continuous and enthusiastic support.

Just as the social and political movements of the 1960s critiqued Apollo in an effort to make it more responsive, so, too, did numerous colleagues provide instructive criticism that has made *Apollo in the Age of Aquarius* a much better book. Such feedback began with space historians who not only taught me about rockets and satellites but also welcomed me onto their conference panels, invited me to speak at their institutions, and included me in their social life. Here I'd like to thank especially Roger Launius, Martin Collins, and Margaret Weitekamp at the National Air and Space Museum, as well as Teasel Muir-Harmony, Eric Conway, Patrick McCray, Peter Westwick, Asif Siddiqi, Alexander Geppert, Steve Garber, Ruth Rand, and especially John Krige, who has become an invaluable mentor. While known mostly for their incredible space technology, both the National Air and Space Museum and NASA have amazingly rich historical collections, and the librarians and archivists at both of these institutions are some of the best in the business. Here I would like to thank in particular Colin Fries, at the NASA Headquarters library and archive, who never once failed to track down material for me. Arlene Balkansky at the Library of Congress also went above and beyond to locate images for the book.

While the space history community introduced me to the stars, my environmental history colleagues kept my feet, and head, planted on terra firma during the long hard work that went into *Apollo in the Age of Aquarius*. I first want to thank Cindy Ott, Paul Sutter, and Thomas Andrews, who over the years have become my great friends who also just happen to be environmental historians; the three of them have supported me more than they will ever know. I also want to thank Mark Fiege, who over the past few years has been a "partner in crime," of sorts, in our effort to promote an environmental history that engages more mainstream history writ large. Additional colleagues and friends from the field who have helped along the way include Peter Alagona, Brian Black, Ben Cohen, Kip Curtis, Sterling Evans, Emily Greenwald, Sarah Gregg, Josh Howe, Karl Jacoby, Ari Kelman, Matt Klingle, Nancy Langston, Gregg Mitman, Kathy Morse, Linda Nash, Sarah Phillips, Jenny Price, Sara Pritchard, Michael Reidy, Adam Rome, Kendra Smith-Howard, Ellen Stroud, Jay Taylor, Brett Walker, Louis Warren, Marsha Weisiger, Bob Wilson, and Frank Zelko. As always,

Don Worster continues to guide my thinking, and my spirit, in all things environmental history.

This book also benefited from an expert team of readers and editors. Adam Rome and John Krige read every page of the manuscript and provided invaluable suggestions. My fellow fellows at the Charles Warren Center, including Cathy Gudis, Nick Howe, Sarah Luria, Kathy Morse, Aaron Sachs, Liz Mesok, Max Kenner, and especially Cindy Ott, along with the center's faculty conveners, Robin Kelsey, Joyce Chaplin, and Larry Buell, allowed me to workshop several chapters and in each instance provided perceptive direction followed by warm socializing; my year in Cambridge, because of these people, was one of the best of my professional career. I would also like to thank Sarah Flynn and her deep historical knowledge of, and personal connections to, the social and political movements of the 1960s, for guiding *Apollo in the Age of Aquarius* to a new level. My wonderful editor, Kathleen McDermott, has been an enthusiastic champion of this project from the start and went to bat to make the book better throughout the publishing process. Her colleagues at Harvard University Press, including Katrina Vassallo, Louise Robbins, and Stephanie Vyce, as well as production editor Melody Negron at Westchester Publishing Services, are equally talented. Finally, although she has yet to read much of this book, my former graduate advisor, Lizabeth Cohen, shaped its research and writing; I just couldn't shake her voice in my head, thankfully, as I pored over documents, organized outlines, and wrote draft upon draft upon draft.

I want also to acknowledge my family and nonprofessional friends who have endured, for far too long, my long-winded rants about rockets and the 1960s revolution. I thank Keric Brown, Randy Hibbits, Shep Kopp, Tom McDonald, Tom Quinn, and Mike Zinn for keeping me sane with camping and surf trips over the years, and also because I'm tired of hearing Shep complain that I failed to acknowledge them in my first book. The Elfers, Bowes, Healy, Tulacro, and Greco clans have always been supportive of the historian in our tribe, as has Mitchel Dobbs. My mother and brother, along with his wife, Stephanie, and their two beautiful daughters, Charlotte and Olivia, have lived this book with me every step of the way, and their continuous encouragement and love remind me of that expressed by our family's original history buff, my father. I wish he could read this one, too.

But most important I want to thank my wife and best friend, Stacy. We met at the very outset of this book project, and she has supported it and me ever since. I should have known back then, when after dating for only a few months she agreed to travel with me to Cape Canaveral and wait for five hours under Florida's scorching sun to watch a thirty-second space shuttle liftoff, that she was no ordinary woman. Since then we have launched a family of our own, and it is because of her that when I read our sons bedtime stories about going to the moon I think instead about our wonderful, happy life here on Earth.

Illustration Credits

Fig. 1.1: Cartoon by Chester Commodore, courtesy of the *Chicago Defender*

Fig. 1.2: A 1969 Herblock Cartoon, © The Herb Block Foundation

Fig. 1.3: Courtesy of the cartoonist, Bill Sanders

Fig. 1.4: Cartoon by Hugh Haynie, courtesy of the Louisville *Courier-Journal*

Fig. 1.5: National Institute of Standards and Technology

Fig. 1.6: Associated Press

Fig. 1.7: Photo by Elaine Tomlin

Fig. 1.8: Bettmann / Getty Images

Fig. 2.1: Courtesy of the cartoonist, Gene Basset

Fig. 2.2: Cartoon by John Fischetti, courtesy of the John Fischetti Estate

Fig. 2.3: Cartoon by Bruce Shanks, courtesy of *The Buffalo News*

Figs. 2.4, 3.1, 3.8, 3.9, 3.10, 3.12, 4.1: NASA (National Aeronautics and Space Administration)

Figs. 3.2, 5.5: Author's collection

Figs. 3.3, 3.4, 3.7, 5.6, 5.7, 5.8: NASA, from author's collection

Fig. 3.5: Cartoon by Scott Long, courtesy of the Minneapolis *Star Tribune*

Fig. 3.6: Cartoon by L. D. Warren

Fig. 3.11: Reprinted by permission from Macmillan Publishers Ltd: J. C. Farman, B. G. Gardiner, and J. D. Shanklin, "Large Losses of Total Ozone in Antarctica

Reveal Seasonal ClO_x / NO_x Interaction," *Nature* 315 (no. 6016): 207–210, 16 May 1985. © 1985 by Nature Publishing Group

Fig. 4.2: Courtesy of ILC Dover, LP

Fig. 4.3: U.S. Geological Survey

Fig. 4.4: Cartoon by Guernsey LePelley from *Christian Science Monitor*, July 26, 1969. © 1969 Christian Science Monitor. All rights reserved. Used by permission and protected by the Copyright Laws of the United States. The printing, copying, redistribution, or retransmission of this Content without express written permission is prohibited.

Fig. 4.5: Cartoon by Jim Lange, courtesy of *The Oklahoman*

Figs. 4.6, 4.7, 4.8: Courtesy of Schlesinger Library, Radcliffe Institute, Harvard University

Fig. 4.9: Cartoon by Ken Alexander, courtesy of the *San Francisco Examiner*

Fig. 5.1: Cartoon by Charles G. Brooks, courtesy of the Charles G. Brooks Estate

Fig. 5.2: Cartoon by Robert Crumb, courtesy of Lora Fountain Literary Agency, Paris, France

Fig. 5.3: Technology Utilization Division / NASA

Fig. 5.4: Courtesy of The Green Center, Inc., Hatchville, MA

C.1: Cartoon by Franklin Morse. Copyright © 1969, *Los Angeles Times*. Reprinted with permission.

Index

Abernathy, Ralph, 11–13, 20, 48–49, 50, 51, 102, 235

Abzug, Bella, 168

Africa, Landsat technology used in, 80, 82

Air-Delivered Seismic Intrusion Detector (ADSID), 64

air pollution, 39–40, 221

Aldrin, Buzz, 35, 104, 146

Alsop, Joseph, 67–68

American flag, 192–193, 194, 199, 228–229

American Friends Service Committee, 73

"American Know-How," 32

American Physical Society, 73

Ames Research Center studies, 72–73, 174–175, 205

Anders, William, 125, 186

Anderson, Clinton, 166

animal experiments, 147

anti-war activism, 65–74, 87–88, 272–273n66, 282n155

Apollo (god), 187–188

Apollo 8, 127, 183–189, 202, 224–225

Apollo 11: environment and ecology of, 14–20; and flag placed on moon, 192–193, 194, 199; Furgurson on, 188; landing of, 230; launch of, 3–4, 50–51; media coverage of protests against, 30–32; music inspired by, 198; pollution caused by, 104–108; postcards depicting, 117, 295–296n95,96; protests against, 11–14, 20–21, 25–27, 48–49; Rand on, 190; scientific exploration through, 110; significance of, 233; simulators for, 152–155, 157; and war on poverty, 35–36

Apollo 14, 25, 27

Apollo 17, 94–96, 155–156, 169. See also *Whole Earth*

Apollo Lunar Surface Experiments Package (ASLEP), 64, 110, 271n52

Apollo program: and conservative politics, 189, 193–196; scientific "brain drain" experienced by, 110–111; scientific exploration through, 110; seismometers and, 271n52; significance of, 233; support for, 33–35, 242n5, 255n94; timeframe of, 243–244n14

Apollo-Soyuz Test Project (ASTP), 74–75

Apollo spacesuit, 148–152, 190

Application of Aerospace Technologies to Urban Community Problems, 39

Applications Technology Satellites (ATS), 63

"arks," 201–202, 207

Armstrong, Neil, 146, 147, 152, 157–158

A7L spacesuit, 148–152, 163–164

Asia, Landsat technology used in, 82–83. *See also* Vietnam; Vietnam War

Asimov, Isaac, 161

astronaut bodies: as cultural symbols, 143–147; and extreme outer space environment, 147–148; simulators and, 152–158; spacesuits and, 148–152. *See also* female body; women astronauts; women's movement

Barefoot and Pregnant Ball, 167

Barry, Marion, 25–26

beauty pageant, 169–170, 182

Berrigan, Daniel, 230

Berrigan, Philip, 73

Berry, Charles, 148
Berry, Wendell, 198–199
Beyer, Robert, 16–17
"Beyond Solar Suburbia" (Calthorpe), 210
Black, Shirley Temple, 80
"blue marble," 93, 94, 284n4. See also *Whole Earth*
Bombeck, Erma, 102–103
Bookchin, Murray, 4, 112, 304n166
Borgstedt, Douglas, 104
Borman, Frank, 177, 186
Bowie, David, 198
Brand, Stewart, 92–96, 124, 134–136, 197, 201, 235, 331n73
Brazil, 81, 167
British Antarctic Survey, 129
Brooks, Charles, 193, 194
Brower, David, 124
Brown, Allan H., 111, 112
Brown, Frank A., 147
Buckley, William F. Jr., 183–184, 185, 189, 202, 224–225
Bulletin of the Atomic Scientists, 111–112
Bunge, Paul, 34, 235
Bush, George H. W., 236
Business Week magazine, 109

Calthorpe, Peter, 210, 335n115
cameras, carried by astronauts, 125
Canaveral National Seashore, 115–116
Cape Canaveral: creation of Merritt Island National Wildlife Refuge at, 114; environment of, 4, 116; lunar simulations at, 156; technological makeover of, 97–98, 117
carnations, 137, 305n10
Carpenter, Scott, 144–145, 154
Carson, Rachel, 112, 136
Cashin, Joan, 222
Cashin, John, 222
centrifuge, 153, 174–175
Century Plaza protest, 26–27, 253n76
Challenger, 137, 158
chauvinism, 159–163, 167. *See also* second-wave feminism; women's movement
Church, Thomas, 212
Ciaravella, Stacy, 11
civil rights: and media coverage of demonstrations against space race, 27–35, 254n80; space race and, 11–13, 20–27, 48–53, 245n12; and urban housing crisis, 47–48
Clear Creek High School, Texas, 224
Clear Lake City, Texas, 218, 220

climate change, 129–136
Cloud Camera, 63, 65, 273n75
Cobb, Jerri, 164–165, 166, 168–169, 177
Collins, Michael, 153
Columbia University, 69
Commercial Space Launch Amendments Act, 236
Commoner, Barry, 112
communes, 200–202, 330n68
Community Action Program, 37
Conrad, Pete, 167, 169
conservatives. *See* New Right
Cooke, Dennis, 17–18, 249n42
Cooper, J. N., 233
CORONA, 61
corporate flight, 211–213, 218–219
corporate parks, 211–213
counterculture: author's early experiences with, 230–231; battle between New Right and, 2, 183–186, 202; communes, 200–202, 330n68; scholarship on, 325n7; vision regarding space technology and natural environment, 225–226
Cousteau, Jacques, 122–123
Crippen, Robert, 139, 140
Crumb, R., 196–197
Cuban Missile Crisis, 74

Davis, Rennie, 69
Debus, Kurt, 114
Deese, Jim, 97
defoliation, 59–60, 86–87, 268n28
Department of Housing and Urban Development (HUD), 30, 37, 41–48, 209, 269n139
détente, 74–75, 83–87, 275n94
Diamond, Edwin, 101
difference feminists, 177–178
Dionysus, 187–188
Discoverer 1, 61
Disney, Walt, 205
diverse goals strategy, 111, 112
Donnelly, John, 114
drought, 82, 220
Dwight, Edward J., 21

Earth, photographs of, 124–134. See also *Whole Earth*
Earth Day, 127–128, 300n140, 301n143
Earth Resources Observation Satellite (EROS), 119
Earth resources satellites, 76–83, 88–90. *See also* remote-sensing technology

Earth Resources Survey Symposium, 80
Earthrise, 125, 300n139
Earth Science Storytelling Project, 133
Economic Opportunity Act, 37
editorial cartoons, 162, 170, 180–181, 193, 233. *See also* political cartoons
Ehrlich, Paul, 112
Eisele, Donn, 145
Eisenhower, Dwight, 213
Elder, John, 223
employment, of minorities, 21, 37, 250nn50,53, 321n166
environment: of Apollo spaceship, 15–20; application of NASA technology to, 113–118; NASA technology's effect on, 103–113; and relationship between racial politics and space race, 13–14; scholarship on history of, 5–6; suburbs' impact on, 220–224; technology and natural, 4–5, 189–190, 198–199, 225–226. *See also* natural resources; pollution; space environment
Environmental Action, 300n140
environmental activism / environmental movement, 9, 92–96, 103–125, 134–136
Environmental Control Subsystem, 19, 42
Environmental Protection Agency (EPA), 40
Equal Employment Opportunity (EEO) Program, 167–168, 173–174, 176
equality feminists, 177–178, 324n190
equal opportunity employment, 37, 167–168, 173–174, 176
"Establishment of the Florida National Seashore" memorandum, 293n85

Faas, Horst, 58
Fair, Stanley D., 59
Fauntroy, Walter, 22
female body: and feminist movement, 140, 305n10; resiliency of, 142–143; and space exploration, 141. *See also* women astronauts; women's movement
feminist movement. *See* second-wave feminism; women's movement
Ferguson, James, 60
Finger, Harold, 41
First Indochina War, 58
Fischetti, John, 71
Flaningan, Peter, 51
Fletcher, James, 121–122, 167–168, 173, 206
Flickinger, Don, 162–163, 318n134
flight nurses, 174, 322n169
flowers, refused by Ride, 137, 140, 305n10

Forbes, Ray, 98
Forbes magazine, 125
Fortune magazine, 211, 212
Friedan, Betty, 166–167; *The Feminine Mystique,* 143
Friedman, Milton, 343n6
Fuller, Buckminster, 4
Fulton, James, 318n122
Furguson, Ernest B., 188, 326n17

Gagarin, Yuri, 146
Gandhi, Indira, 73
Gemini 12, 104
gender roles, suburban, 222–223. *See also* women's movement
General Development Corporation, 218
General Dynamics, 219
General Motors Technical Center, 212–213, 336n125
Gerstenmaier, William, 237
ghettos, 23–35
Gifford, Mrs. Michael, 228–229
Ginsberg, Allen, 69
Glenn, John, 125, 144, 145, 159, 165–166, 181
Glennan, T. Keith, 143–144
Global Atmospheric Research Program (GARP), 129
global warming, 129–134
Global Weather Experiment, 129
Goddard Institute for Space Studies, 129
Goddard Space Flight Center, 129–130, 132–133, 214–215, 216, 279n135
Goldwater, Barry, 188
Gorbatko, Victor, 54, 86, 87, 91
Gordon, Frederick, 89
Graves, Dale, 175
Greenbelt, Maryland, solar energy demonstration, 209
Grigsby, Charles, 34
Grissom, Virgil, 22

Hair affair, 1, 9–10, 228–229
Hamilton, John, 30
Hansen, James, 129–130, 302n152
Harris, Hugh, 138
Harris, Louis, 291–292n75
Harris, Ruth Bates, 250n53, 321n166
Hart, Jane, 164–169, 177, 178
"Have You Seen the Saucers" (Jefferson Airplane), 198
Hawley, Steve, 137
Hayden, Tom, 272–273n66

Hays, Ted, 42
Heat Transport Section, 19
Heide, Wilma Scott, 167–168
Herblock, 30
Hess, Wilmont, 110
hippies. *See* counterculture
Hise, Greg, 218–219
Ho Chi Minh Trail, 58–59, 271n55
Hoffman, Abbie, 195–196, 199, 200, 235
House That NASA Built, The, 334n103
"Housewives in space tests," 174–175
HUD (Department of Housing and Urban Development), 30, 37, 41–48, 209, 269n139
Hudson, Robert, 120
Humphrey, Hubert, 38, 256n104
Huntsville, Alabama, 218, 222. *See also* Marshall Space Flight Center

Infiltration Surveillance Center, 64
"Infinity" visual display, 154
inner space, 199–201, 202, 210
Instrumentation Laboratory, 67–68
Integrated Thermal Micrometeoroid Garment, 150
Intercosmos Council, 83–85
International Geophysical Year (IGY), 128
International Latex Corporation, 151–152, 313n78
Irwin, James, 155, 158

Jackson, J. B., 215
Jasanoff, Sheila, 304n165
JASON Defense Advisory Group, 69
Jefferson Airplane / Jefferson Starship, 198
jet pilots, 158–159
Johnson, Katherine, 250n50
Johnson, Lady Bird, 35–36, 166, 278n122
Johnson, Lyndon, 10, 35–38, 164, 237, 256n111, 319n137
Johnson Space Center, 80, 168–170, 178, 182, 214, 221
Jones, Bernice, 23

Kantner, Paul, 198
Keeling, Charles, 129
Kendall, E. A., 192
Kennedy, John F., 146–147, 191
Kennedy, Ted, 34–35, 255n99
Kennedy Space Center (KSC), 100–104, 114–116, 156–157, 217
Khrushchev, Nikita, 142–143
King, Elbert, 110

King, James, 120
King, Martin Luther Jr., 22–23, 24
Kistiakowsky, George, 109
Koch, Ed, 34
Kozakoff, Emily, 164
Kraft, Chris, 170, 176–177
Kunen, James, 183–184, 185, 194–196, 202, 224–225, 325nn1,4

Landsat, 76–83, 88–91, 119, 279n135, 283n162
Langley Research Center, 39, 62, 118, 206–207
Large Area Crop Inventory Experiment (LACIE), 77–78, 80
Lassiter, Matthew D., 223, 326n9
Launch Complex 39, 98
Leary, Timothy, 200–201
Lewis, John, 22, 250n55
Life magazine: on dangers of outer space, 148; depiction of astronauts in, 145–146; Mercury Seven wives appear in, 158; on scuba training, 154; on spacesuits, 151; on spirit of space exploration, 147; and suburban development, 213, 218; on women astronauts, 142, 143
Limb Irradiance Monitor of the Stratosphere (LIMS), 122
Limited Warfare Committee, 60–64
Lindsay, John, 27
Liquid Cooling Garment, 150
"Little Boxes" (Reynolds), 221, 341n166
Little Rock, Arkansas, 22
"living machines," 201–202
Long, Scott, 104
"Loose Happening by Spaced-Out Free People, A," 199
Lord, Chip, 197–198
Lord, M. G., 222–223
Lovelace, Randolph, 144, 165
Lovelace Clinic, 144, 168, 177, 321n159
Lovell, James, 1, 152, 167, 186, 228–229, 233
Lovelock, James, 20
Low, George, 43, 77, 113–114, 122, 163
Luce, Clare Boothe, 143, 177
Luckman, Charles, 214, 219
Lunar Extravehicular Visor Assembly, 150–151
Lunar Landing Festival, 169–170
lunar landscape, simulated, 153–158, 316n103
Lunar Module Mission Simulator, 154–155
lunar orbiter program, 315n94
lunar seismometers, 64, 68

MacLeish, Archibald, 125–127, 136, 232
Mailer, Norman, 3–4, 145, 195
Malcolm X, 23
Manned Spacecraft Center, 214, 215
"March against Moon Rocks," 25
Marionetti, Gene, 228
Marshall Space Flight Center, 206, 215, 216
Massachusetts Institute of Technology
  (MIT), 67–68
Mauldin, Bill, 159–160
Mauna Kea, Hawaii, 155–156
McDivitt, James, 253n78
McGirr, Lisa, 223
Mekong Committee, 83, 89–91
Mercury Seven, 143–146, 153
Merritt Island National Wildlife Refuge, 114,
  116, 118
meteorology, as military aid, 62–63
Miller, David, 230
Mills College, 169
Mission to Planet Earth, 235–236
MIUS Integration and Subsystem Test
  (MIST) lab, 43–44
MKF-6 multizone camera, 85–86, 88–89
Modular Integrated Utility System (MIUS),
  43–46, 260n139
moon: pollution of, 104–105, 107–108;
  scientific exploration of, 110; U.S. flag on,
  192–193, 194, 199
moon rocks, 25
Morse, Franklin, 233
Ms. magazine, 139, 168–169, 179, 321n159
Mueller, George E., 63, 115
Myers, Jack, 15–16

Nader, Ralph, 109
NASA: and aerospace suburbs, 217–224;
  budget for, 113, 203, 269n41, 277nn110,113;
  and Cape Canaveral construction, 97–98;
  competing interests in, 7–8; criticized as
  square, 193–202; demonstrations against,
  11–14, 20–25, 27–33, 51–53, 254n80; ecological research of, 16–19; environmental
  activism of, 113–124; environmental
  movement and, 95–96, 109, 135–136; Equal
  Employment Opportunity (EEO) Program,
  167–168, 173–174, 176; feminism's impact
  on, 173–177, 180–182; feminist opposition to, 164–173; global data collected by,
  125–134; lunar orbiter program, 315n94;
  misses ozone hole, 302n151; and New
  Left, 66–74, 87–88, 234–236; physical

training at, 143–147; postcards of, 101,
  116–117, 287–288nn32–34, 295–296n95,96;
  promoted as square, 186–193; promotion
  of technology of, 98–103, 152; public image
  and funding of, 77; racial discrimination in, 21, 250nn50,53, 321n166; and
  remote-sensing technology, 76–83, 88–91;
  research parks, 210–217; Ride's publicity
  tour, 179–180; Rocket Gardens built by,
  100, 287n30; scientific "brain drain" experienced by, 110–111, 123; sexism in, 141,
  158–167; sides with New Right, 203–210;
  simulations done by, 152–158, 316n105;
  social and political movements transformed by, 231–234; Southern strategy of,
  224–227; studies of flight nurses, 322n169;
  and suburban sprawl, 217–220, 340n159;
  support for, 3, 33–35; technology of, as
  environmental hazard, 103–113; tests space
  environment, 147–148; tours of, 100–103;
  transforms second-wave feminism,
  177–179; urban renewal undertaken by,
  41–48; and Vietnam War, 56, 60–68; war
  on poverty, 35–41
National Day of Participation, 25–26
National Oceanic and Atmospheric Administration (NOAA), 78
National Organization of Women (NOW),
  139, 166–174, 304n6
National Welfare Rights Organization, 24
natural resources: analysis of Earth's, 76–83;
  Vietnam's wartime use of, 58–59. *See also*
  environment
Nelson, Gaylord, 73
New Alchemists, 201–202, 207, 208, 209
New Buffalo commune, 201
New Frontier, 190–192, 198
New Left, 65–74, 87–88, 234–236, 266n12,
  282n155
New Right: and aerospace suburbs, 223–224;
  battle between hippie counterculture and,
  183–186, 202; NASA criticized as square,
  193–202; NASA promoted as square, 186–
  193; and NASA research parks, 210–217;
  NASA sides with, 203–210; use of term,
  326n10; vision regarding space technology
  and natural environment, 225–227
Newsweek magazine, 20, 30, 47, 67, 101, 106,
  107, 125, 145, 200, 201, 217
Nguyen Van Hieu, 86
Nichols, Nichelle, 176
Niger, 82

Nimbus Experiment Team (NET), 123
Nimbus Observation Processing System
  (NOPS), 123
Nimbus satellites, 120, 122–123, 128–129
Nixon, Richard: and demonstrations against
  NASA, 25–27; and NASA technology, 78, 190;
  and New Frontier rhetoric, 191–192; and
  "silent majority," 1–2, 185, 188; on solar
  energy, 203–204; and Southern strategy,
  226–227, 326n9; and Vietnam War, 274n86
nuclear fallout, 112
Nye, David, 327n27

Odum, Eugene, 16–18, 20, 249n42
Odum, Howard, 16, 247n23
Office of Space Science Applications (OSSA),
  110
O'Leary, Brian, 111
O'Neill, Gerard, 205
"On to Mars!," 31–32
Operation Breakthrough, 42–48, 263n162
Operation Igloo White, 64
Orians, G. H., 268n28
Our Synthetic Environment (Bookchin),
  304n166
oxygen, in space capsule pressurization,
  248n35
ozone layer, 120–121, 129, 131–133, 136, 302n151

Paine, Thomas O.: on aerospace suburbs,
  219–220; on American space program, 225;
  and battle between New Right and coun-
  terculture, 186–188; on hippies' view of
  NASA technology, 197; on mastering space
  environment, 190; on NASA budget cuts,
  35, 203; and urban housing crisis, 11–12,
  41, 48–49, 50–51
Palme, Olof, 73
Pereira, William, 219
Perry, Alfred, 44–45
Petrov, Boris, 84, 85
Pfeiffer, E. W., 268n28
Philpott, Gladys, 164
photographs of Earth. See *Earthrise*; remote-
  sensing technology; *Whole Earth*
Picayune, Mississippi, 218
Pickering, William, 292n80
Piezoelectric crystals, 271n55
pioneers. See New Frontier
Playtex, 151–152, 313n78
political cartoons, 30–32, 69–71, 104–105,
  107–108. See also editorial cartoons

pollution, 39–40, 104–109, 113, 118, 220–223,
  291–292n75
Poor People's Campaign, 11–14, 20–21,
  48–49
Poppoff, I. G., 120
Portable Life Support System, 150
Port Malabar, 218, 222
postcards, 101, 116–117, 287–288nn32–34,
  295–296n95,96
potatoes, 147
poverty, war on, 35–41
privatization of space exploration, 236–237
Project Able, 63–64, 65, 74
Project Independence, 204
Proxmire, William, 35, 68, 106
Pupin Physics Laboratory, 69

race relations. See civil rights
Rand, Ayn, 187–188, 189–190, 192, 217
Reagan, Ronald, 139
Remote Measurement of Pollution, 118, 131
remote-sensing technology, 76–83, 88–90,
  118–120, 122–124, 128–131, 277n117
research parks, 210–217
Reynolds, Malvina, 221, 341n166
Ride, Sally, 137–140, 158, 176, 178–181,
  235–236, 324n196
Rocket Gardens, 100, 287n30
Rodgers, Daniel, 343n6
Rome, Adam, 340n163
Romney, George, 30, 41, 42

Saarinen, Eero, 212–213, 214
Salinas, Gayla, 173
Salyut space station, 54–55, 84, 88
Sandler, Harold, 175–176
Saturn V, 14–15, 102, 184, 189–190
Savio, Mario, 65–66
Schumacher, E. F., 4
Science Action Coordinating Committee, 67
"Science and Urban Development" work-
  shop, 41
scientific exploration, 109–111
Scientific Visualization Studio, 132–133,
  134
Scott, Ann, 173
Scott, David, 155
Scott, James, 267n22
Scott, Joan, 182
Scott-Heron, Gil, 29–30
SEALAB, 154
Seasat, 119–120

second-wave feminism, 138–140, 177–182, 304n6
Seeger, Pete, 341n166
segregation, 222, 253n78
seismometers, 64, 271n52
sewage treatment, 40–41, 50, 221
sexism, 158–163, 167. *See also* second-wave feminism; women's movement
Shanks, Bruce, 71
Shepard, Alan, 155, 250n55
Shoemaker, Eugene, 110
"Shuttle Re-Entry Acceleration Tolerance in Male and Female Subjects Before and After Bedrest," 174–175
Siddiqi, Asif, 6
Silent Spring, 112, 136
simulators / simulations, 152–158, 165, 174–176, 314n84, 316n103, 316n105
Slayton, Deke, 111
Slick, Grace, 198
Snyder, Gary, 199
"social movement spillover," 344n14
soil erosion, 220, 340n163
solar energy, 203–210
Solar Energy Panel (SEP), 203
Solar Heating and Cooling Demonstration Act (1974), 209
"solar ponds," 202
Southern strategy, 226–227, 326n9
Soviet Union: alliance with Vietnam, 54–57; Cold War propaganda of, 54–55; gender equality in, 142–143; portrayal of astronauts from, 146; space diplomacy of, 88; technology of, 64–65, 83–89
Soyuz 37, 54, 85–87, 91
"Space, Science, and Urban Life" conference, 39
"Space Biology: Ecological Aspects" symposium, 15–16, 247n23
space colonies, 198, 204–206
space environment, 147–148, 152–163, 174–176, 177
space junk, 104
"Space Oddity" (Bowie), 198
spacesuits, 148–152, 161–164, 190, 313n78
Stafford, Tom, 75
Stanford Torus, 205–206
Star Electricity, 65
Steel, Allen, 206
Steinem, Gloria, 168, 182, 235
Strategic Arms Limitation Agreement (SALT I), 74

Stratosphere Research Program, 121
Stratospheric Aerosol Measurement (SAMS-II), 122
STS-7, 137, 139, 180, 324n196
Students for a Democratic Society (SDS), 66
Study of Man's Impact on Climate, 131
suburban sprawl, 220–222, 340n159, 341n166
suburbs: corporate migration toward, 211–212; NASA and aerospace, 217–220; and NASA research parks, 212–217; political ecology of aerospace, 220–224
Summit Plaza, 44–46
Sunset Crater, 156
supersonic transport (SST), 106, 120–121
Suri, Jeremi, 275n94
Swain, Charles, 208
Swain, Elaine, 208
Swigert, John, 1, 228–229

Tech House, 206–208, 210
technology: of Apollo spaceship, 15, 18–19; counterculture communalists' relationship to, 201; environment and, 4–5, 189–190, 198–199, 225–226; promotion of, 98–103; relationship between political ideology and, 54–55; and relationship between racial politics and space race, 13–14; scholarship on history of, 6; of Soviet Union, 64–65; and war on poverty, 38–41
Technology Utilization Program, 38, 41
television, promotion of space technology via, 101
Tereshkova, Valentina, 141–143
Thematic Mapper, 76
Thompson, Wayne, 39
Time magazine, 2, 20, 98, 153, 174, 188, 192, 204
TIROS-1 (Television Infrared Observing Satellite), 61
Todd, John, 201–202, 331n76, 335n115
total energy systems, 43–46, 48, 50, 260n139
Total Ozone Mapping Spectrometer (TOMS), 122
"Transported" (Herblock), 30
Tuan, Pham, 54, 55–57, 84–89, 91
22727. See *Whole Earth*

Udall, Stuart, 109, 115, 119
underwater training, 154
U.N. flag, 192
unionization, 341n173

United States: Cold War propaganda of, 54–55; debate regarding national purpose of, 232–233; ecological impact of war efforts in Vietnam, 73; and energy self-dependence, 204; Tuan on aggression of, during Vietnam War, 55–56; visibility problem during Vietnam War, 59–60
University of South Dakota, 169
urban development, 23–35, 41–48
Urban Systems Project Office, 41–42
urban waste, 40–41
U.S. flag, 192–193, 194, 199, 228–229
U.S. Geological Survey (USGS), 155–156

Vehicle Assembly Building (VAB), 98, 101
Vietnam: alliance with Soviet Union, 54–57; and Intercosmos program, 84–87; Lower Mekong Landsat experiment in, 83, 89–91; natural environment of, 55, 58–59
Vietnam War: ecological impact of, 73; NASA and, 56, 60–68; and New Left, 65–74, 266n12; use of natural environment in, 58–59; visibility technology in, 59–60, 87, 268n28
von Braun, Wernher, 35, 160–161, 181, 191, 205, 332n92

Wainwright, Loudon, 145
Waldman, Anne, 199
war on poverty, 35–41
Warren, L. D., 107–108
waste removal, 40–41, 46, 162
Water Immersion Facility, 154
water treatment, 39, 220–221
Watts riots, 24
Webb, James, 36, 38, 39, 60, 62, 115, 159, 166
weightlessness, 153–154, 162, 174–175
Western settlement, space exploration compared to, 190–192
Where Today Meets Tomorrow, 213
White, Edward, 145, 253n78
"Whitey on the Moon" (Scott-Heron), 29–30

Whole Earth, 92–95, 124–125, 134–136, 232, 304n165
"Why Haven't We Seen a Photograph of the Whole Earth Yet?" buttons, 92–93, 95, 134–135
Williams, Hosea, 20–21, 51–52
Wilmarth, Verl, 79
Wilson, Richard, 30
Winter, David, 174, 175
Wise, Donald, 110
wives, in suburbia, 222–223
Woman's National Democratic Club (WNDC), 166
women, in suburbia, 222–223
women astronauts: feminist support for, 164–173, 180–181; language in debates over, 323n188; and space environment, 158–163, 174–176, 177. *See also* female body; women's movement
Women in Space Program, 165–166, 168–169
Women's Equality Day, 170–171
women's movement: and astronaut bodies as cultural symbols, 143–147; and difference feminists, 177; and equality feminists, 177, 324n190; and female body, 305n10; manned spaceflight and, 137–141; and support for female astronauts, 164–173, 180–181; Tereshkova and, 141–143; and transformation of NASA, 173–177. *See also* women astronauts
Woodstock, 2, 187–188, 195–196, 199–200, 230, 235
Worcester Polytechnic Institute (WPI), 187
World Bank, 82–83
World's Fair (1964), 98–99

Young, Andrew, 22
Young, John, 22
Young, Whitney, 23

"ZAP action" campaign, 170–172
Zenit, 65
zero gravity, 153–154, 162, 174–175